Studies in Computational Intelligence

Volume 505

Series Editor

Janusz Kacprzyk, Warsaw, Poland

For further volumes:
http://www.springer.com/series/7092

A. Şima Etaner-Uyar · Ender Özcan
Neil Urquhart
Editors

Automated Scheduling and Planning

From Theory to Practice

Editors
A. Şima Etaner-Uyar
Elektrik-Elektronik Fakultesi
Istanbul Teknik Universitesi
Istanbul
Turkey

Neil Urquhart
Centre for Emergent Computing
Edinburgh Napier University
Edinburgh
UK

Ender Özcan
School of Computer Science
University of Nottingham
Nottingham
UK

ISSN 1860-949X ISSN 1860-9503 (electronic)
ISBN 978-3-642-43596-6 ISBN 978-3-642-39304-4 (eBook)
DOI 10.1007/978-3-642-39304-4
Springer Heidelberg New York Dordrecht London

Printed on acid-free paper

Springer is part of Springer Science+Business Media (www.springer.com)

To Pudra and Alice - Ş.E.U.
To Ayla and Burhan - E.Ö.
To Siân- N.U.

Foreword

I was delighted and honoured to be asked by the editors of this book to write a short foreword to help set the scene. The goal of the book is to provide introductions to search methodologies and their applications to real world scheduling problems. I think that this is very much a worthwhile aim that resonates with the international scientific research agenda in scheduling research. The goal of closing the gap between real world practice and scientific theory in this research field plays a prominent role in that agenda. Scheduling problems are ubiquitous. They appear in many different forms across industry, leisure and the public sector. All of these sectors are represented here. Indeed, the breadth of the application areas is one of the particularly impressive features of this volume. This book brings together a selection of world leading authors from across a wide range of disciplines and scientific backgrounds. The editors have carefully constructed a volume which not only introduces modern search methodologies for the selected application areas, but it also provides insightful case studies which illustrate the effectiveness of some of these techniques. The book reflects a variety of important methodologies for a broad spectrum of challenging application areas.

The automation of scheduling problems across all of these important application areas represents a major challenge and it also represents significant potential impact. Intelligent decision support systems offer the potential to generate significant environmental, financial and social benefits. Some of the example application areas presented in this book provide compelling evidence for this claim. More effective radiotherapy scheduling has the potential to save patients lives. More efficient personnel scheduling can lead to a happier and more productive workforce. High quality airport scheduling could lead to lower levels of aircraft fuel burn. Factory floor scheduling can lead to improvements in production. Search methods can underpin the engines of intelligent decision support systems and this book provides an insight into how search methods can address challenging scheduling problems.

I have enjoyed reading through the chapters of this book. I would like to congratulate the editors on putting together such an interesting and informative volume. I am sure that this will provide a valuable resource to the scientific community and to practitioners for many years to come. I hope that you enjoy reading it as much as I have.

March 2013 Edmund Burke

Preface

This book was conceived as a result of the EvoStim (Nature-inspired Techniques in Scheduling, Planning and Timetabling) tracks held in Turin in 2011 and Malaga in 2012, as part of EvoStar: The Leading European Event on Bio-Insipired Computation. This book encompasses a wide range of research areas that fall under the generic title of automated scheduling, including healthcare, aviation, timetabling, manufacturing and computing. A very deliberate emphasis is placed on real-world applications.

We would like to offer our gratitude to all our distinguished authors for their valuable contributions and their diligence, without whom this book would not have been possible. They have met our deadlines and then patiently awaited this book to appear in print. We would also like to thank Edmund Burke for writing the foreword and providing invaluable advice. Finally, special thanks go to the staff at Springer, in particular Holger Schäpe, for their support.

We hope that you enjoy reading this book.

Istanbul, Nottingham, Edinburgh,
February 2013

A. Şima Etaner-Uyar
Ender Özcan
Neil Urquhart

Preface

Contents

List of Contributors

Bahriye Akay
Dept of Computer Engineering, Erciyes University, 38039, Melikgazi, Kayseri, Turkey,
e-mail: bahriye@erciyes.edu.tr

María Arsuaga-Ríos
Beams Department, European Organization for Nuclear Research, CERN, CH-1211, Geneva 23, Switzerland,
e-mail: maria.arsuaga.rios@cern.ch

Jason Atkin
School of Computer Science, Jubilee Campus, The University of Nottingham, NG8 1B, UK,
e-mail: jason.atkin@nottingham.ac.uk

Burak Bilgin
CODeS, KAHO Sint-Lieven Gebr. De Smetstraat 1, 9000 Gent, Belgium,
Tel: +32 9-265.86.10
e-mail: burak.bilgin@kahosl.be

Elkin Castro,
ASAP research group, School of Computer Science, University of Nottingham, Jubilee Campus, Wollaton Road, Nottingham, NG8 1BB, UK,
e-mail: edc@cs.nott.ac.uk

Patrick De Causmaecker
KU Leuven Campus Kortrijk, Department of Computer Science E. Sabbelaan 53, 8500 Kortrijk, Belgium, Tel: +32 56-28.28.73
e-mail: patrick.decausmaecker@kuleuven-kortrijk.be

Johannes Gärtner
XIMES GmbH, Austria,
e-mail: gaertner@ximes.com

Luca Di Gaspero
DIEGM, Università degli Studi di Udine, Italy,
e-mail: l.digaspero@uniud.it

İsa Güney
Faculty of Engineering and Architecture, Yeditepe University,
Istanbul,Turkey,

Mark Johnston
Victoria University of Wellington, New Zealand,
e-mail: mark.johnston@msor.vuw.ac. nz

Truword Kapamara
Faculty of Engineering and Computing, Coventry University, Priory Street,
Coventry, CV15FB, UK,
e-mail: d.petrovic@coventry.ac.uk

Graham Kendall
University of Nottingham, Nottingham, UK and Malaysia,
e-mail: graham. kendall@nottingham.ac.uk

Jeffrey H. Kingston
School of Information Technologies, University of Sydney, Australia,
e-mail: jeff@it.usyd.edu.au

Gürhan Küçük
Faculty of Engineering and Architecture, Yeditepe University, Istanbul, Turkey,
e-mail: gkucuk@cse.yeditepe.edu.tr

Michalis Mavrovouniotis
Centre for Computational Intelligence (CCI), School of Computer Science and Informatics, De Montfort University, The Gateway, Leicester LE1 9BH, U.K.,
e-mail: mmavrovouniotis@dmu.ac.uk

Nysret Musliu
DBAI, Technische Universität Wien, Austria,
e-mail: musliu@dbai.tuwien.ac.at

Su Nguyen
Victoria University of Wellington, New Zealand,
e-mail: su.nguyen@ecs.vuw.ac.nz

Dobrila Petrovic
Faculty of Engineering and Computing, Coventry University, Priory Street,
Coventry, CV15FB, UK,
e-mail: d.petrovic@coventry.ac.uk

Sanja Petrovic
Division of Operations Management and Information Systems,
Nottingham University Business School, Jubilee Campus, Wollaton Road,
Nottingham NG8 1BB, UK,
e-mail: sanja.petrovic@nottingham.ac.uk

Dmitry Ponomarev
SUNY Binghamton,
e-mail: dima@cs.binghamton.edu

Andrea Schaerf
DIEGM, Università degli Studi di Udine, Italy,
e-mail: schaerf@uniud.it

Werner Schafhauser
XIMES GmbH, Austria,
e-mail: schafhauser@ximes.com

Pieter Smet
CODeS, KAHO Sint-Lieven Gebr. De Smetstraat 1, 9000 Gent, Belgium,
Tel: +32 9-265.86.10
e-mail: pieter.smet@kahosl.be

Wolfgang Slany
IST, Technische Universität Graz, Austria,
e-mail: wolfgang.slany@tugraz.at

Kay Chen Tan
National University of Singapore, Singapore,
e-mail: eletankc@nus.edu.sg

Greet Vanden Berghe
CODeS, KAHO Sint-Lieven Gebr. De Smetstraat 1, 9000 Gent, Belgium,
Tel: +32 9-265.86.10
e-mail: greet.vandenberghe@kahosl.be

Miguel A. Vega-Rodríguez
ARCO Research Group, University of Extremadura, Dept. Technologies of
Computers and Communications, Escuela Politécnica, Cáceres, Spain,
e-mail: mavega@unex.es

Stephan Westphal
Institute for Numerical and Applied Mathematics,
Georg-August University, Germany,
e-mail: s.westphal@math.uni-goettingen.de

Shengxiang Yang
Centre for Computational Intelligence (CCI), School of Computer Science and
Informatics, De Montfort University, The Gateway, Leicester LE1 9BH, U.K.,
e-mail: syang@dmu.ac.uk

Xin Yao
Center of Excellence for Research in Computational Intelligence and Applications, School of Computer Science, University of Birmingham, Birmingham B15 2TT, U.K.,
e-mail: x.yao@cs.bham.ac.uk

Mengjie Zhang
Victoria University of Wellington, New Zealand,
e-mail: mengjie.zhang@ecs.vuw.ac.nz

Airport Airside Optimisation Problems

Jason A.D. Atkin

1 Introduction and Problem Context

This chapter aims to give the reader an accessible overview of airside airport operational research problems, with a particular focus upon runway scheduling, which is the subject of the case study. A number of problems are described, highlighting the direction of the research in each area and pointing the reader towards key publications where more information can be gained. Some of the surrounding problems are also outlined, to better understand the airport context. A case study is then provided, describing a system which was developed to aid runway controllers at Heathrow. Importantly, this considers a combination of two separate problems and the way in which these are simultaneously handled by the solution method. Results are provided for the presented case study, showing the potential benefits of decision support in that area. The chapter ends with a discussion of the likely ongoing importance of considering increasingly realistic objectives and constraints, of combining problems, and of targeting the environmental challenge at airports.

Airports and the airspace which connect them, together form the framework for the air transportation system worldwide. There has been increasing interest in air transportation optimisation and efficiency over the last few years, partly due to the SESAR[1] and NextGen[2] initiatives, and partly because the level of technology and computational speeds have now reached the point where improvements are actually possible. However, the field of air transportation is huge, and could in no way

Jason Atkin
School of Computer Science, Jubilee Campus, The University of Nottingham, NG8 1B, UK
e-mail: jason.atkin@nottingham.ac.uk

[1] Single European Sky ATM Research, a joint undertaking by Eurocontrol, the European Commission and Industrial bodies to build the future of the air traffic management system, see http://www.eurocontrol.int/content/sesar-and-research,

[2] A US program to design and develop the Next Generation of Air Transportation Systems, see http://www.faa.gov/nextgen/ for more information.

A.Ş. Etaner-Uyar et al. (eds.), *Automated Scheduling and Planning*,
Studies in Computational Intelligence 505,
DOI: 10.1007/978-3-642-39304-4_1, © Springer-Verlag Berlin Heidelberg 2013

be encompassed by a single chapter of a book. Consequently, this chapter concentrates upon airside airport operations, and particularly upon the runway sequencing problem.

It is useful to consider the airside optimisation and sequencing problems at an airport in the order in which they will be experienced by an aircraft: A runway-sequencing problem has to be handled first, to get aircraft on the ground safely; the aim being to determine the order in which the aircraft will land. This is explained in Section 7. Aircraft then have to taxi from the runway to their allocated stands, where they will be unloaded, refueled and re-loaded with passengers and baggage. This ground movement problem is considered in Section 6, determining how the aircraft will taxi around the airport and which aircraft will take priority at any points of contention (e.g. taxiway intersections). A stand allocation (or gate allocation) problem also has to be solved, to determine which stand (or gate) each aircraft will be allocated to. This will usually determine one of the end points for each taxi operation and the relative loads which are put on different stands/gates and is discussed in Section 5. A number of resource allocation problems then need to be resolved at the stands, so that refueling facilities, baggage handling facilities and ground crews are available. In general, these have the same format as the stand/gate allocation problem, and are discussed in Section 5.3. Once loaded with passengers, cargo and fuel, an aircraft will then taxi out again (so the ground movement problem has to be solved once more) and queue near to a departure runway, awaiting its position in the take-off sequence. This will be determined by the solution of another runway sequencing problem. The runway sequencing problem is an important one, since the runways often form bottlenecks for the entire arrival/departure system. In addition to the description in Section 7, the case study in Section 8 also considers the take-off sequencing problem for London Heathrow.

In addition to the aforementioned optimisation problems, this chapter also considers some of the structure of the local airspace, including down-stream constraints upon the airport (e.g. congestion on departure routes), and summarises some of the issues which airlines have to consider when building their schedules, along with the effects that these decisions have upon airports. Further, but less recent, information about these problems can be found in Wu and Caves (2002) and Yu (1998), which review the air traffic management research at the times of the publications. Of course many logistics problems (e.g. supplying shops), flow problems (e.g. passenger flow through terminals) and staffing/rostering problems (e.g. security staff rostering) also have to be considered by airports, but these are beyond the scope of this chapter.

1.1 The Usual Problem Decomposition

The airside operations are perhaps best understood by considering the air transportation system as a cyclic process, rather than focussing upon only a single airport. Safety considerations are always paramount within any of these processes, and constraints upon one part of the process may be a result of ensuring safety or controlling workload in a later part. For example, en-route congestion may be controlled by metering departures from airports, resulting in increased delays at those airports. Similarly, passenger preferences (and the resulting financial considerations in such a competitive marketplace) may result in higher delays for all concerned, as airlines compete for the more lucrative timeslots. The effects of these interactions will be highlighted at various points in this chapter, since these constraints can cause detrimental effects upon the throughput of other parts of the system.

The overall airport optimisation problem could be considered one of attempting to utilise scarce airport resources, such as the stands, taxiways and runway(s), in as efficient a manner as possible. Unfortunately, each of the various problems is usually large, has to be solved (or re-solved) quickly (within a few seconds in some cases) as the situation develops over time (especially when things do not go to plan) and often has multiple, possibly conflicting, objectives which should be met. It will be obvious from even a cursory glance at the existing literature, that the optimisation problem for even a single airport is usually decomposed into a number of constituent subproblems, and that each airport is usually considered independently. This is at least partly an attempt to simplify the individual problems, making them more tractable for different solution methods. This is especially important when real-time solutions are required for these problems, to account for the dynamic, constantly changing, nature of the airport. However the decomposition is also a symptom of the fact that different problems tend to be handled by different companies or organisations, who may (and often do) have differing objectives. This decomposition can mean that even apparently optimal solutions for individual sub-problems can result in suboptimal utilisation of the airport as a whole, whereas cooperation could potentially have greater benefits for all.

1.2 More Recent Initiatives

Until recently there were few attempts to link together the various stages and stakeholders, although this is starting to change as information sharing and collaborative decision making is becoming more evident. The NextGen program has involved significant investment in infrastructure, aircraft design and process development in order to bring cost, performance and environmental improvements for the US air transportation industry. Significant funding has become available for companies and researchers in related areas in order to achieve these improvements. System-Wide Information Management (SWIM) is an important part of NextGen, aiming to ensure that information is made available in a more timely manner, to an improved accuracy, to all interested stakeholders. Improving information availability is an important requirement of increasing the quantity and efficiency of automation and decision support tools in airports; since it means that important information about likely decisions is already available electronically, so can be more easily shared.

SWIM is also an important element of the SESAR Joint Undertaking and airport-CDM is perhaps the most important element for the purpose of this chapter. Airport Collaborative Decision Making (CDM) involves the various stakeholders at an airport sharing information so as to make improved global decisions. With so many different stakeholders involved, including the airport staff, airlines, air traffic controllers and ground handlers, each with their own computer systems and methods of operation, this is not necessarily a simple task, but is one which is being successfully faced at a number of European airports. With better information about the progress of aircraft through the system, each stakeholder should be able to better allocate their resources so as to reduce delays and conflicts, determining the appropriate allocation of resources when they may be required in multiple places at almost the same time. These systems also allow a better coordination with the rest of the air transportation system, beyond the airport, by providing downstream systems with better predictions for take-off times for aircraft, and accepting improved estimations for arrival times for incoming aircraft, distributing the information appropriately throughout the airport. In this way, resources (such as gates, refueling trucks or tugs) can be ready and waiting for an aircraft when it arrives, or potentially reallocated to other aircraft as soon as it is known that an aircraft will be late.

2 Solution Methods

The following sections describe a number of operational research problems at airports. Each has one of more objectives and a number of hard constraints which must be satisfied. Both exact and heuristic methods have been utilised for all of these problems, so this section provides an overview of many of these methods, directing the reader to sources from which to find out more, to avoid repetition in the following sections.

Firstly, it is relatively common to formulate the problems as Mixed Integer Linear Programs (MILPs), especially where exact solution methods will be applied. A good introduction to formulating problems in this way has been provided in the book by Williams (1999). The underlying theory behind the solution of Linear Programs and MILP models, in addition to that of many other solution techniques is covered in Hillier and Lieberman (2010). These formulations require that all objectives and constraints are linear combinations of variables (although some solvers are increasingly able to cope with quadratic terms). There are common ways for linearising constraints or objectives, for example by introducing additional variables, or by finding piece-wise linear approximations for functions. In theory, such linear models can be passed straight to a MILP solver (such as CPLEX[3] or Gurobi[4]) to solve the problem. If the models contain only continuous variables, the solution is usually trivial, however when variables can only take discrete (e.g. integer) values the resulting models can be too large or complex to solve in a reasonable time. Models may have many thousands of variables and constraints under these formulations, which can often cause even machines with huge amounts of disk space and

[3] See for example: http://www.ibm.com/software/integration/optimization/cplex-optimizer/

[4] http://www.gurobi.com/

memory to run out of space for the model, or to take prohibitively long (sometimes potentially years or more) to solve these problems.

When a MILP solver is not used, or not solely used, two particular exact solution methods are very common for these types of problems. Firstly, dynamic programming is often used, where a problem is divided into a number of discrete stages, each of which can be optimally solved in turn, producing one or more optimal states to pass to the next stage. Secondly, hand-crafted branch and bound or branch and cut approaches can be used (perhaps in combination with a MILP solver), which will recursively divide the problem into smaller and smaller parts, discarding parts which are provably sub-optimal. (The MILP solvers tend to do this themselves, but hand-crafted approaches can sometimes utilise information which the solvers cannot automatically infer, or can more easily deal with non-linearities.) These approaches are described in Hillier and Lieberman (2010), or many other Operational Research books.

Heuristic methods will usually be used when a problem structure or size is problematic for exact solution methods, or when insufficient solution time is available to apply an exact method. Some heuristic approaches will use problem-specific information to make decisions, knowing that such a decision is likely to be good, or at least not be 'too' bad. The path allocation heuristic in Section 8.4 is an example of such a problem-specific approach. Other approaches, such as local search or metaheuristics, may be guided less by the problem and more by the value of potential solutions, gradually homing in on possible good solutions.

A local search algorithm will start at a known potential solution and investigate other potential solutions which are similar to it. 'Similarity' will usually be defined in a problem-specific way, by generating a 'neighbourhood' of solutions to investigate at each step. A neighbourhood is usually defined in terms of the 'moves' which will be used to make a change to the current solution, for example to swap the positions of two parts of a solution. The aim is usually to make incremental changes, moving to better solutions and ignoring worse solutions, eventually moving to better and better solutions over time, in the hope of finding an optimal solution. Eventually the search will find a 'local optimum': a solution (or set of identically valued solutions) such that all solutions in its neighbourhood are worst than it. Many problems have local optima which are not globally optimal (i.e. better valued solutions exist, but these are not in the local neighbourhood), resulting in an inability for the basic algorithm to guarantee that it will find a globally optimal solution.

Metaheuristics utilise the idea of some higher level guidance on top of a low level heuristic algorithm. Commonly used strategies for avoiding or escaping local optima are: to alter the neighbourhood dynamically; to include a memory of where the search has been, or of the characteristics of good or bad solutions which have already been considered so as to prefer or avoid solutions; to restart from another solution when the search gets stuck; or to consider multiple solutions simultaneously (population-based approaches), potentially interchanging information between these solutions. Useful surveys and explanations of metaheuristic methods can be found in Glover and Kochenberger (2002); Blum and Roli (2003); Gendreau and Potvin (2005); Talbi (2009). The case study in Section 8 utilises a tabu search meta-heuristic, which is discussed further in Section 8.4. Further information about tabu search, as explained by its inventor, can be found in Glover (1989, 1990); Glover and Laguna (1997).

Many solution methods for the problems described in this chapter utilise Genetic Algorithms. These are population-based approaches which use a mechanism based upon natural selection to interchange parts of the solutions between them, generating further solutions which are hybrids of the 'parents'. Further information about Genetic Algorithms can be found in Goldberg (1989); Sastry et al (2005). A good introductory tutorial to problem solving using genetic algorithms can be found in Michalewicz and Fogel (2000).

3 The Airport Layout and Some Definitions

Airports vary greatly in their layouts, however there are some common factors. Airports will have one or more '*terminals*', where the passengers check in and where passenger facilities are located. '*Gates*' will be situated around these terminals and passengers will wait at a gate to leave a terminal in order to board an aircraft. The term '*gate*' is used here to denote the exit point from the terminal to the airside, where the aircraft are located. In common with the usual meanings at UK airports, the term '*stand*' will be used in this chapter to denote a parking position for an aircraft. Stands will usually be either at gates, so that passengers can embark straight onto the aircraft, or will be remote stands, requiring a walk or a short bus journey to be taken by passengers to get from the gate to the remote stand. It is common, especially in the US, to refer to the area close to the terminal, on which stands may be located, as the Apron.

In order to increase the number of gates at a terminal, piers are often built out from the terminals. These piers can be relatively long passages for passengers to walk along, often with gates on either side. Aircraft will then park along the sides of the piers. It is common to see multiple parallel piers extruding from terminals. Cul-de-sacs, or alleys, will be formed between the piers, such that aircraft may have to taxi past other stands in order to reach an allocated stand at the end of a cul-de-sac.

An airport will have one or more runways. Each runway could be used as an arrival runway (for landings only), a departure runway (for take-offs only) or in mixed mode (for both arrivals and take-offs). It is common for runways to be usable in either direction, allowing aircraft to ensure that they take off or land into the wind to maximise their lift. When an airport has parallel runways which are sufficiently far apart, these runways can be treated independently. When runways are not parallel, or are close together, some dependencies may be formed between the runway sequencing for the runways.

Taxiways usually run between runways and stands. The layout of the taxiways can vary greatly between airports. Sometimes taxiways are only used in one direction, but they are often bidirectional, adding complexity for the ground movement controller (GMC) who has to direct the aircraft. Sometimes bottlenecks form on taxiways, for example to go around or between buildings. Again, these will complicate the problem for the GMC. Sometimes runways are positioned such that aircraft have to cross one runway in order to travel between another runway and the allocated stand. In order to do this, a gap has to be added to the runway sequence, for the runway crossing. This can complicate both the ground movement problem and the runway sequencing problem, introducing interdependencies between them.

4 The Airline Scheduling Problem

It is useful to understand something about airline scheduling, at least as far as the effects that it has upon airports are concerned. A more detailed overview of the various problems involved in airline scheduling can be found in Barnhart et al (2003) or Qi et al (2004).

Airlines will usually start planning their routes many months in advance, the aim usually being to maximise their income from passengers in a competitive marketplace. Importantly, they will not usually consider the effects of their decisions upon an airport, since doing so may put them at a disadvantage against competitors who do not do so. The planning problem is usually decomposed into a number of distinct stages, first determining the flight legs (trips from one airport to another) to include in the schedule, and at what times the flights should take place, then allocating aircraft types to legs. Specific aircraft of the correct types will then be allocated to flight legs at a later date, taking into consideration issues such as maintenance requirements. Finally, crew will often be allocated to aircraft in an even later stage. The timing of flights can have considerable effects upon the demand for the flights, with specific types of passengers (e.g. long-haul vs short-haul, eastward vs westward bound, business vs holiday) preferring different days, or times of day, for journeys. Consequently, airlines will not always have the same flights on every day of the week and will not usually evenly space out their flights across the day. These preferences will naturally ensure that the demand upon the airport resources is not steady across a day or week and can cause congestion both at airports and in the en-route airspace at certain times of the day.

Many airlines run hub and spoke networks, where they have a central base (the hub airport) at which connections are made, and a number of spokes to many other airports. Journeys are then possible between any two of the spoke airports by making a single connection at the hub airport. Airline schedules will, therefore, attempt to facilitate passenger connections at the hub by timing flights so that a number of flights will arrive at similar times, be on the ground at similar times, while the connections are made, and take off at similar times.

There has been increasing interest in creating more robust schedules (e.g. Burke et al (2010)), especially since any modern optimisation methods which assume deterministic environments tend to produce more efficient schedules by removing slack which is perceived to be unnecessary, but which would have absorbed delays. However, even with reasonable levels of slack, some delays will occur and airlines have a number of ways in which to recover from these delays, see

Filar et al (2001); Clausen et al (2010). One recovery method which is used is to switch aircraft between flight legs, effectively moving the delay between aircraft and possibly moving it to aircraft where there is more slack in the schedule to absorb the delay, or to recover later. With this flexibility, it is potentially possible to strategically allocate slack through a schedule rather than allowing slack for all flights, reducing the wasted (i.e. slack) time while still allowing for recovery. More swap opportunities are available when there are more aircraft available on the ground at the same time, which is another reason for grouping multiple arrivals and departures together.

In summary, the benefit for the airlines from having multiple aircraft arrive and depart around the same time is another cause for uneven load on airports across the days. Indeed, peaks and troughs in load are usually obvious from a consideration of the frequency of take-offs and landings over the day. This uneven load will contribute to congestion on the ground, with more aircraft wanting to taxi at the same time (affecting the ground movement problem), increased demand for stands (more aircraft requiring simultaneous usage of parking places), and increased delay at the runway (as aircraft queue to use the limited runway throughput). This can result in significant problems at some times of day for even relatively quiet airports.

Other interesting side effects are that small delays are relatively common and, conversely, the slack that is often added to a schedule can sometimes mean that aircraft are ready to depart from gates earlier than expected. This means that it is hard to predict in advance exactly (i.e. within a few minutes) when each aircraft is likely to want to pushback. In addition, when the aircraft which is allocated to a flight is changed by an airline, the stand/gate for the flight will often change to that at which the new aircraft is already located. This can significantly alter the planned departure time of that aircraft, with only limited warning. Together, these issues mean that it is complex for any party other than the airline itself to obtain relatively accurate pushback times for aircraft. Airport-CDM aims to avoid this problem by having the airline do this prediction and publish the information to other airport partners.

5 The Gate/Stand Allocation Problem

The Gate Allocation Problem (GAP) involves allocating gates or stands to aircraft so as to meet airline and airport preferences, as well as to fulfil any hard constraints such as ensuring that an aircraft will actually fit on its allocated stand. There has been considerable research into the gate allocation problem and a good review of the research and open problems at the time can be found in Dorndorf et al (2007a). Both heuristic (e.g. Genetic Algorithms, Bolat (2000a)) and exact (e.g. Dynamic Programming, Jaehn (2010)) methods have been utilised in the past, and the ideal solution method will probably depend upon the problem size and the precise objectives used. An overview of the basic problem will be provided in this section, along with a discussion of the effects that gate allocation has upon the other airside operations.

5.1 *Primary Objective and Hard Constraints*

The key objective is usually to allocate as many flights to stands as possible, then to meet as many of the soft constraints as possible, subject to ensuring that all hard constraints are met. At most airports, it is hoped that sufficient stands are available, if they are correctly allocated, however, this is not always the case, see Ding et al (2005). The result of solving this problem will be a set of aircraft-stand pairs, specifying which stand each aircraft will be allocated to. In most formulations, the on-stand (arrival time of the aircraft at the stand) and off-stand (departure time of the aircraft from the stand) times are fixed, however, in some versions of the problem some flexibility is permitted in these timings and the on-stand and off-stand times would also be outputs of the solution method, Lim et al (2005).

The key hard constraints which usually have to be considered can be summarised as follows:

No Stand Can Be Simultaneously Occupied by Two Aircraft: The on-stand and off-stand times for aircraft are usually specified in the input problem, in which case this constraint usually involves ensuring that no two flights are allocated to the same stand if their on-stand times overlap. When on-stand and off-stand times are an output of the model, these constraints are more complex and may need to be modelled as objectives in the extreme case, since an aircraft pair could always be allocated to the same stand if the second aircraft could be delayed long enough that the times no longer overlapped. However, there would be a cost associated with this delay.

An Aircraft Can Only Be Allocated to an Appropriate Stand: It is uncommon for every aircraft to be able to use every stand, so there will usually be a number of aircraft-stand pairings which are disallowed. For example, it is necessary to prevent an aircraft from being allocated to a stand which is too small for it, or has inappropriate facilities. In addition, airport agreements or facility availability may restrict certain aircraft to certain sets of stands, either because an airline needs specific facilities which are only available at certain stands, because of legal agreements between the airlines and airport, or because their ground handlers have resources which are only available at certain stands. It is also necessary to prevent an aircraft from being allocated to a stand with inappropriate security measures. For example, domestic or Schengen flights do not require the same level of security measures that are required by international flights. Obviously, it is inappropriate to allocate an international flight to a domestic/Schengen gate. Certain destinations (for example international departures to the US) or sources (for arrivals) may require even more security, which may only be available at certain gates, so these flights will be even more constrained. All of these constraints can usually be easily enforced within a model by reducing the potential range of stand values for each flight, e.g. forcing specific aircraft-stand pairing variables to zero.

Prohibited Combinations of Aircraft-Stand Pairings: The final common hard constraint is often termed the shadowing constraint, see Dorndorf et al (2007a). This constraint usually applies to pairs of aircraft and stands, specifying that if specific aircraft are allocated to one of the stands, then either the other stand cannot be used or it can only be used by specific (types of) aircraft. A common case is where large stands can be divided into two sides (often named 'left' and 'right'), for use by two small aircraft, or kept as a whole for use by a single large aircraft. This can be modelled as three stands, such that the larger stand cannot be used if either of the smaller ones is used, with appropriate size restrictions for which aircraft can use which stands. Another case is where the use of a stand by a large aircraft prevents the use of one or more adjacent stands due to its size. In fact both of these cases can be modelled identically since the former case could be modelled by saying that if one side is being used by a large aircraft then the other side cannot be used. Again, it is usually relatively easy to add constraints to a model to enforce these hard constraints, adding constraints between aircraft-stand pairings preventing any two inconsistent pairings from being made simultaneously.

5.2 Soft Constraints and Objectives

The soft constraints tend to vary far more widely than the hard constraints, depending upon the objectives of the airport or the group who are attempting to solve the problem. A realistic objective function for this problem could well include a number of different factors, perhaps weighted together, or perhaps handled using multi-objective methods. The most common factors considered are:

Gate/Stand Preferences: Airlines may have preferred stands such that they could use other stands but would prefer not to, or may prefer to allocate certain stands first in preference to others when there is spare capacity. There may also be preferences from the airport side for utilising some stands rather than others, such as to reduce congestion at the airport, or to keep taxi times lower for larger aircraft. These objectives can be modelled by applying a cost for aircraft-stand pairings, for example to penalise any usage of a stand.

Flight-Stand Pair Preferences: Airlines may prefer to put specific flights on specific stands, or to avoid putting some flights on some stands. For example, when considering allocations across a week or longer, it may be useful to have the same allocations across multiple days, to increase familiarity for staff and frequent travellers, thus airlines may express such preferences. Similarly, there may be advantages to having a regular schedule, so that allocations are relatively similar on each Monday, for example. A reference schedule may be available (e.g. the schedule for the previous week or day) and there may be a benefit for staying close to it (see Dorndorf et al (2012)), penalising flight-stand pairs which deviate from the reference schedule and potentially increasing the penalty for larger deviations. These objectives could be modelled by applying a cost for specific aircraft-stand pairings.

Walking Distance Reduction: The tendency to position gates along piers means that some gates may involve a lot more passenger walking distance than others. This is particularly important for transfer passengers, who may be moving from one pier to another, or even from one terminal to another, often passing security one or more times (especially when transferring from an international to a domestic flight) with a deadline for making the transfer (the departure time of the second flight). One of the more commonly studied objectives has been to reduce passenger walking distance, Kim et al (2010). The common model for this problem is to consider the number of passengers transferring between each pair of flights, along with the number of passengers who originate from the current airport and the number of passengers who are terminating their journey there. Given a distance measure for each walk, from the airport entrance/exit to each gate, and between each pair of gates, the objective is to reduce the total walking distance, in terms of the product of the distance of the walk and the number of passengers making it. One practical difficulty for this objective is in determining the number of transfer passengers for flight pairs, so estimations of passenger numbers (e.g. based on historical data) may be needed for this purpose. This objective can cause problems when there are cul-de-sacs between piers (see Section 3), since this formulation will tend to position the larger aircraft (which will have more passengers) nearer to the terminals (reducing their distance to security/the exit), and hence further down the cul-de-sacs. This will often increase the taxi distance for the aircraft and also makes them more likely to be delayed by other users of the cul-de-sacs, both of which will increase the fuel burn. Furthermore, walking distance did not seem to be an important consideration for the airports with whom we have spoken, and environmental and financial considerations seem to be more important at the moment.

Spacing Preferences: It may be desirable to accumulate flights into a small physical area of the airport, perhaps to reduce time for ground crews to move between aircraft, or to reduce passenger walking distance, as previously discussed. Alternatively, it may be better to spread flights out more widely, so as to spread the load upon scarce resources or to reduce the load or congestion on the taxiways around the stands, e.g. Kim et al (2009, 2010). These objectives can be modelled by penalising combinations of multiple stand-flight pairs, for example such that a penalty is applied when multiple flights with similar planned departure times are allocated to stands on the same cul-de-sac.

Ensure Buffer Times between Uses: Since arrival and departure times for flights are not entirely deterministic, some deviation from the planned times is usually expected. These disruptions can cause problems and must be handled, Dorndorf et al (2007b). Good gate/stand allocations will be robust to small changes (see Bolat (2000b); Dorndorf et al (2008); Kim and Feron (2011)), so that assignments are less likely to overlap even when there are small delays for one of the aircraft. A penalty could, therefore, be applied to any schedule where the gaps between allocations are below a certain value, even when they do not actually overlap, with larger penalties for smaller gaps. Alternatively, one of the stands may need to be reallocated (see Bolat (1999)), in which case the aim will usually be to recover a feasible allocation while keeping as many stand allocations as possible unchanged.

5.3 Problem Variants and Other Resource Allocation Problems

When there is a shortage of stands, it may be sensible to move aircraft which have long ground times away from their stand after unloading, to park them at a remote location, then move them back again prior to loading. In this case, the stand to which they return may differ from the stand at which they were unloaded and two stand allocations are necessary - one for unloading and one for loading. In other airports, it may be necessary to move aircraft between gates between unloading and loading, for example if an aircraft arrives as a domestic flight and departs as an international flight, or vice versa, requiring a gate with different security measures in place. In this case the towing operations are extremely important since the tugs are expensive resources at an airport and thus likely to be limited resources. Additional objectives for this problem include reducing the number of tugs required or reducing the total towing distance. In this case, the unloading and loading would usually be modelled as separate stand usages, perhaps modelling the aircraft as two separate flights (an arrival and a departure). A cost would then be associated with the towing operations, whereby the allocated stands for each activity would determine the costs involved, in terms of time and distance, using a model which is very similar to that for passenger walking distance.

Although gate allocation is the academically most widely studied of the airport resource allocation problems, there are many other similar resource allocation problems which need to be considered near the gates. Ground handling operations, which consist of operations such as refueling and re-provisioning the aircraft, offloading baggage, loading the new baggage, attaching steps or walkways (so that passengers can get in or out of the aircraft), and cleaning the aircraft all require resources, often consisting of both equipment to perform the operations and crew to use the equipment. At many airports these operations will be performed by either the airport operating company or separate ground handling agents rather than the airlines themselves. It is common for the equipment and crews for these operations to be used across multiple gates, and allocating the appropriate equipment and crews to aircraft is another resource allocation problem. For example, the baggage sorting station problem (see Abdelghany et al (2006); Ascó et al (2011)) involves allocating baggage sorting stations (the airside positions at which baggage is accumulated for loading into aircraft, and where offloaded baggage is placed to feed it into the airport baggage system) to flights. Many of these problems have a similar structure to the gate allocation problem, although the details of the objectives and constraints can differ slightly.

Other resource allocation problems may need to include a scheduling element and involve not only assigning items but also determining the times at which item assignment will start and end. For example, there may be time windows while an aircraft is on a gate within which it must be refueled. Ensuring that refueling occurs within this time window will require consideration of the locations of refueling facilities and crew at any time and of allowing sufficient time not only for the refueling operations but also for equipment and crew to be moved between aircraft.

6 The Ground Movement Problem

The airport ground movement problem involves moving all aircraft from their starting positions (usually at stands or runways) to their final positions (usually at other stands or runways) within some specified timeframes (earliest and latest times for the start and end of each move). Objectives may differ for different aircraft. For example, the objective for aircraft which have just landed would usually be to arrive at their allocated stand as soon as possible (i.e. to reduce the taxi time). Depending upon whether a take-off time has already been planned for an aircraft or not, the objective for aircraft taxiing from their stands to a take-off runway would usually be one of the following: to arrive at the runway as early as possible; to arrive on time for a predicted take-off time; or, more recently, as fuel burn and environmental considerations become increasingly important, to arrive on time for take-off but to start the engines (i.e. commence the movement) as late as possible. A recent review of previous research into ground movement is available in Atkin et al (2010b).

6.1 Models of the Ground Movement Problem

Many different models have been utilised for this problem, however they are usually conceptually very similar. The airport is usually modelled as a directed graph, with taxiways being modelled as a set of arcs, and a set of nodes being used to model the intersection points of taxiways. One difference between the models is whether aircraft are conceptually located at nodes or on arcs. In the former case, additional nodes will usually be added to split long arcs according to how many aircraft can fit on them. In the latter case, arcs may have capacity constraints. However, in general, the choice of arcs or nodes to contain the aircraft makes little difference to the conceptual model. Very small examples of ground movement graphs are given in Figures 2 and 3 (which are discussed in Section 8), showing the ways in which aircraft could move through the taxiway structures near to the runways of Heathrow around 2004.

The output of the ground movement problem will usually be a route (a sequence of arcs or nodes) for each aircraft, and either times at which each intersection point should be passed, or an order in which aircraft should pass each intersection. Solution approaches tend to consider either the sequence in which aircraft should pass different points, and use this to determine the times at which points are passed, or to directly consider the timings for aircraft. The basic constraints in the model are that aircraft must maintain at least a specified minimum safe distance between them, and that maximum taxi speeds for aircraft will limit how quickly the arcs can be traversed. Genetic Algorithms have been used successfully (e.g. Gotteland et al (2001), Deau et al (2008)), but the majority of research has tended to use Mixed Integer Linear Programming approaches, e.g. Smeltink et al (2004), Roling and Visser (2008), Rathinam et al (2008), Keith and Richards (2008), Clare and Richards (2011), Marín (2006), Marín and Codina (2008).

6.2 *Taxi Time Prediction and Environmental Effects*

Taxi time prediction is a closely related problem to ground movement, but at a more abstract level. The aim is to predict how long an aircraft will take to complete a taxi operation, without necessarily having to plan the route or sequence of interactions. Various statistical methods have been utilised for this, with Multiple Linear Regression being popular due to its firm mathematical foundations, see Ravizza et al (2012a), Idris et al (2002). Taxi-out time (from stand to runway) usually includes queueing time at the runway. At busy airports, this queueing time can be a considerable proportion of the entire taxi time, to the point where taxi time estimations can be produced by considering the runway queue size, e.g. Idris et al (2002); Simaiakis and Balakrishnan (2009). Other research has identified that aircraft taxi faster overall if they have to make fewer turns (e.g. Rappaport et al (2009)), and that this speed decrease seems to be quantifiable (e.g. Ravizza et al (2012a)). Other approaches have used reinforcement learning methods (e.g. Balakrishna et al (2010)), or fuzzy rule based systems (e.g. Chen et al (2011)), and there are some indications that the non-linearity of the fuzzy rule based systems has benefits for taxi time prediction.

6.3 *Recent Work*

Taxi speed information tends to be an input to the ground movement problem, and many systems assume constant taxi speeds. However taxi speed can depend upon the route which is taken around the airport, as shown by Ravizza et al (2012a). Route-specific taxi times could be utilised in a ground movement algorithm, improving the travel time predictions and the accuracy of the ground movement models. Recent work has started to consider the combination of taxi time prediction and ground movement research, e.g. Ravizza et al (2013).

There has also been an increasing interest in the environmental effects of taxiing (see Nikoleris et al (2011)), and particularly upon the impact that runway queues have upon this, for example Simaiakis and Balakrishnan (2010). Various queueing models have been developed to try to predict waiting times and pushback metering has been used to hold aircraft upon the stands for longer, for example Andersson et al (2000), Simaiakis and Balakrishnan (2009), Simaiakis et al (2011).

Even more recently, researchers have started to look at the environmental advantages of slower taxiing, reducing the engine power and thus the fuel burn, for example Ravizza et al (2012b). Along similar lines, the benefits of single-engine taxi operations have been considered by airlines, and some airlines are using these already, although this is often more practical for taxi operations after landing rather than prior to take-off, since it is better not to leave engine start-up, with its associated checks, until too late before take-off.

7 Runway Sequencing

Since the runways are often the limiting factor for the capacity of the arrival and departure systems, runway capacity has had considerable research for many years, from theoretical models such as that used by Newell (1979) to simulation-based methods such as were used in Bazargan et al (2002). An insufficient landing capacity will cause delays for arrivals, potentially requiring these to be 'parked' in stacks near the airports, as discussed in Section 7.7. Similarly, an insufficient take-off capacity can result in long queues at the runways or the application of stand holds to hold some aircraft at the stands/gates for longer.

The basic aim for runway sequencing is to determine a runway sequence (consisting of landings, take-offs or both) and landing/take-off times for each aircraft, so that all constraints are met and the value of some objective function (which depends upon the landing/take-off times) is improved. Sequence dependent separations must be applied between any two aircraft which use the runway, thus the runway capacity can be highly dependent upon the take-off or landing sequence. Mixed mode operations (where the runway is used for both arrivals and departures) are usually more efficient than segregated mode (see Newell (1979)), since the required gaps in the departure stream may only need to be slightly widened to fit an arrival between departures, and vice versa. It is also usually possible to trade off arrival throughput against departure throughput in this circumstance, see Gilbo (2003).

A recent review of runway scheduling research can be found in Bennell et al (2011). An earlier paper by Fahle et al (2003) compares some of the formulations for the arrival problem and Mesgarpour et al (2010) provides an overview of some of the solution approaches which have been applied. Various solution methods have been utilised, including exact methods such as dynamic programming (e.g. Psaraftis (1980); Trivizas (1998); Chandran and Balakrishnan (2007); Balakrishnan and Chandran (2010)), or branch and bound (e.g. Beasley et al (2000); Ernst et al (1999)) and heuristic methods such as genetic algorithms, especially when a non-linear objective function is used (e.g. Beasley et al (2001)).

Since runway sequencing is key for the case study in this chapter, a mathematical model of the problem is presented in this section, alongside the discussion of the problem. The exact objective function and constraints in operation will depend upon the airport, but the various elements are explained here and the specific elements which are important for London Heathrow are discussed in the case study.

7.1 Decision Variables and Constants

Since this model applies to both arrivals and departures, the concept of a runway time is used to denote either the landing time (for an arrival) or the take-off time (for a departure). The variables and constants can be summarised as follows, where the aim is to find values for s_j and d_j for each aircraft j:

s_j	Decision variable. The position of j in the planned runway sequence.
d_j	Decision variable. The predicted runway time of aircraft j.
et_j	Constant. The earliest runway time for aircraft j.
ht_j	Constant. A hard constraint upon the latest runway time for aircraft j, such that j MUST take-off or land before that time.
lt_j	Constant. A soft constraint upon the latest runway time for aircraft j such that schedules where it takes off or lands after this time are penalised.
bt_j	Constant. A base time for aircraft j from which the delay can be calculated. This will often be equal to et_j.
RS_{ij}	Constant. The minimum required separation between the runway times of aircraft i and j, when i uses the runway before j.
a_j	Constant. The position of j in a first-come-first-served sequence, for example by ordering the aircraft in increasing order of bt_j.

7.2 *Time Window Constraints*

Time windows may be either hard or soft constraints, indicating mandatory or preferred compliance. Hard time windows are potentially useful for reducing the problem complexity since they may implicitly specify a partial ordering for aircraft (e.g. if time windows do not sufficiently overlap), but soft time windows are not so helpful. When time windows are formulated as soft constraints they should be met when possible, but this may not always be achievable. When they are not met, it may be useful to minimise the amount by which they are missed.

7.2.1 Hard Constraints

Aircraft will usually have an earliest take-off or landing time, et_i, before which they will not be available, and this will usually be modelled as a hard constraint. These may be due to physical limitations upon how early an aircraft can reach the runway, due to flying time for arrivals or taxi time and line-up time for departures. Departures will sometimes perform final take-off preparations only once the taxi operation has commenced, and take-off cannot occur until these have been performed. There may also be latest runway time constraints, for example due to safe fuel considerations, on-board emergencies for arrivals (e.g. a seriously ill passenger), or regulations, although latest runway times are less common for departures than arrivals.

Issues of ensuring fairness in delay distribution between aircraft are usually far more common in preventing long delays for aircraft than fuel concerns. Due to suggested fuel loads, it should be rare for a properly prepared aircraft to be forced to land due to fuel shortages, and this is even more rare with take-offs, since the fuel-burn is far lower when idling on the ground. In general, however, equity/fairness considerations tend to be soft constraints, which prefer rather than enforce more equitable/fair sequences, although hard constraints are sometimes also applied, to limit either the latest runway times or the positional deviation from a first-come-first-served sequence.

An important consideration for European airports is the Calculated Take-Off Time (CTOT). Since European airspace is in high demand, restrictions may be applied at source airports to limit the load upon en-route sectors. The CFMU (Central Flow Management Unit) at Eurocontrol uses heuristic methods to estimate the load on sectors, considering the known flight plans of all aircraft passing the sectors. A calculation will be performed to ensure that the predicted load on sectors does not exceed an acceptable limit. In the case of likely congestion, the shortest take-off delay which would remove the capacity problem will be calculated and a ground delay will be applied to an aircraft by allocating it a Calculated Take Off Time (CTOT), as discussed in de Matosa and Ormerod (2000) and Eurocontrol (2012). Aircraft are permitted to take off up to five minutes before this time, or ten minutes after it, thus it applies both an earliest and latest runway time constraint upon these aircraft. Although the airspace in the US tends to be less restricted, bad weather can still require ground holds to be applied (see Hoffman and Ball (2000)), limiting the earliest take-off times for aircraft in a very similar way.

Together the earliest and latest landing/take-off time (if it exists) will imply a window around the potential runway times. Inequality 1 ensures that any hard time-window constraints are met.

$$et_j \leq d_j \leq ht_j \tag{1}$$

7.2.2 Soft Time Slots

At some airports the end of the CTOT time may actually be modelled as a soft constraint, since it may be possible for controllers to obtain a small extension for a time window. Since missing a time window is intrinsically bad, a large fixed penalty can be applied for each miss, preventing the adoption of such schedules where possible. It may also be useful to add an additional penalty related to the amount by which it is missed. In this way the aim is to reduce both the total number of misses and the amounts by which CTOTs are missed. Extending this, there could also be multiple penalty periods in some cases, whereby a small miss incurs a cost, but a miss by more than a certain time (e.g. more than a possible extension) incurs an extremely large cost.

Let lt_i and $l2_i$ denote the ends of two different time periods, whereby the aircraft should be scheduled before lt_i, but if that is impossible then it should be scheduled before $l2_i$. Let ω_2 and ω_4 be constant penalties which are applied for missing the lt_i and $l2_i$ deadlines, respectively. Let ω_1 and ω_3 be constant costs per second for missing the lt_i and $l2_i$ deadlines, respectively. The cost of CTOT compliance could then be expressed by Formula 2, which is a simplified version of the objective function used in Atkin (2008); Atkin et al (2007, 2012). Using a value of $\beta = 1$ will give a linearly increasing cost for misses. Values of $\beta > 1$ could be used to avoid larger misses, for example encouraging the adoption of two shorter misses rather than one larger miss. It is worth noting that Formula 2 is not convex, thus use of it as (part of) an objective function will be problematic for some solution methods.

$$E = \begin{cases} 0 & \text{if } et_i \leq d_i \leq lt_i \text{ (i)} \\ \omega_1 (d_i - lt_i)^\beta + \omega_2 & \text{if } lt_i < d_i \leq l2_i \text{ (ii)} \\ \omega_3((d_i - l_i)^\beta) + \omega_4 & \text{if } l2_i < d_i \quad \text{(iii)} \end{cases} \tag{2}$$

7.3 Separation Rules

Gaps are required between aircraft taking off, for a number of reasons:

Wake Vortex Separations: The act of producing lift from the aerofoil of a wing will result in wake vortices being produced behind the trailing edge of the wing. For safety reasons, minimum separations are always required between aircraft using the runway, in order to ensure that the wake vortices from the previous aircraft have had time to dissipate prior to the following aircraft passing that point. Aircraft are all assigned a weight class, according to their size, and the mandated separation times are larger when the leading aircraft is of a larger weight class than the following aircraft, since larger aircraft will produce larger wake vortices, but be affected less by wake vortices. Although the separation requirements vary between the arrival and take-off problems, both problems have these asymmetric separations, so the runway sequence can affect the throughput. Where only wake vortex separations apply (e.g. for landings), it is usually a good idea to group aircraft of the same weight class together in the sequence, thus reducing the number of larger separations due to a smaller weight class aircraft following a larger weight class aircraft.

Route and Speed Separations: Aircraft which take off from an airport will usually follow a SID (Standard Instrument Departure) route. Each SID mandates a specific conflict-free departure route, including details of when to make turns and the exit points for the TMA (the local airspace around the airport), so that aircraft need only an overview by a controller rather than approval for each turn or flight level change. This greatly reduces the workload and communication required of the controller. If SIDs are divergent, aircraft may be able to depart soon after each other, however if SIDs are close together, a longer gap may be required at take-off in order to ensure in-air separations and to control the workload for the downstream controllers to whom the SIDs will deliver aircraft. These SID separations may be further modified by the speed groups of the two aircraft, to allow for the gap closing or expanding in flight. The required separations for any pair of SIDs and speed groups can be determined using look-up tables. SID separations mean that it is common to alternate departure directions, and at some airports the SID separations can be much more constraining upon the problem than the wake vortex separations, as shown in Atkin et al (2009) for Heathrow. Importantly, route/speed-based separations mean that take-offs cannot usually be ordered by weight class, and do not always obey the triangle inequality, so it is not always possible to consider only the separations between adjacent take-offs.

Increased Separations: When local airspace becomes congested, or the local sector controller is at risk of having too high a workload, a short-term flow control measure can be applied at the runway by temporarily increasing the mandated separation for certain SIDs. For example, a separation which is normally two minutes could be increased to three or more minutes (or to as high as ten or more minutes in extreme cases), by applying a Minimum Departure Interval (MDI) restriction. SIDs could even be temporarily closed if the airspace becomes extremely busy or unavailable, for example due to an emergency flight or adverse weather.

Runway Occupancy Time: The runway occupancy time must also be considered since only one aircraft is permitted to be using the runway at any time. This can enforce a minimum separation between consecutive users of the runway even when other separations may not apply (e.g. between a take-off and a landing).

Modelling the Separation Rules

Regardless of the cause of the separation requirement, these constraints can be modelled using a single minimum separation value, RS_{ij} which gives the required separation between aircraft i and j if i uses the runway before j. RS_{ij} must be the maximum of all required separation values, including wake vortex separations, route and speed separations and runway occupancy times, to ensure that all are met. The separation time constraint can then be modelling as the disjunction in Formula 3, and the runway sequence can be determined from the order of the d_i or d_j values.

$$d_j \geq d_i + RS_{ij} \quad \text{OR} \quad d_i \geq d_j + RS_{ji} \quad \forall i \neq j \tag{3}$$

Many formulations of the disjunction will introduce auxiliary variables to handle the disjunction. Sequence variables, s_j, are obvious candidates, allowing the disjunction of Formula 3 to be formulated as Inequality 4. Alternatively, binary variables could be used to specify whether i or j comes first in the sequence (e.g. a variable k_{ij} for each pair of aircraft i and j, which is 1 if $d_i < d_j$ and 0 otherwise).

$$d_j \geq d_i + RS_{ij} \quad \forall i, j \text{ s.t. } s_i < s_j \tag{4}$$

It is worth noting that this model assumes that separations are constant over time. If this is not the case then RS_{ij} may actually have a dependency upon d_i, greatly complicating the model. In this case, take-off times must be predicted for aircraft in runway sequence order, so that d_i will already be known for all previous take-offs/landings, allowing RS_{ij} to be determined.

7.4 Runway Sequencing Objectives

The objectives for the runway sequencing problem can take various forms and will depend upon the airport and whether it is arrivals or departures which are being sequenced. The objectives can be summarised as aiming to minimise some combination of one or more of the following values, A, B, C or D:

$$A = max_j \, d_j \tag{5}$$

$$B = \sum_j d_j \tag{6}$$

$$C = \sum_j (s_j - a_j)^2 \tag{7}$$

$$D = \sum_j (d_j - bt_j)^\alpha \tag{8}$$

Maximise Throughput: Runway throughput (i.e. the number of aircraft per hour) is often assessed by considering the difference between the runway time of the first aircraft in the sequence and the runway time of the last aircraft in the sequence. Thus, the objective is usually to complete the landing/take-off of the last of the set of aircraft as early as possible, as expressed by Equation 5. Throughput can often be increased by re-sequencing, to reduce the total separations between aircraft.

Since Formula 5 only considers the landing/take-off time of the last aircraft, time windows can cause a problem for this objective: late availability of the final aircraft can greatly affect the runway time of that aircraft and the value of a sequence can become highly dependent upon the last aircraft rather than the sequencing of the preceding aircraft. This can hide any poor sequencing of the earlier aircraft, potentially delaying them unnecessarily as long as such delays do not delay the last take-off/landing. This is especially a problem in the dynamic case, where new aircraft can become available over time. Better sequencing of earlier aircraft may allow a new arrival to be slipped into a sequence ahead of the last aircraft, whereas poor sequencing may prohibit this. This is discussed further in Atkin (2008).

Reduce Delay: An alternative objective is to consider the total delay for aircraft, rather than considering the runway time of only the last aircraft. This formulation considers the times for all aircraft, aiming to reduce the sum of these times, as expressed by Equation 6. However, this formulation can introduce another problem. Certain types of aircraft (e.g. particularly slow or heavy aircraft) will have separation rules associated with them such that as soon as they are scheduled a larger separation is required. This objective will tend to move aircraft with larger separations later in the sequence, so that fewer aircraft will be affected by the increased delay. If not controlled, this can limit the fairness of the resulting sequences.

Avoid Inequity/Unfairness: Re-sequencing is vital for reducing delay, but it must be controlled to avoid unfairness. An obvious solution is to minimise positional delay, as expressed by Equation 7. Indeed, some previous research (e.g. Psaraftis (1980); Dear and Sherif (1989, 1991); Trivizas (1998); Balakrishnan and Chandran (2010)) has worked with a maximum position shift constraint, which applies a hard constraint to the positional shifts which any aircraft can have. This can be expressed as Inequality 9, where MPS is a constant for the the maximum allowed position shift:

$$a_j - MPS \le s_j \le a_j + MPS \tag{9}$$

This is an excellent idea for the purpose of limiting inequity and can greatly reduce the complexity of the problem which has to be solved. It appears to work well for arrivals in some cases, but does not work so well for departures, since the delays can vary greatly across departure routes, so it is important to allow arbitrarily large relative positional shifts between different departure routes. Similarly, when time-windows are involved, conformance with these can require large relative position shifts in the sequence, which must not be prohibited.

An alternative approach is to use a non-linear factor of delay in the objective function, to account for equity of delay as well as overall delay reduction, as expressed by Formula 8, where $\alpha > 1$. In this case, the value of α will determine the trade-off between delay reduction and equity. For obvious reasons, linear programming models have tended to avoid non-linear factors in the objective functions (although quadratic objective functions can now be handled more easily by some solvers than was previously possible), instead utilising hard constraints in terms of maximal delays or maximum positional delays to limit the inequity. However, previous experiments with non-linear delay costs have had good results in terms of controlling the inequity of delays, e.g. Beasley et al (2001); Atkin et al (2010a). A trade-off between equity of delay and overall delay was observed in Atkin et al (2010a), such that the overall delay increased as the pressure for more equitable delay was increased. A power of 2 (i.e. squaring the delay) appears to be popular for this (e.g. Beasley et al (2001)), however it has been suggested that this may be too high for departure sequencing (see Atkin et al (2010a)), since it may result in unnecessarily high overall delays.

Penalise Deviation from a Reference Schedule: Similar approaches could also be used to find results which are similar to a reference schedule, or to previous decisions in a dynamic problem (see Beasley et al (2004)), explicitly penalising positional shifts or changes in runway times.

Meet Any Timeslots: As previously mentioned, the cost of missing any timeslot which is modelled as a soft constraint would need to be included into the objective function. For example, in the form of Inequality 2.

7.5 Similarities with Other Problems

There has been significant research into the arrival sequencing problem, perhaps due to the similarities between the single runway sequencing problem and the Travelling Salesman Problem (TSP), which has itself been well studied for many years (see Bellmore and Nemhauser (1968), Laporte (1992)). The TSP is equivalent to the single machine scheduling problem with sequence-dependent processing or set-up times, see Bianco et al (1988). The main requirement for problem equivalence is that separations must require looking back only to the immediately preceding aircraft, thus the arrival sequencing problem (where this is usually the case) is much more similar to it than the departure sequencing problem is (where separations often do not obey the triangle inequality). However, since reducing delay is usually a more appropriate objective function than directly maximising the throughput, the problem is actually closer to a cumulative travelling salesman problem, see Bianco et al (1993), Bianco et al (1999).

7.6 *Multiple Runways and Runway Allocation*

When there are multiple arrival or departure runways, a runway allocation problem must be considered in addition to the runway sequencing problem for each runway. In some cases aircraft will have already been allocated to runways according to some other criterion, for instance allocating each aircraft to a runway according to its arrival or departure route (avoiding aircraft paths crossing, with the consequent interactions between runways) or according to the terminal at which the aircraft is parked (to reduce taxi times, de-conflict traffic to simplify the taxi operations, or to avoid runway crossings, reducing inter-runway dependencies). In this case, if there are no other dependencies between the runways, each single-runway sequencing problem can be solved independently. Where there are dependencies between runways, but runway allocations are known, the entire problem can be solved as single sequencing problem, with the RS_{ij} values considering the separations which are required between two aircraft given the runways to which they have been allocated, taking into account the inter-runway dependencies.

If runway allocation also has to be performed, the problem is more complex. In general, the multiple runway problem with runway allocation is analogous to a Vehicle Routing Problem in the same way that the single runway problem is analogous to the Travelling Salesman Problem, since runway allocation is analogous to allocating customers to vehicles. Approaches which add one aircraft at a time to a sequence, utilising sub-sequences could dynamically adapt the RS_{ij} values according to the allocated runways for each aircraft as they are added, and branch upon the options for the runway for the next aircraft. This includes any method which already utilises a pruned tree-search, branching upon the options for next aircraft, such as Atkin et al (2012). However, the increase in problem size will ultimately be exponential in the number of aircraft, so good pruning would probably be needed for larger problems. Some previous work has successfully considered the runway allocation problem for arrivals (e.g. Ernst et al (1999), where a Mixed Integer Linear Programming model was used), however the majority of work has considered only single runway or fixed allocation problems.

7.7 *Arrival Scheduling and Stacks*

Arrival runway throughput may not always keep up with the arrival rate at busy airports. Unlike on the ground, aircraft cannot simply be parked stationary in the air to await a landing time, so instead they circle near beacons, in stacks, one above the other. Each level of a stack will be dedicated to a specific aircraft, so that there can be no conflicts between aircraft regardless of their airspeed (they maintain a vertical separation). Controllers will usually take aircraft from the bottom of a stack, or

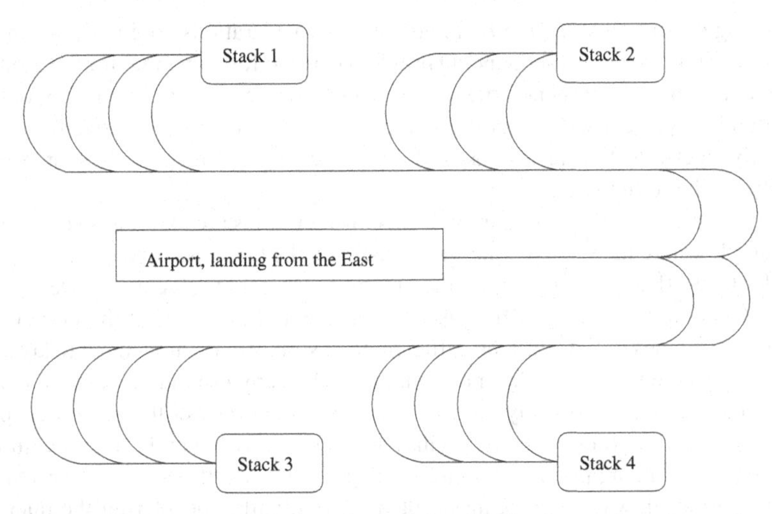

Fig. 1 Stylised diagram of arrival routes from stacks, showing how trajectories can be varied in order to achieve desired separations at the runway

occasionally from the second or higher level if it is simple to do so and there are obvious benefits. The aircraft will then follow merging trajectories to the runway, to slot into the landing sequence. The remaining aircraft in the stacks will then spiral down to fill the vacant levels. An example, stylised, diagram of a stack structure is shown in Figure 1, showing how aircraft can make turns at different points in order to shorten or lengthen trajectories as required (see Smith (1998)), in addition to being able to accelerate or decelerate to achieve the correct spacing.

The arrival stacks add extra constraints to the sequencing problem for arrivals, which may help or hinder solution approaches, depending upon their flexibility. If only the bottom aircraft can be taken, these can potentially simplify the problem significantly, since it could become a problem of interleaving/merging queues, with some restrictions upon the frequency with which each queue can be used, to allow aircraft time to spiral down to fill gaps. However the ability to take aircraft from other levels when it is useful can add significant complexity to the problem. Although it currently appears that significant throughput increases could result from taking aircraft from higher levels, it is extremely important to understand the effects of this kind of decision, in terms of equity of delay and increased workload for controllers before this could be assumed.

7.8 Combined Runway Sequencing and Ground Movement Problems

For arrival sequencing, the landing times tend to feed into the ground movement problem as the times at which aircraft enter the ground movement system. However, it would be rare for ground movement considerations to have an effect upon the landing sequence, since the fuel and equity considerations tend to be far more important. Thus the arrival sequencing problem tends to form an input for the ground movement problem. This is not true for the take-off sequencing problem. Obviously, the ground movement will occur prior to take-off, thus the ground movement will implicitly determine the earliest take-off time for each aircraft (i.e. how soon it will actually reach the runway).

At airports where the ground movement considerations are extremely important, for example where there are distinct bottlenecks on the taxiways, it may be useful to consider take-off sequencing within the ground movement problem, e.g. Deau et al (2008), Clare and Richards (2011). At other airports, the runway is the bottleneck and runway queues will usually be sufficient to absorb ground movement delays, so the runway sequence may be far more important than any ground movement. However, in these cases the runway queues themselves may restrict the re-sequencing. In some cases, the order of aircraft within a runway queue may be fixed, with re-sequencing only being possible by interleaving the queues. It may then be possible to simplify the runway sequencing problem by explicitly considering the queues, e.g. Bolender and Slater (2000); Bolander (2000). The case study in the next section considers a less restrictive runway queue problem, where the runway queues affect the potential take-off sequences, but are insufficiently constraining to simplify the problem.

Finally, at some airports it may be necessary to cross a runway in order to reach another runway. In these cases the runway crossing time may need to be considered in the landing or take-off sequencing, allowing sufficient gaps to perform the crossings (see Anagnostakis and Clarke (2002)), introducing dependencies between the ground movement and runway sequencing problems.

8 Case Study: Heathrow Departure Sequencing at the Runway

This case study considers a system which was developed to potentially provide decision support for departure runway controllers at London Heathrow. It considers a combined runway sequencing and local ground movement problem, and utilises a hybrid solution method to solve the overall problem. This research was funded by NATS (formerly National Air Traffic Services) and the Engineering and Physical Sciences Research Council (EPSRC) through the Smith Institute for Industrial Mathematics and Systems Engineering. Two questions were posed: "Could a decision support system be developed to give 'real time' advice to the runway controller, given the complexity of the task?" and "Is there likely to be any benefit from using such a system?"

London Heathrow is an extremely popular and busy two-runway airport. Its location, close to London, means that aircraft fly over built-up areas, and thus it is important to consider the noise for residents near the flight paths. A noise-control policy means that each of the runways may only be used for arrivals at certain times of the day, with the other runway being used for departures. The runways are, therefore, normally used in segregated mode and the departure runway controller has the task of attempting to find the best runway sequence from the aircraft which are available to him/her.

Since taxi times have rarely been entirely predictable, with delays around the stands being especially hard to predict unless they are explicitly modelled (see Atkin et al (2012)), the operational method at the time was to release aircraft from the stands as soon as possible, unless there was obvious congestion already, taxi them to a holding area near the end of the current departure runway and have them wait there for the controller to re-sequence them for take-off. The controller would normally consider only the aircraft which were available at the time, along with the locations of these aircraft in the holding area, and would attempt to build a good take-off sequence, directing aircraft through the holding area to achieve the sequence.

The holding areas are normally a set of one or more queues, with possibilities to move aircraft between queues, in order to overtake or be overtaken. Directed graph models for two of the holding areas are provided in Figures 2 and 3. Each node in the graph is assumed to be able to hold exactly one aircraft, and the arcs show the normal paths through the holding areas. Re-sequencing will be successful if aircraft can be moved from their current positions onto the runway, in the take-off order, without ever requiring multiple occupancy of any node. Importantly, not all good take-off sequences will be achievable, due to the limited re-sequencing which is possible within the holding area, although many sequences will be achievable in multiple ways. The controller has to consider the work involved in achieving the sequence as well as the feasibility of performing the re-sequencing. Any decision support tool also has to consider whether the re-sequencing will appear to be sensible to the controller (i.e. will be likely to be adopted). In general, these factors will depend upon both the absolute paths which are taken through the holding area (avoiding difficult paths), and the relative paths which are taken (ensuring that the longer paths are taken by aircraft which have the time available to do so).

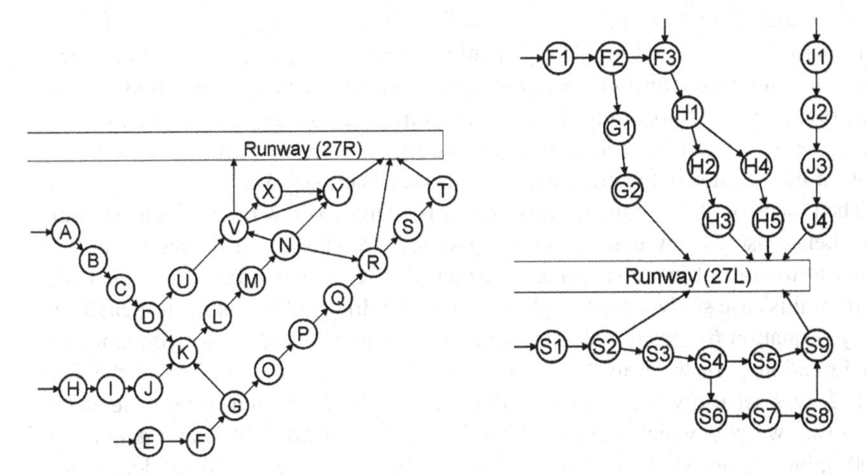

Fig. 2 Diagram of the 27R holding area

Fig. 3 Diagram of the 27L holding area

This problem can be considered to be a combination of a runway sequencing problem plus a small ground movement problem. The individual problems are discussed in Sections 8.1 and 8.2, respectively, before the combined problem is discussed in Section 8.3. The main question is how best to solve the combined problem and the developed solution method will be discussed and justified in Section 8.4. Some results are then presented and discussed in Section 8.5, evaluating the potential benefits of such a system. A more detailed discussion of the holding area sequencing problem, the modelling of the holding areas and the reasons for selecting the chosen solution method, can be found in Atkin (2008) and Atkin et al (2007).

8.1 The Runway Sequencing Sub-problem

The runway sequencing problem at Heathrow involves finding the best take-off sequence, given the available aircraft and their current positions on the airport surface. It is a variant of the general runway sequencing problem which was described in Section 7. The key elements which apply to Heathrow are explained in this section.

Take off separations apply, depending upon the weight class, speed group and departure route of the aircraft (see Section 7.3). These do not obey the triangle inequality and are not symmetric.

Given the position of each aircraft in the holding area, it is possible to determine how long it would take it to reach the runway, line up and take off. It is also relatively easy to estimate taxi times for aircraft which are taxiing along the taxiways towards the runway, once they have left the cul-de-sacs, and to predict earliest take-off times for these aircraft. These determine the earliest take-off time, et_j, for any aircraft j.

CTOTs (Calculated Take-Off Times) are extremely important, as discussed in Section 7.2, and specify a fifteen minute take-off window. The earliest take-off time was enforced as a hard constraint in this work, by modifying the earliest take-off time (et_j) accordingly. As discussed in Section 7.2, the end time was modelled as a soft constraint in this work, since some CTOTs may be tight and the controllers were permitted a limited number of 5 minute extensions, to use only where necessary to smooth the airport operations and avoid the need for last minute CTOT renegotiations. Therefore, hard constraints on the latest take-off times are assumed not to apply to aircraft (so $ht_i = \infty$ in Inequality 1), since aircraft may have a long wait for busy routes in the presence of MDIs (see Section 7.3).

The objective of the sequencing is primarily to miss as few time-slots as possible (and hence use as few extensions as possible), secondarily to reduce delay, and tertiarily to control inequity in the sequencing. The objective function which was used for this case study was a weighted sum of the different components, penalising delay (Equation 6), positional delay (Equation 7) and CTOT misses (Equation 2). For Equation 2, lt_i is set to the CTOT time plus ten minutes (the end of the slot) and $l2_i$ was set to five minutes later than that (i.e. the duration of the extension). The exact weights which were used can be found in Atkin (2008) and Atkin et al (2007), but are not vital for this discussion. A linear cost was used for delay since it was discovered that, with the positional delay penalty, the holding area structure itself enforced a degree of equity upon the sequences, making a non-linear cost unnecessary. However, in the absence of the sequencing constraints enforced by the holding area structure, a power for delay of 1.5 was found to be more appropriate for runway sequencing, see Atkin et al (2010a).

8.2 The Ground Movement Sub-problems

The ground movement element of the problem introduces a number of other objectives, which were identified from discussion with controllers and can be summarised as follows:

1. Ensure that the re-sequencing which is required is possible within the holding area structure. This is a hard constraint, which can be tested using the directed graph model of the holding area.
2. Ensure that the re-sequencing is simple to achieve. This objective turned out to be related to the choice of paths which aircraft use to traverse the holding area, and can be considered to mean that certain paths would never be used (e.g. due to the number of turns which would have to be made) and other paths would only ever be used if there was no way to use a simpler path. The acceptable routes were identified by discussing the problem with the controllers, and from identifying the routes which were actually used in practice, and the circumstances under which they would be used, by considering the playback of historic ground positional data.
3. Ensure that re-sequencing is sensible to the controllers. In addition to the route elimination in point 2, which will help for this, it is also important to consider the routes which are allocated in relation to the re-sequencing which is being performed. For example, if one aircraft is to overtake another in the holding area, the overtaking aircraft will obviously be in the holding area for less time than the overtaken aircraft, thus the overtaking aircraft should be assigned to the shorter routes, reducing the probability of delay. In effect, this means that overtaken aircraft are held out of the way, rather than the overtaking aircraft having to go around them.

8.3 The Combined Problem

Considering the combined problem, a number of observations can be made:

1. The solution space for the sequencing problem alone is of size $n!$ for n aircraft.
2. The solution space for the ground movement problem, with the flexible holding areas shown in Figures 2 and 3 was even larger: most feasible sequences could be achieved in many different ways, greatly outweighing the fact that only a fraction of the potential sequences are feasible.
3. Many sequences are unachievable given the structure of the holding area. In particular, there are often limitations upon how many other aircraft any aircraft

can overtake. Overtaking other aircraft can also limit the relative re-sequencing which can be performed upon the overtaken aircraft.

4. There will usually be a single preferred method in which any specified re-sequencing should be performed (i.e. a specific solution for the ground movement problem), which will minimise the workload for the controllers and pilots.
5. Most objectives are related to the take-off sequence rather than the means by which it was achieved. The ground movement problem effectively determines whether a sequence can be achieved easily, and if so how best to achieve it.

There was a requirement for a successful decision support system to return results 'instantly', which was measured as 'within a second'. Heuristic methods were considered for that reason and local search algorithms were chosen.

8.4 Solution Method

Despite the success of research into the combined problem where the ground movement was the main constraint, as discussed in Section 7.8, experiments with using a local search algorithm to manipulate the ground movement sequencing and evaluate the resulting take-off sequences, were not very successful. Various neighbourhoods were investigated, consisting of changing the allocated path for an aircraft, changing the relative sequence in which aircraft pass points, and changing part of an allocated route from a decision point onwards (i.e. changing the arc which an aircraft uses to leave a node, along with later arcs), however various problems were discovered:

- It is not simple to make small isolated changes to the movement. A single change of order, or path, at one point in the holding area could require many subsequent changes, since aircraft may no longer be in appropriate positions for the previous subsequent movement.
- It was not easy to guide the search using the ground movement. The runway sequence identifies the cost of the solution, however, relatively small changes in the ground movement could have far-reaching effects upon the consequent take-off sequence, due to the enforced changes upon subsequent holding area movement.
- Even when good solutions were found, the method of achieving the sequence was often not simple. Many theoretically good sequences required far too much manoeuvring in the holding area, and hence had to be rejected under that criterion.

The sequencing and routing problems are intrinsically linked. The ground movement objective (objective 3 in Section 8.2) indicates that the appropriate routing will be highly dependent upon the re-sequencing which is performed. A decision was made to solve the sequencing problem as the master problem, and to determine the feasibility of the ground movement as a subordinate problem. A tabu search algorithm was developed which had the following properties:

- The search was repeated for 200 iterations.
- In each iteration, 50 neighbouring solutions were randomly generated, by randomly selecting a move type then randomly generating an alternative solution using that move type. Three move types were utilised:
  - Swap two aircraft: select two random aircraft and swap their positions in the sequence. This was used 30% of the time.
  - Shift aircraft: select a consecutive sequence of between 2 and 5 aircraft and, maintaining their relative order, shift them to another position in the sequence. This was used 50% of the time.
  - Randomise order: select a consecutive sequence of between 2 to 5 aircraft and randomly re-order them. This was used 20% of the time.
- Whenever a move was made, the old positions of each aircraft which moved were recorded. Any move within the next ten iterations which put all of these aircraft back into the original positions was prohibited, by declaring it tabu, unless it improved upon the best solution found so far.
- Each remaining solution was tested for feasibility using the ground movement model, and infeasible solutions were rejected from further consideration.
- The best solution found in each iteration was adopted for the next iteration.
- The best solution found during the search was returned at the end.

Full details of the ground movement element, including the path allocation and routing algorithms, can be found in Atkin (2008). It can be summarised as follows:

1. Separately consider the set of aircraft for each of the holding area entrances. Heuristically allocate paths to these aircraft such that their relative re-sequencing can be performed sensibly. e.g. if one has to overtake another, then allocate the overtaken aircraft a path which moves it out of the way.
2. Move aircraft through the holding area graph, one arc at a time. The next node is known for each aircraft (based upon the allocated path), so the only decision which has to be made is the order in which aircraft should enter a node.
3. The input sequence of aircraft at the holding area entrances is fixed. The target take-off sequence fixes the relative sequence for all aircraft at the holding area exits. At intermediate points, all aircraft with a common initial path must maintain their relative entrance sequence and all aircraft with a common set of nodes at the end of their path must maintain their relative exit sequence. These common sub-paths are identified and used by the algorithm.
4. A look-ahead algorithm is used, counting the number of empty nodes later on each path (i.e. vacant nodes which an aircraft could be moved into) to determine whether moving an aircraft will block another aircraft from exiting to the runway at the correct time. Using the look-ahead count and the partial sequences from (3) above, an algorithm is applied to move each aircraft in turn so long as doing so does not block other aircraft. Pre-processing ensures that this algorithm is very fast; fast enough to execute within each iteration of the tabu search.

Experiments showed that the chosen moves were able to move from one good solution to another very easily. A local search algorithm, without the tabu list, also performed very well, showing that this neighbourhood was very good for this problem. However, adding the tabu list made the search perform (statistically) significantly better. The success relied upon a fast and successful solution for the ground movement sub-problem, and upon the search only needing to move within the feasible region of the search space (i.e. it ignores solutions which cannot be achieved), which required that the feasible search space was connected.

8.5 Summary of Results

The developed algorithm was executed using historic data (ten half-day datasets) and each of the three commonly used holding area structures; the two previously shown and a much simpler structure for the 09R runway. A simple simulation was built, which assumed that the system was executed once a minute (for less than a second each time), considering all of the aircraft which it would have been aware of at the time. An assumption was made that the sequences would be adopted by the controller, and that the movement through the holding area which the algorithm suggested would actually be performed. Each problem therefore had the aircraft from the previous iteration, plus any new aircraft which entered the system in the last minute, less any aircraft which would have taken off by that time.

Figure 4 shows the percentage of delay which the system was able to achieve, in comparison to the real delay on the day, for each of ten datasets and each of three holding area structures. These results assume that the system could be given

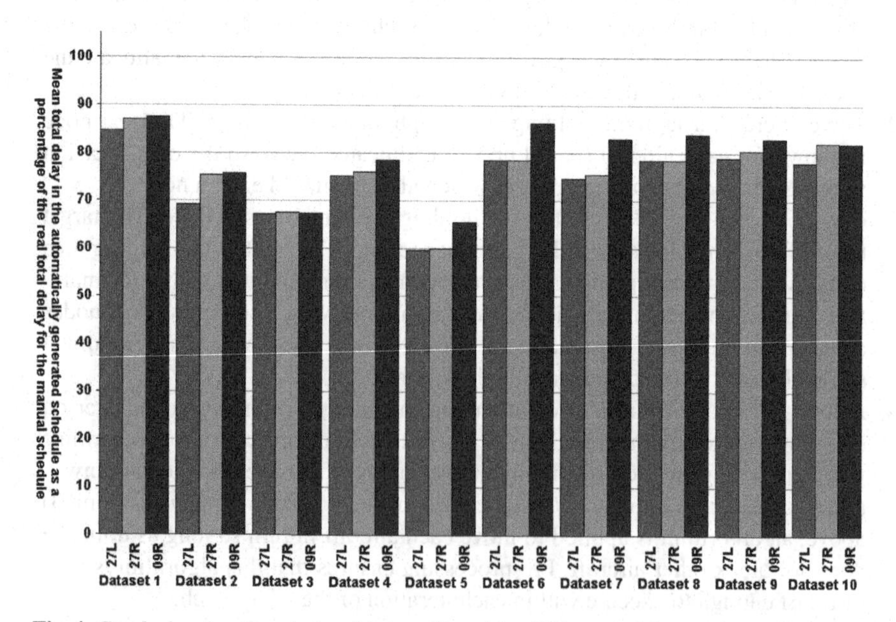

Fig. 4 Graph showing the relative delays achieved for different holding area structures across ten different datasets

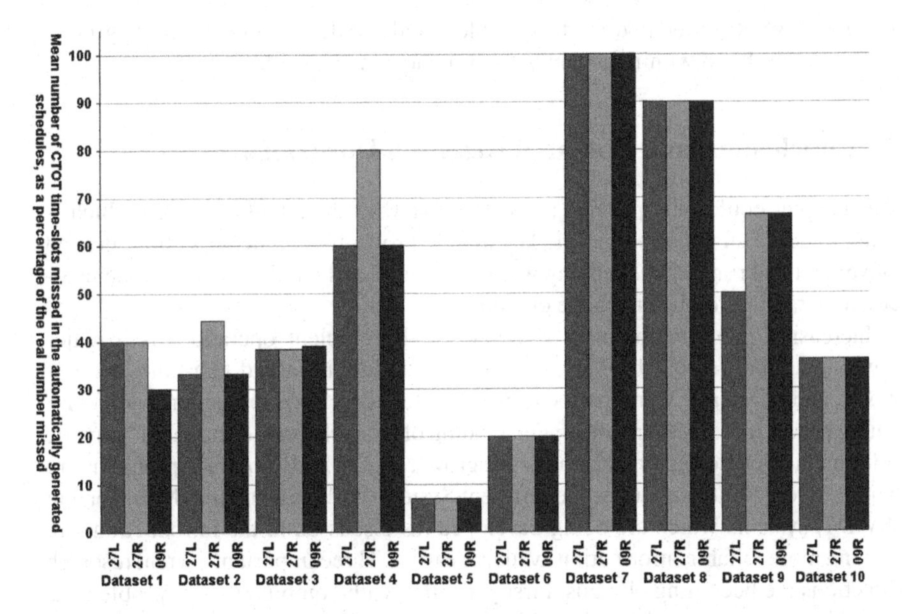

Fig. 5 Graph showing the relative delays achieved for different holding area structures across ten different datasets

knowledge of taxiing aircraft 15 minutes before they arrived at the holding area, or at pushback time if the taxi time was less than 15 minutes, since pushback times were thought to be highly unpredictable until they actually occurred. In all cases, the system found sequences which had lower delays than the controller-generated sequences, indicating that there was at least a potential for a developed system to help the controllers to improve delays.

Further research showed that greater knowledge of taxiing aircraft increased the performance of the system, as discussed in Atkin et al (2006). Results with no knowledge of taxiing aircraft were roughly equivalent to those which controllers produced, indicating that the controllers were performing very well and that the benefits came from considering the aircraft which were still taxiing at the time. As long as the system was made aware of aircraft, there were significant benefits even when only approximate taxi time predictions were available, as shown in Atkin et al (2008), although increased prediction accuracy improved the benefits. Taxi time prediction research is of particular interest for this reason, and current research (see Ravizza et al (2012a)) using data from Heathrow indicates that taxi times can be predicted relatively accurately if sufficient factors are taken into account.

Figure 5 shows the percentage of CTOTs which were missed (i.e. for which extensions were required) compared with the number that were actually missed by the real controllers. This shows that the system was able to achieve the lower delays by using no more CTOT extensions than the controllers, and usually far fewer.

In summary, these results show that the system was able to perform extremely well, even with the tight execution time constraints. They justify both the choice of a heuristic solution method for this problem and the decision to solve the ground movement problem within the tabu search sequencing algorithm.

9 Conclusions and Potential Research Directions

This chapter could only provide a short overview of the airside research which is being undertaken for airports. The key airside problems of gate allocation, ground movement and runway sequencing were considered and the differences in the models, constraints and objectives are obvious.

Increasing pressure for more environmentally efficient operations is a major driver for airport research. Reductions in delays, fuel burn and taxi times are obvious candidates, and were the focus of the system described in Section 8. As computing power increases, and the understanding of exact, heuristic and meta-heuristic solution method design and performance grows, increasingly complex problems are being solved. Simultaneously, the different System-Wide Information Management (SWIM) systems which are being developed (as discussed in the introduction) are making more reliable information available earlier. Three important current research directions are becoming obvious. Firstly, dealing with combinations of problems is becoming increasingly feasible and important, where the solution of one problem at least considers the effects upon other elements of the system, to obtain better global solutions and/or consider the objectives of multiple stake holders. Secondly, considering complex real-world constraints rather than academic simplifications is becoming more common, such as the consideration of non-linear functions of delay rather than makespan reduction, or multiple objectives, perhaps weighted together. Thirdly, it is becoming increasingly necessary to deal with the uncertainty in real world dynamic problems, since predicted data usually has some inaccuracy associated with it which needs to be handled. As models become more complex and accurate, they are likely to be of increasing use to airports, with fewer and fewer special cases which are not handed by the systems having to be considered.

This chapter included a case study which described a solution method that combines a ground movement and runway sequencing algorithm. Further research, detailed in Atkin et al (2012), which considered the task of predicting take-off times while aircraft are still at the stands and using this to allocate stand holds, has already been successfully implemented at Heathrow. A particularly interesting extension would be to integrate together the algorithms from the case study and Atkin et al (2012), potentially extending the ground movement model to span the area between the two. The combination of larger ground movement problems with runway sequencing has already been considered for airports where the runway sequencing is less sensitive to the selection of aircraft than at Heathrow (for example when mixed mode runways are used), as discussed in Section 7.8, and developing methods to do this for airports such as Heathrow has obvious benefits.

When dealing with uncertain data, reducing the uncertainty by improving prediction methods can help. This could involve introducing factors which may have previously been ignored, such as those which influence taxi speeds (see Ravizza et al (2012a)) or the modelling of the contention in the cul-de-sacs (see Atkin et al (2012)). However, decision support tools will need to explicitly deal with the remaining imprecision and unreliability of data, so deterministic models are unlikely to be sufficient for long. Similarly, the ability to not only have slack to absorb delays, but also to have alternative recovery solutions, with minimal changes, has obvious value, such as the airline practice of swapping aircraft between flights.

In summary, there is still significant potential for researchers to make an impact on these real world problems, building upon the existing research.

Acknowledgements. I wish to thank EPSRC and NATS (formerly National Air Traffic Services) for funding the research which was discussed in the case study, and the Smith Institute for Industrial Mathematics and System Engineering, for facilitating the original work as well as the later work on arrival scheduling. I would also like to thank Manchester Airport Group and Flughafen Zürich AG for providing the knowledge and data to help various students with airport scheduling and optimisation problems. I would particularly like to thank John Greenwood (NATS) for his constant help and feedback in my considerations of airport operations.

References

Abdelghany, A., Abdelghany, K., Narasimhan, R.: Scheduling baggage-handling facilities in congested airports. Journal of Air Transport Management 12, 76–81 (2006)

Anagnostakis, I., Clarke, J.P.: Runway operations planning, a two-stage heuristic algorithm. In: Proceedings of the AIAA Aircraft, Technology, Integration and Operations Forum, Los Angeles, CA, report AIAA-2002-5886 (2002)

Andersson, K., Carr, F., Feron, E., Hall, W.D.: Analysis and modelling of ground operations at hub airports. In: Proceedings of the 3rd USA/Europe Air Traffic Management R&D Seminar, Napoli, Italy (2000)

Ascó, A., Atkin, J.A.D., Burke, E.K.: The airport baggage sorting station allocation problem. In: Proceedings of the 5th Multidisciplinary International Scheduling Conference (MISTA 2011), Phoenix, Arizona, USA (2011)

Atkin, J.A.D.: On-line decision support for take-off runway scheduling at london heathrow airport. PhD thesis, The University of Nottingham (2008)

Atkin, J.A.D., Burke, E.K., Greenwood, J.S., Reeson, D.: The effect of the planning horizon and freezing time on take-off sequencing. In: Proceedings of the 2nd International Conference on Research in Air Transportation (ICRAT 2006), Belgrade, Serbia, Montenegro (2006)

Atkin, J.A.D., Burke, E.K., Greenwood, J.S., Reeson, D.: Hybrid meta-heuristics to aid runway scheduling at London Heathrow airport. Transportation Science 41(1), 90–106 (2007)

Atkin, J.A.D., Burke, E.K., Greenwood, J.S., Reeson, D.: On-line decision support for take-off runway scheduling with uncertain taxi times at London Heathrow airport. The Journal of Scheduling 11(5), 323–346 (2008)

Atkin, J.A.D., Burke, E.K., Greenwood, J.S., Reeson, D.: An examination of take-off scheduling constraints at London Heathrow Airport. Public Transport 1, 169–187 (2009)

Atkin, J.A.D., Burke, E.K., Greenwood, J.: TSAT allocation at London Heathrow: the relationship between slot compliance, throughput and equity. Public Transport 2, 173–198 (2010a)

Atkin, J.A.D., Burke, E.K., Ravizza, S.: The airport ground movement problem: Past and current research and future directions. In: Proceedings of the 4th International Conference on Research in Air Transportation (ICRAT 2010), Budapest, Hungary, pp. 131–138 (2010)

Atkin, J.A.D., Burke, E.K., Greenwood, J.S., Reeson, D.: Addressing the Pushback Time Allocation Problem at Heathrow airport. Transportation Science (2012) (to appear), `http://dx.doi.org/10.1287/trsc.1120.0446`

Balakrishna, P., Ganesan, R., Sherry, L.: Accuracy of reinforcement learning algorithms for predicting aircraft taxi-out times: A case-study of tampa bay departures. Transportation Research Part C: Emerging Technologies 18(6), 950–962 (2010)

Balakrishnan, H., Chandran, B.G.: Algorithms for scheduling runway operations under constrained position shifting. Operations Research 58(6), 1650–1665 (2010)

Barnhart, C., Belobaba, P., Odoni, A.R.: Applications of operations research in the air transport industry. Transportation Science 37(4), 368–391 (2003)

Bazargan, M., Fleming, K., Subramanian, P.: A simulation study to investigate runway capacity using taam. In: Proceedings of the Winter Simulation Conference, San Diego, California, USA, vol. 2, pp. 1235–1243 (2002)

Beasley, J.E., Krishnamoorthy, M., Sharaiha, Y.M., Abramson, D.: Scheduling aircraft landings - the static case. Transportation Science 34, 180–197 (2000)

Beasley, J.E., Sonander, J., Havelock, P.: Scheduling aircraft landings at London Heathrow using a population heuristic. Journal of the Operational Research Society 52, 483–493 (2001)

Beasley, J.E., Krishnamoorthy, M., Sharaiha, Y.M., Abramson, D.: Displacement problem and dynamically scheduling aircraft landings. Journal of the Operational Research Society 55(1), 54–64 (2004)

Bellmore, M., Nemhauser, G.L.: The traveling salesman problem: A survey. Operations Research 16, 538–558 (1968)

Bennell, J., Mesgarpour, M., Potts, C.: Airport runway scheduling. 4OR: A Quarterly Journal of Operations Research 9, 115–138 (2011)

Bianco, L., Ricciardelli, S., Rinaldi, G., Sassano, A.: Scheduling tasks with sequence-dependent processing times. Naval Research Logistics 35, 177–184 (1988)

Bianco, L., Mingozzi, A., Ricciardelli, S.: The travelling salesman problem with cumulative costs. Networks 23, 81–91 (1993)

Bianco, L., Dell'Olmo, P., Giordani, S.: Minimizing total completion time subject to release dates and sequence-dependent processing times. Annals of Operations Research 86, 393–415 (1999)

Blum, C., Roli, A.: Metaheuristics in combinatorial optimization: Overview and conceptual comparison. ACM Computing Surveys 35(3), 268–308 (2003)

Bolander, M.A.: Scheduling and control strategies for the departure problem in air traffic control. PhD thesis, University of Cincinnati (2000)

Bolat, A.: Assigning arriving flights at an airport to the available gates. Journal of the Operational Research Society 50(1), 23–34 (1999)

Bolat, A.: Models and a genetic algorithm for static aircraft-gate assignment problem. Journal of the Operational Research Society 52(10), 1107–1120 (2000a)

Bolat, A.: Procedures for providing robust gate assignments for arriving aircrafts. European Journal of Operational Research 120(1), 63–80 (2000b)

Bolender, M.A., Slater, G.L.: Analysis and optimization of departure sequences. In: Proceedings of the AIAA Guidance, Navigation and Control Conference and Exhibit, Denver, CO, pp. 1672–1683 (2000)

Burke, E.K., Causmaecker, P.D., Maere, G.D., Mulder, J., Paelinck, M., Berghe, G.V.: A multi-objective approach for robust airline scheduling. Computers & Operations Research 37(5), 822–832 (2010)

Chandran, B., Balakrishnan, H.: A dynamic programming algorithm for robust runway scheduling. In: Proceedings of the American Control Conference, New York, USA, pp. 1161–1166 (2007)

Chen, J., Ravizza, S., Atkin, J.A.D., Stewart, P.: On the utilisation of fuzzy rule-based systems for taxi time estimations at airports. In: Proceedings of the 11th Workshop on Algorithmic Approaches for Transportation Modelling, Optimization, and Systems (ATMOS 2011), Saarbrücken, Germany. OpenAccess Series in Informatics (OASIcs), vol. 20, pp. 134–145 (2011)

Clare, G.L., Richards, A.G.: Optimization of taxiway routing and runway scheduling. IEEE Transactions on Intelligent Transportation Systems 12(4), 1000–1013 (2011)

Clausen, J., Larsen, A., Larsen, J., Rezanova, N.J.: Disruption management in the airline industry–concepts, models and methods. Computers & Operations Research 37(5), 809–821 (2010)

Dear, R.G., Sherif, Y.S.: The dynamic scheduling of aircraft in high density terminal areas. Microelectronics and Reliability 29(5), 743–749 (1989)

Dear, R.G., Sherif, Y.S.: An algorithm for computer assisted sequencing and scheduling of terminal area operations. Transportation Research Part A, Policy and Practive 25, 129–139 (1991)

Deau, R., Gotteland, J.B., Durand, N.: Runways sequences and ground traffic optimisation. In: Proceedings of the 3nd International Conference on Research in Air Transportation (ICRAT 2008), Fairfax, VA, USA (2008)

Ding, H., Lim, A., Rodrigues, B., Zhu, Y.: The over-constrained airport gate assignment problem. Computers & Operations Research 32(7), 1867–1880 (2005)

Dorndorf, U., Drexl, A., Nikulin, Y., Pesch, E.: Flight gate scheduling: state-of-the-art and recent developments. Omega 35(3), 326–334 (2007a)

Dorndorf, U., Jaehn, F., Chen, L., Hui, M., Pesch, E.: Disruption management in flight gate scheduling. Statistica Neerlandica 61(1), 92–114 (2007b)

Dorndorf, U., Jaehn, F., Pesch, E.: Modelling robust flight-gate scheduling as a clique partitioning problem. Transportation Science 42(3), 292–301 (2008)

Dorndorf, U., Jaehn, F., Pesch, E.: Flight gate scheduling with respect to a reference schedule. Annals of Operations Research 194(1), 177–187 (2012)

Ernst, A.T., Krishnamoorthy, M., Storer, R.H.: Heuristic and exact algorithms for scheduling aircraft landings. Networks 34, 229–241 (1999)

Eurocontrol. Air traffic flow & capacity management operations, ATFCM Users Manual, edition 16.0 (2012)

Fahle, T., Feldmann, R., Götz, S., Grothklags, S., Monien, B.: The aircraft sequencing problem. In: Klein, R., Six, H.-W., Wegner, L. (eds.) Computer Science in Perspective. LNCS, vol. 2598, pp. 152–166. Springer, Heidelberg (2003)

Filar, J.A., Manyem, P., White, K.: How airlines and airports recover from schedule perturbations: a survey. Annals of Operations Research 108, 315–333 (2001)

Gendreau, M., Potvin, J.Y.: Metaheuristics in combinatorial optimization. Annals of Operations Research 140, 189–213 (2005)

Gilbo, E.P.: Arrival/departure tradeoff optimisation: a case study at the st. louis lambert international airport (stl). In: Proceedings of the 5th USA/Europe Air Traffic Management R&D Seminar, Budapest, Hungary (2003)

Glover, F.: Tabu search - part i. ORSA Journal on Computing 1(3), 190–206 (1989)

Glover, F.: Tabu search - part ii. ORSA Journal on Computing 2(1), 4–32 (1990)
Glover, F., Kochenberger, G. (eds.): Handbook of Metaheuristics. Kluwer Academic Publishers (2002)
Glover, F., Laguna, M.: Tabu Search. Kluwer Academic Publishers (1997)
Goldberg, D.E.: Genetic Algorithms in Search Optimization and Machine Learning. Addison Wesley (1989)
Gotteland, J.B., Durand, N., Alliot, J.M., Page, E.: Aircraft ground traffic optimization. In: Proceedings of the 4th USA/Europe Air Traffic Management Research and Development Seminar, Santa Fe, NM, USA (2001)
Hillier, F.S., Lieberman, G.J.: Introduction to Operations Research, 9th edn. McGraw-Hill (2010)
Hoffman, R., Ball, M.: A comparison of formulations for the single-airport ground-holding problem with banking constraints. Operations Research 48, 578–590 (2000)
Idris, H.R., Clarke, J.P., Bhuva, R., Kang, L.: Queuing model for taxi-out time estimation. Air Traffic Control Quarterly 10(1), 1–22 (2002)
Jaehn, F.: Solving the flight gate assignment problem using dynamic programming. Zeitschrift fr Betriebswirtschaft 80, 1027–1039 (2010)
Keith, G., Richards, A.: Optimization of taxiway routing and runway scheduling. In: Proceedings of the AIAA Guidance, Navigation and Control Conference, Honolulu, Hawaii, USA (2008)
Kim, S., Feron, E.: Robust gate assignment. In: Proceedings of the AIAA Guidance, Navigation, and Control Conference, pp. 2991–3002 (2011)
Kim, S.H., Feron, E., Clarke, J.P.: Assigning gates by resolving physical conflicts. In: Proceedings of the AIAA Guidance, Navigation and Control Conference, Chicago, USA (2009)
Kim, S.H., Feron, E., Clarke, J.P.: Airport gate assignment that minimizes passenger flow in terminals and aircraft congestion on ramps. In: Proceedings of the AIAA Guidance, Navigation, and Control Conference, Toronto, Canada, vol. 2, pp. 1226–1238 (2010)
Laporte, G.: The traveling salesman problem: An overview of exact and approximate algorithms. European Journal of Operational Research 59(2), 231–247 (1992)
Lim, A., Rodrigues, B., Zhu, Y.: Airport gate scheduling with time windows. Artificial Intelligence Review 24(1), 5–31 (2005)
Marín, Á.: Airport management: Taxi planning. Annals of Operations Research 143(1), 191–202 (2006)
Marín, Á., Codina, E.: Network design: Taxi planning. Annals of Operations Research 157(1), 135–151 (2008)
de Matosa, P.L., Ormerod, R.: The application of operational research to european air traffic flow management understanding the context. European Journal of Operational Research 123(1), 125–144 (2000)
Mesgarpour, M., Potts, C.N., Bennell, J.A.: Models for aircraft landing optimization. In: 4th International Conference on Research in Air Transportation (ICRAT 2010), Budapest, Hungary, pp. 529–532 (2010)
Michalewicz, Z., Fogel, D.B.: How to solve it: Modern metaheuristics, 2nd edn. Springer, Heidelberg (2000)
Newell, G.: Airport capacity and delays. Transportation Science 13(3), 201–240 (1979)
Nikoleris, T., Gupta, G., Kistler, M.: Detailed estimation of fuel consumption and emissions during aircraft taxi operations at Dallas/Fort Worth International Airport. Transportation Research Part D: Transport and Environment 16(4), 302–308 (2011)
Psaraftis, H.N.: A dynamic programming approach for sequencing groups of identical jobs. Operations Research 28(6), 1347–1359 (1980)

Qi, X., Yang, J., Yu, G.: Scheduling problems in the airline industry. In: Handbook of Scheduling - Algorithms, Models and Performance Analysis, pp. 50.1–50.15. Chapman & Hall/CRC (2004)

Rappaport, D.B., Yu, P., Griffin, K., Daviau, C.: Quantitative analysis of uncertainty in airport surface operations. In: Proceedings of the AIAA Aviation Technology, Integration, and Operations Conference (2009)

Rathinam, S., Montoya, J., Jung, Y.: An optimization model for reducing aircraft taxi times at the Dallas Fort Worth International Airport. In: Proceedings of the 26th International Congress of the Aeronautical Sciences, ICAS 2008 (2008)

Ravizza, S., Atkin, J.A.D., Maathuis, M.H., Burke, E.K.: A combined statistical approach and ground movement model for improving taxi time estimations at airports. Journal of the Operational Research Society (2012a) (to appear), http://dx.doi.org/10.1057/jors.2012.123

Ravizza, S., Chen, J., Atkin, J.A.D., Burke, E.K., Stewart, P.: The trade-off between taxi time and fuel consumption in airport ground movement. In: Proceedings of the Conference on Advanced Systems for Public Transport (CASPT 2012), Santiago, Chile (2012b)

Ravizza, S., Atkin, J.A.D., Burke, E.K.: A more realistic approach for airport ground movement optimisation with stand holding. Journal of Scheduling (to appear, 2013), http://dx.doi.org/10.1007/s10951-013-0323-3

Roling, P.C., Visser, H.G.: Optimal airport surface traffic planning using mixed-integer linear programming. International Journal of Aerospace Engineering 2008(1), 1–11 (2008)

Sastry, K., Goldberg, D., Kendall, G.: Genetic algorithms. In: Burke, E.K., Kendall, G. (eds.) Search Methodologies, pp. 97–125. Springer (2005)

Simaiakis, I., Balakrishnan, H.: Queuing models of airport departure processes for emissions reduction. In: Proceedings of the AIAA Guidance, Navigation, and Control Conference (2009)

Simaiakis, I., Balakrishnan, H.: Impact of congestion on taxi times, fuel burn, and emissions at major airports. Transportation Research Record: Journal of the Transportation Research Board 2184, 22–30 (2010)

Simaiakis, I., Khadilkar, H., Balakrishnan, H., Reynolds, T.G., Hansman, R.J., Reilly, B., Urlass, S.: Demonstration of reduced airport congestion through pushback rate control. In: Proceedings of the 9th USA/Europe Air Traffic Management Research and Development Seminar, Berlin, Germany (2011)

Smeltink, J.W., Soomer, M.J., de Waal, P.R., van der Mei, R.D.: An optimisation model for airport taxi scheduling. In: Proceedings of the INFORMS Annual Meeting, Denver, Colorado, USA (2004)

Smith, C.: Final approach spacing tool. In: Proceedings of the 2nd USA/Europe Air Traffic Management R&D Seminar, Orlando, USA (1998)

Talbi, E.G.: Metaheuristics. In: From Design to Implementation, 1st edn. John Wiley & Sons, Inc. (2009)

Trivizas, D.A.: Optimal scheduling with maximum position shift (MPS) constraints: A runway scheduling application. Journal of Navigation 51, 250–266 (1998)

Williams, H.P.: Model Building in Mathematical Programming, 4th edn. John Wiley & Sons, Ltd. (1999)

Wu, C., Caves, R.: Research review of air traffic management. Transport Reviews 22, 115–132 (2002)

Yu, G. (ed.): Operations Research in the Airline Industry. International Series in Operations Research & Management Science. Kluwer Academic Publishers (1998)

Instruction Scheduling in Microprocessors

Gürhan Küçük, İsa Güney, and Dmitry Ponomarev

Abstract. The Central Processing Unit (CPU) in a microprocessor is responsible for running machine instructions as fast as possible so that the machine performance is at its maximum level. While simple in design, in-order execution processors provide sub-optimal performance, because any delay in instruction processing blocks the entire instruction stream. To overcome this limitation, modern high-performance designs use out-of-order (OoO) instruction scheduling to better exploit available Instruction-Level Parallelism (ILP), and both static (compiler-assisted) and dynamic (hardware-assisted) scheduling solutions are possible. The hardware-assisted scheduling integrates an OoO core that requires a complex dynamic instruction scheduler and additional datapath structures are utilized to hold the in-flight instructions in program order to support the reconstruction of precise program state. The logic becomes even more complex when superscalar (those capable of executing multiple instructions every clock cycle) designs are used. This chapter gives a brief introduction to instruction scheduling on pipelined superscalar architectures, and, then, explains some of the keystone static and dynamic instruction scheduling algorithms.

1 Introduction

The processor performance is always considered to be one of the major criteria for evaluating microprocessors. This chapter mainly focuses on instruction scheduling techniques that target performance metric in such processors. The processor performance is best measured by the CPU execution time formula given in Equation 1.

$$T_{exec} = N \cdot CPI \cdot \tau \tag{1}$$

Gürhan Küçük · İsa Güney
Yeditepe University

Dmitry Ponomarev
SUNY Binghamton

A.Ş. Etaner-Uyar et al. (eds.), *Automated Scheduling and Planning*,
Studies in Computational Intelligence 505,
DOI: 10.1007/978-3-642-39304-4_2, © Springer-Verlag Berlin Heidelberg 2013

In this equation, T_{exec} refers to the execution time of an application running on a processor. T_{exec} is directly proportional to N which refers to the number of instructions that are executed. *CPI* is an average value that refers to *Cycles Per Instruction*, which is specific to both the organization of the machine and the applications running on the machine. Finally, clock cycle time, τ, indicates the time spent on running a single processor clock cycle. The clock cycle time is inversely proportional to the clock rate (or the clock frequency), f.

In a scalar processor, there are three major algorithmic steps running in a cyclic fashion to execute machine instructions: *Fetch*, *Decode* and *Execute*. The *Fetch* step retrieves the next instruction to be executed from the program memory in its binary form. Then, the *Decode* step decodes the instruction and extracts its operation code and operands, and, consequently, reads operand values from a register storage unit which is known as the Register File. Finally, the *Execute* step moves the instruction to an available *Arithmetic Logic Unit* (ALU) that calculates the result of an operation. At the end of the *Execute* step, the CPU returns to the *Fetch* step and continues executing instructions one after another. Figure 1 shows the execution of five consecutive instructions in a scalar processor. When execution time formula is considered, the *CPI* value is always 1 and it never changes. Besides, the clock cycle time, τ, is chosen to be as large as possible to make sure that all execution steps are completed within a single clock cycle. This makes the clock frequency, f, to be chosen as small as possible.

instructions \ cycles	1	2	3	4	5
inst.#1	F D E				
inst.#2		F D E			
inst.#3			F D E		
inst.#4				F D E	
inst.#5					F D E

Fig. 1 Execution of five consecutive instructions in a scalar processor

Instruction pipelining is a technique targeting improved processor throughput by creating an assembly line for instructions inside the processor. In this technique, the execution steps of an instruction are completed in different processor cycles, and, therefore, the clock cycle time is much shorter compared to the clock cycle time of a scalar processor. In a traditional pipelined datapath, there are five stages: *Fetch, Decode, Execute, Memory* and *Writeback*. The first three stages are identical to the ones that reside in a scalar processor. The *Memory* stage is dedicated to memory instructions, such as *loads* and *stores*, which can directly access the memory. The final stage, *Writeback*, is the stage where the destination register of an instruction is updated with the output values generated at the end of either the *Execute* or *Memory* stages. Figure 2 depicts a pipelined execution with five consecutive instructions; and, there are several things that worth to mention in this Figure:

- The clock cycle time is much shorter, since a pipelined processor runs a fraction of the total Fetch-Decode-Execute steps within a single clock cycle. As a result, a pipelined processor usually has a higher clock frequency.
- Note that, after cycle 5, the pipeline becomes full, and each of five instructions is served in different stages of pipeline, in parallel.
- Assuming that five consecutive instructions are totally independent from each other, once the pipeline becomes full, a pipelined processor may still promise a *CPI* value of 1.

instructions \ cycles	1	2	3	4	5	6	7	8	9	10	11	12
inst. #1	F	D	E	M	W							
inst. #2		F	D	E	M	W						
inst. #3			F	D	E	M	W					
inst. #4				F	D	E	M	W				
inst. #5					F	D	E	M	W			

Fig. 2 Execution of five consecutive instructions in a pipelined scalar processor

Unfortunately, the instructions are usually dependent on each other. For example, an instruction may be the producer of a value, and there may be one or many consumer instructions waiting for that value as an input so that they can start their execution. Such instruction dependencies are hazardous to pipeline performance and must be avoided as much as possible. Actually, there are three categories of pipeline hazards: *structural, control* and *data.*

The structural hazards are due to shared pipeline resources. For instance, when the memory provides a single read/write port, the example schedule in Figure 2 becomes impossible. Specifically, the problem is due to two instructions that require access to the memory unit competing for the same memory port within the fourth clock cycle. The first instruction may be a *load* instruction that initiates its *Memory* stage, meanwhile, the fourth instruction goes into the *Fetch* stage, again requiring access to the same resource. The simplest and the cheapest solution is to stall the youngest instruction (i.e. the fourth instruction) in that cycle. As a result, the stalled instruction retries the *Fetch* stage in the next cycle. But, of course, this solution degrades the processor performance since all the consumer instructions of a stalled instruction are also indirectly stalled. Other solutions, such as pipelining the resource or replicating the resource, require more complex hardware, but they may bring the performance impact of a structural hazard to its minimum.

The control hazards are other type of pipeline hazards due to, as the name implies, *control* instructions. A control instruction changes the program flow depending on its given condition. For instance, consider the example instruction sequence in Figure 3. A MIPS[1] assembly branch instruction, `BEQ R1, R2, L`, jumps to the *L* label when the value of *R1* register is equal to the value of *R2* register. When the branch condition is *true*, the next instruction to be fetched is `inst.#9`. But, when the branch condition is *false*, the next instruction to be fetched becomes `inst.#3`.

```
          inst.#1
          BEQ R1, R2, L
          inst.#3
          ...
L:
          inst.#9
```

Fig. 3 An example code for demonstrating a control hazard

The problem here is that the outcome of a branch condition cannot be known until the branch instruction is executed. Figure 4 presents the corresponding pipeline chart for the code example given above. It shows that the branch instruction is executed at cycle 4 and its result is known at cycle 5. However, the next instruction, which is either `inst.#3 or inst.#9` depending on the branch outcome, should be allowed to enter the pipeline at cycle 3. This introduces a two-cycle stall, which is also known as the *delay slot*, after each branch instruction as shown in pipeline chart given in Figure 4. Any types of pipeline stalls have to be taken very seriously since they can quickly degrade the processor performance by increasing the average *CPI* value of Equation 1.

instructions \ cycles	1	2	3	4	5	6	7	8	9
inst. #1	F	D	E	M	W				
BEQ R1, R2, L		F	D	E	M	W			
STALL			■	■	■	■	■		
STALL					■	■	■	■	
inst. #3 or inst. #9					F	D	E	M	W

(The two STALL rows are bracketed as "delay slot".)

Fig. 4 A control hazard stalls the pipeline for two cycles

[1] MIPS is an acronym for Microprocessor without Interlocked Pipeline Stages.

The delay slot in Figure 4 may be filled with two independent instructions that come before the branch instruction in program order. This type of instruction scheduling is known as static scheduling, since the instructions are reordered at compilation time. It is compiler's job to find two independent instructions that come before the branch instruction for filling each delay slot. If there are less than two instructions that can be scheduled to a delay slot, the compiler may issue No Operation (NOP) instructions instead. This way the hardware becomes unaware of the fact that the instructions are being executed out of program order, and no additional circuitry is required.

The third type of pipeline hazards is due to data dependencies among instructions. In an ideal pipeline, no dependency among tasks is assumed, and an ideal schedule similar to the one shown in Figure 2 is expected. Unfortunately, instruction dependencies are unavoidable in any type of application. Specifically, there are two types of data dependencies: *True* and *False*. True dependencies are also known as Read-after-Write (RAW) data dependencies, and they cannot be totally avoided. When an instruction writes to a register, there must be at least one instruction reading that register[2]. True dependencies are the building stones of any type of algorithm. In contrast, false dependencies exist among instructions due to the use of a common name (or register), and, they only matter when the Out-of-Order (OoO) scheduling is considered. Once, those instructions with false data dependencies start using unique names, this type of dependencies may be totally avoided. False dependencies can be examined under two subcategories: *Anti* and *Output*. Anti dependencies are also known as Write-after-Read (WAR), and Output dependencies are known as Write-after-Write (WAW) data dependencies.

Figure 5 depicts a code sequence containing various types of data dependencies and its corresponding data flow graph. In this example, there is a true dependency on R1 between inst. #1 and inst. #2. This is shown with a directed arc between these two instructions. Anti dependencies are depicted using directed arcs with an attached slash (/) symbol, and output dependencies are represented with arcs containing a circle (o) symbol. Generally, static scheduling algorithms rely on such data flow graphs for compiler optimizations that aim to reduce the penalty of all type of pipeline hazards.

For further improving the processor performance, two types of processor architecture is considered: Very Long Instruction Word (VLIW) and SuperScalar (SS) architectures. The main difference between these processors is that VLIW architecture focuses on static scheduling and suggests using a simple hardware whereas the SS architecture chooses the other direction and proposes a dynamic scheduler with a simpler compiler. Figure 7 shows an example schedule for fifteen consecutive instructions with the data dependencies as given in Figure 6. Here, the width of an instruction word is three, and the VLIW compiler finds and schedules three independent instructions that can execute out of program order within the same

[2] Otherwise, if there is no consumer instruction, the producer instruction is treated as a dead code and eliminated by the compiler.

```
inst. #1: ADD R1, R2, R3    /* R1 ← R2 + R3 */
inst. #2: SUB R3, R1, R5    /* R3 ← R1 - R5 */
inst. #3: MUL R1, R4, R6    /* R1 ← R4 * R6 */
```

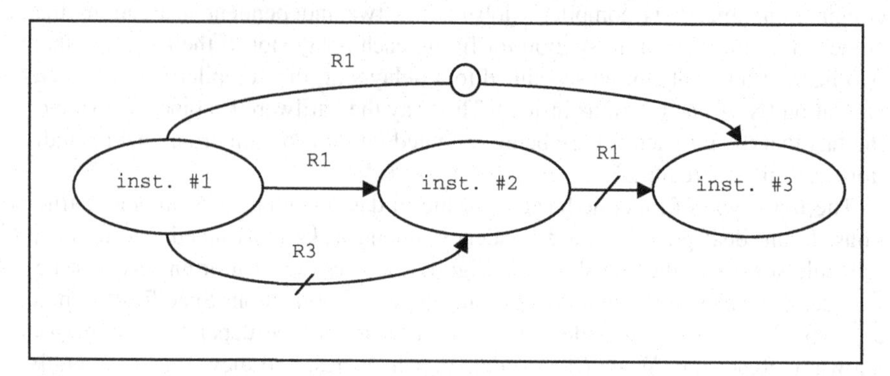

Fig. 5 An example code for demonstrating various data dependencies among instructions

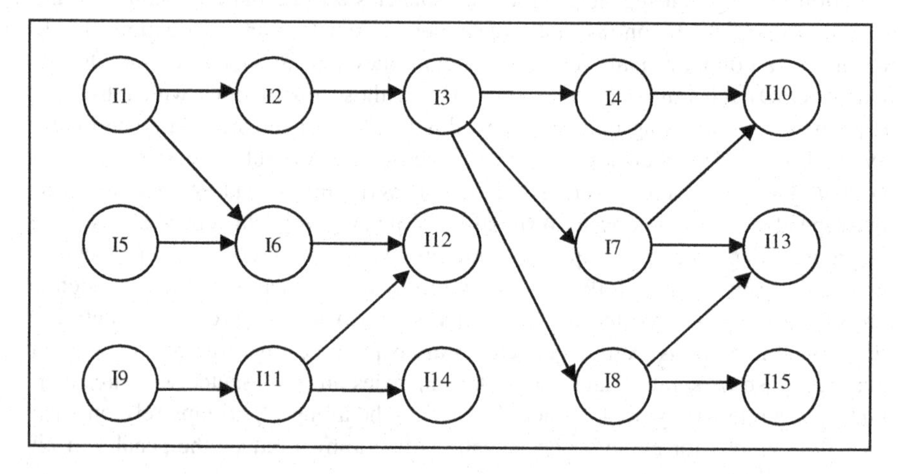

Fig. 6 An example data flowgraph showing true data dependencies among 15 consecutive instructions

clock cycle. On the hardware side, a VLIW machine fetches a very long instruction word, decodes it into three instructions and executes those instructions, in parallel. VLIW architecture heavily depends on the abilities of the compiler for high performance. However, there are several things that are unknown at compilation time, and those give hard times to VLIW compilers. For instance, memory address dependencies among memory instructions are not known until runtime of an application, and the compiler should usually generate pessimistic schedules since there is always a chance that the address of a *store* instruction that writes to a memory coinciding with the address of a later *load* instruction that reads from the memory.

cycles / instructions	1	2	3	4	5	6	7	8	9
I1, I5, I9	F	D	E	M	W				
I2, I6, I11		F	D	E	M	W			
I3, I2, I14			F	D	E	M	W		
I4, I7, I8				F	D	E	M	W	
I10, I13, I15					F	D	E	M	W

Fig. 7 A VLIW processor running instructions from I1 to I15 in an OoO fashion

The SS architecture, on the other hand, relies on dynamic instruction scheduling and is much more popular in the commercial market. In an n-way SS machine, the hardware of a pipeline scalar machine is modified so that it can fetch, decode and execute n instructions within the same clock cycle. The idea is very similar to that of the VLIW architecture. However, instead of the compiler, now the hardware is made responsible for finding n independent instructions that can execute in parallel. These processors come in two different flavors: *in-order* and *out-of-order*. In in-order SS architecture, the program order among instructions is preserved during each stage of the pipeline. Since the dynamic scheduler is only responsible for selecting n instructions in program order, the hardware complexity required to implement in-order processors is minimal. Figure 8 presents a pipeline chart for an out-of-order SS processor with a similar schedule given in Figure 7. Here, the hardware fetches three instructions per cycle, and, in the beginning, it will not be able to reach *I5* and *I9*. However, from Figure 7, it is shown that the VLIW compiler can manage to bundle those three instructions into an instruction word, since it can generate the data flowgraph and extract dependency information in a much wider context. Note that, in cycle 7, *I13*, *I14* and *I15* are not scheduled for execution, since a 3-way processor can schedule up to three instructions in a single clock cycle, and within that cycle, the processor schedules *I10*, *I11* and *I12* for execution assuming the oldest-instruction-first scheduling policy is being used. The following section describes the state of the art in both static and dynamic instruction scheduling.

cycles / instructions	1	2	3	4	5	6	7	8	9	10
I1	F	D	E	M	W					
I2	F	D	stall	E	M	W				
I3	F	D	stall	stall	E	M	W			
I4		F	D	stall	stall	E	M	W		
I5		F	D	E	M	W				
I6		F	D	stall	E	M	W			
I7			F	D	stall	E	M	W		

Fig. 8 A 3-way out-of-order SS processor running instructions from I1 to I15

I8			F	D	stall	E	M	W		
I9			F	D	E	M	W			
I10				F	D	stall	E	M	W	
I11				F	D	stall	E	M	W	
I12				F	D	stall	E	M	W	
I13					F	D	stall	E	M	W
I14					F	D	stall	E	M	W
I15					F	D	stall	E	M	W

Fig. 8 (*continued*)

2 Literature Survey

In contemporary processors, the execution order of machine instructions has a significant impact on the overall processor throughput. There are many studies targeting the compiler-assisted static instruction scheduling. For instance, Bernstein et al. work on different compiler techniques, such as Basic Block Scheduling (BB), Global Scheduling (GL) and Branch Optimizations (BO), which examine ways to extract *Instruction Level Parallelism* (ILP) that is hidden in dynamic instruction streams [1]. Global Scheduling aims to optimize the code by moving instructions beyond basic blocks. Branch Optimizations are divided into three which are *code replication*, *branch swapping* and *gluing*. Code replication deals with the delay caused by closing branches of loops. Branch swapping aims to eliminate delays caused by an unresolved branch instruction followed by a resolved one. Lastly, gluing is used for optimizing if-then-else statements. When obtaining the scheduled stream of instructions, a set of machine independent optimizations are applied followed by the global scheduling. In order to fine-tune the code produced by global scheduling, a pass of basic block scheduling is also applied. After register allocation, another pass of basic block scheduling is applied, this time on real registers, followed by the last step on branch optimizations. The authors conclude that GL and BO improvements overlap, and GL is capable of doing all improvements which BO and BB can do, except for branch swapping.

In [2], Moon and Ebcioglu study the concepts of static scheduling on extracting ILP on nonnumerical codes; such as global scheduling, speculative code motion, scheduling in absence of branch probability, mispredicted speculative code motion, nonspeculative code motion, mildly speculative code motion, trace based and DAG based code motion and modulo scheduling pipelining loops where multiple execution paths have similar probabilities of being taken at execution time. The authors also propose a technique called *selective scheduling* and present its compiler.

Finding the optimal instruction schedule is an NP-complete problem. Mahajan et al. propose the use of genetic programming, which is a branch of machine learning, to generate heuristics for compile-time instruction scheduling [3]. The study focuses on scheduling instructions in superblocks. Scheduling superblocks are harder than scheduling basic blocks, since there are multiple paths that can be taken in a superblock. Genetic programming used in this study includes techniques such as survival of fittest and crossover, which were introduced in its predecessor, genetic algorithms; as well as new attributes such as variable size expressions.

In a historic paper [4], Weiss and Smith works on the trade-offs between two dynamic scheduling mechanisms, the Tomasulo's algorithm [5] and Thornton's Scoreboard mechanism [6], against static code scheduling of CRAY-1 superscomputer and their proposed approach. This work focuses on four topics: clock period, scheduling of instructions, issue logic complexity and hardware cost, and debugging and maintenance. According to Weiss and Smith, the Tomasulo's algorithm shows the best performance in tests, compared to the static code scheduling of CRAY-1. However, the hardware cost for implementing Tomasulo's algorithm is stated as a serious issue. Thornton's algorithm reduces this cost by abandoning the simultaneous tag matching logic, which causes a performance loss. Weiss and Smith propose an algorithm to overcome this problem, which they call Direct Tag Search (DTS). DTS brings tags implemented in Tomasulo's algorithm back, but reduces the hardware cost by restricting the number of reservation stations per tag to 1, and implementing a tag search table. Since Thornton's algorithm and DTS has less out-of-order execution capabilities, their performance are affected more by the order of instructions, which depends on the performance of compiler's static code scheduling.

The selection logic in a dynamic scheduler is responsible of choosing a good candidate instruction so that performance of the machine is maximized. In a study on instruction criticality, Tune et al. presents a framework for comparing previously proposed methods for determining instruction criticality, ranking instruction criticality and investigating the characteristics of critical instructions [7]. This study focuses on the metric *slack* and proposes a new metric called *tautness* to quantify instruction criticality. The slack of an instruction represents the number of cycles the instruction can be delayed without increasing the execution time of a program. Instructions with more than zero cycles of slack are non-critical. Tautness metric is used to distinguish critical instructions, which is a complementary measure to slack. Tautness of an instruction is defined as the number of cycles by which the execution time is reduced when the result of that instruction is made available to other instructions immediately. This is a useful metric as it quantifies the maximum benefit of applying an optimization to an instruction. It also roughly models what might be achieved by value predicting or speculatively precomputing the result of an instruction. The work has shown that majority of static schedules are never critical, but among those critical instructions that are ever critical, criticality varied frequently; very few static instructions are always critical. Thus, predicting exactly the dynamic instances of these static instructions is difficult but important for a highly accurate predictor. The work also shows that critical path predictors must be able to identify patterns of criticality to achieve high coverage and accuracy. It also demonstrates the need for predictors that quantify criticality rather than just produce a binary prediction.

A problem in superscalar processors emerges from the existence of *long latency load* instructions. Such instructions will cause the dependent instructions to stall and Instruction Queue (IQ), which holds the instructions waiting for execution, will fill up quickly. These instructions will prevent other instructions from entering the IQ and being examined for dependency, therefore lowering the IPC and causing the processor to wait in an idle state until the long latency instruction is completed. One solution to this problem would be increasing the IQ size. However, the IQ stays in the critical path of a processor, and an increase in its access latency directly increase the clock cycle. Lebeck et al. suggest a solution where the IQ size is kept the same; but a Waiting Instruction Buffer (WIB) is introduced [8]. The authors propose that all registers have an extra *wait bit* defined, which are set when a cache miss occurs, and reset when such instructions are completed. In this way, it could be said that an instruction related to *waiting registers* are dependent on the long latency instruction, and all registers related to this new instruction are also said to be *waiting*. Therefore, the chain of instructions dependent to a long latency instruction is identified. To prevent IQ from filling up with waiting instructions, such instructions are moved to a larger structure named as WIB. The advantage of the WIB structure is that it does not need a complex wakeup logic as of IQ. For all outstanding long latency instructions, a bit vector is defined to mark all instructions dependent on that instruction. When the long latency instruction is completed, all related instructions are moved back to IQ. If these instructions are also dependent on another long latency instruction, they are moved back to the WIB. This study also shows that using the extra space for WIB instead of a larger cache has a better influence on the overall performance.

Wang and Sangireddy propose a method to mitigate the effects of long latency instructions [9]. They propose the implementation of a sideline buffer for holding long latency instructions, and the instruction chain dependent to it. Therefore, a performance improvement is achieved without increasing the IQ size, by allowing independent instructions more room in IQ. Instructions in the proposed buffers are issued only in-order, so the instruction selection complexity is increased slightly, whereas the instruction wakeup complexity remains the same.

Sharkey and Ponomarev studied the performance issues on pipelining superscalar processors [10]. The study has shown that the performance losses are incurred only due to a small fraction of instructions, which are intolerant to the non-atomic scheduling. The authors propose a Non-Uniform Scheduler – a design that partitions the scheduling logic into two queues, each with dedicated wakeup and selection logic: a small *Fast Issue Queue* to issue critical instructions in the back-to-back cycles, and a large *Slow Issue Queue* to issue the remaining instructions over two cycles with a one cycle bubble between dependent instructions. Finally, several steering mechanisms to effectively distribute instructions between the queues are studied.

In [11], Ernst et al. propose a novel design called Cyclone in which they increase throughput by reducing issue selection logic complexity, which is on the critical path of the processor. The main factor behind reduced issue selection logic complexity is that Cyclone does not contain any global broadcast signals. In Cyclone, once the instructions are fetched, their latency (time needed before source operands of the instruction are ready) is predicted, and the instruction enters a multi-entry queue in a position based on the predicted latency. If the predicted latency of an instruction is wrong, upon reaching execution by shifting towards execution each cycle, Cyclone is capable of replaying only that instruction and the instructions in the dependence chain. The destination register of a replaying instruction is marked as invalid, and thus the dependence chain is identified. Cyclone has lower IPC than traditional designs; however, the throughput is increased due to much faster clock speeds.

A major consumer of microprocessor power is the IQ. In [12], Buyuktosunoglu et al. study different IQ optimization in terms of performance, power consumption and the deviation from the baseline design. Some superscalar processors implement a latch based *compacting* IQ. Compacting queues feed-forwards each entry to fill the holes created by instruction issue. New instructions are always added to the tail of the queue. Therefore, the queue maintains an oldest to youngest program order, which allows a position based selection mechanism. Each time an instruction is issued; all entries are shifted, which causes a large amount of power dissipation. To eliminate this power cost, the IQ can be made *non-compacting*, where holes in IQ are not filled immediately. However in this design, the position based selection mechanism does not give priority to older instructions. To solve this problem, the ROB numbers each instruction. An extra high-order *sorting bit* is added to ROB to eliminate problems arising from the circular nature of ROB. In the proposed design, each instruction is steered to the bank corresponding to the id number of the unavailable operand, which takes advantage of the fact that dispatched instructions rarely have two source operands unready. In case the instruction has two unready source operands, it is stored into *conflict queue*, which is a conventional IQ.

3 Static Instruction Scheduling

The pipeline hazards that are described in prior sections may severely degrade the processor performance. In this section, some of the well-known examples of static instruction scheduling techniques that attack these hazards are discussed.

3.1 List Scheduling

List scheduling algorithm is a greedy, heuristic approach that focuses on reordering of instructions and elimination of false data dependencies inside a *basic block*. A basic block is a block of code which has only one entry point and one exit point, and it is highly susceptible to optimizations. The algorithm consists of four steps:

1. Register renaming: This initial and optional step eliminates false data dependencies among instructions inside a basic block. Each instruction with a destination register gets a new name for its destination register, and data dependent instructions refer to that new given name instead of the original destination register.
2. Building of a dataflow graph: In this step of the algorithm, the basic block is usually traversed bottom up, and its corresponding dataflow graph, which represents the true data dependencies among instructions, is generated. When the first step of the algorithm is bypassed, this graph also contains false dependencies among instructions. During this process, each edge in the graph is also annotated with the latency of each operation.
3. Prioritize instructions: According to the availability of operands of instructions and latency of each operation, each instruction gets a priority value. In one policy, the longest latency-weighted path is stated as the critical path of execution, and all the instructions within this path get the highest priority. This tends toward a depth-first traversal of the graph. In another policy, an instruction gets the highest priority if it has the maximum number of immediate successor (data dependent) instructions, and this tends toward a breadth-first traversal of the graph.
4. Select and schedule an instruction: This final iterative step is responsible for selecting and scheduling ready instructions among all instructions according to their priority determined in the previous step. The algorithm keeps a list known as the *Ready List*, which lists all instruction that may immediately be scheduled for execution. Then, the algorithm updates the Ready List after scheduling those instructions (removing them from the list) and satisfying the need of data dependent instructions (adding them to the list). This greedy algorithm continues as long as there are instructions in the Ready List.

To better explain the algorithm, consider the sample basic block given in Figure 9.

```
I1:   LOAD R1, R3, #100 /* R1 ← mem[R3+100]*/
I2:   LOAD R2, R3, #104 /* R2 ← mem[R3+104]*/
I3:   ADD R2, R2, R1    /* R2 ← R2 + R1 */
I4:   STORE R2, R3,#108 /* mem[R3+108] ← R2 */
I5:   LOAD R1, R3, #112 /* R1 ← mem[R3+112]*/
I6:   LOAD R2, R3, #116 /* R2 ← mem[R3+116]*/
I7:   SUB R2, R2, R1    /* R2 ← R2 - R1 */
I8:   STORE R2, R3,#120 /* mem[R3+120] ← R2 */
```

Fig. 9 A sample code for list scheduling

When, the first step of the algorithm is applied, all the destination registers are renamed, and the basic block is morphed into the one shown in Figure 10. Here, notice that, all false data dependencies are eliminated at the end of this step. However, Figure 11 shows that this schedule is not ideal, and instructions I3, I4, I7 and I8 are stalled due to delays in their producer instructions. For instance, the execution of I3 is delayed since the result of I2 is not available until the end of cycle 5. I4 is also indirectly delayed since its producer I3 is delayed. As a result, the compiler schedules NOP instructions before I3 and I7 for stalling the pipeline.

Then, in the second step of the algorithm, the corresponding dataflow graph, which is shown in Figure 12, is generated. In this example, we assume that memory instructions (*load* and *store*) have two cycles execution latency (one for the effective address calculation at execute stage and one for the memory access at memory stage) whereas other ALU instructions, such as ADD and SUB, take only one cycle to execute.

```
I1:    LOAD R10, R3, #100/* R10 ← mem[R3+100]*/
I2:    LOAD R11, R3, #104/* R11 ← mem[R3+104]*/
I3:    ADD R12, R11, R10 /* R12 ← R11 + R10 */
I4:    STORE R12, R3,#108/* mem[R3+108] ← R12 */
I5:    LOAD R13, R3, #112/* R13 ← mem[R3+112]*/
I6:    LOAD R14, R3, #116/* R14 ← mem[R3+116]*/
I7:    SUB R15, R14, R13 /* R15 ← R14 - R13 */
I8:    STORE R15, R3,#120/* mem[R3+120] ← R15 */
```

Fig. 10 The resulting code after register renaming step is applied

instructions \ cycles	1	2	3	4	5	6	7	8	9	10	11	12	13	14
I1	F	D	E	M	W									
I2		F	D	E	M	W								
I3			F	D	stall	E	M	W						
I4				F	stall	D	E	M	W					
I5						F	D	E	M	W				
I6							F	D	E	M	W			
I7								F	D	stall	E	M	W	
I8									F	stall	D	E	M	W

Fig. 11 The pipeline chart showing the execution of instructions of Figure 10

In the next step of the algorithm, the instructions are prioritized according to the number of their immediate successors, and during the final step of the algorithm the Ready List is generated. The Initial Ready List contains four instructions: I1, I2, I5, I6, and since all of them have the same priority one of them selected randomly for execution[3]. Figure 13 shows an example schedule at the end of the fourth step of the algorithm. As shown in the pipeline chart given in Figure 14, the final schedule is free from any stalls and no additional NOP instruction is needed.

[3] Another tie breaker rule other than random selection may be utilized, as well.

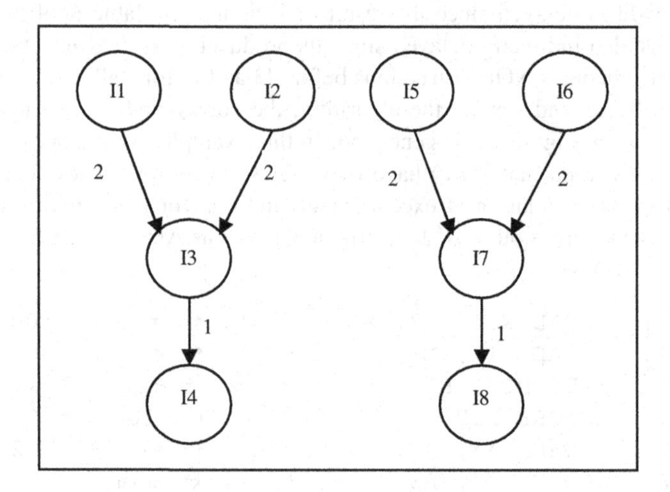

Fig. 12 The corresponding dataflow graph for the code given in Figure 10

```
I1:    LOAD R10, R3, #100/* R10 ← mem[R3+100]*/
I5:    LOAD R13, R3, #112/* R13 ← mem[R3+112]*/
I2:    LOAD R11, R3, #104/* R11 ← mem[R3+104]*/
I6:    LOAD R14, R3, #116/* R14 ← mem[R3+116]*/
I3:    ADD R12, R11, R10 /* R12 ← R11 + R10 */
I7:    SUB R15, R14, R13 /* R15 ← R14 - R13 */
I4:    STORE R12, R3,#108/* mem[R3+108] ← R12 */
I8:    STORE R15, R3,#120/* mem[R3+120] ← R15 */
```

Fig. 13 The final code after list scheduling is applied

instructions \ cycles	1	2	3	4	5	6	7	8	9	10	11	12	13	14
I1	F	D	E	M	W									
I2		F	D	E	M	W								
I3			F	D	E	M	W							
I4				F	D	E	M	W						
I5					F	D	E	M	W					
I6						F	D	E	M	W				
I7							F	D	E	M	W			
I8								F	D	E	M	W		

Fig. 14 The pipeline chart showing the execution of instructions of Figure 13

3.2 *Loop Unrolling*

Loop Unrolling is one of the oldest and well-known loop transformation techniques. In this method, the compiler unrolls a loop by replicating the loop body and rearranging the logic that controls the number of loop iterations. The main aim of the method is to get rid of as many control instructions as possible so that pipeline stalls due to control hazards are minimized. For instance, the loop body, which is shown in Figure 15, may be unrolled as many times as possible to remove branch instructions from the dynamic instruction stream.

```
I1:   L:   LOAD R1, R3, #100 /* R1 ← mem[R3+100]*/
I2:        LOAD R2, R3, #400 /* R2 ← mem[R3+400]*/
I3:        ADD R2, R2, R1    /* R2 ← R2 + R1 */
I4:        STORE R2, R3,#800 /* mem[R3+800] ← R2 */
I5:        ADDI R3, R3, #4   /* R3 ← R3 + 4 */
I6:        BLT R3, R5, L     /* if R3 < R5 jump L */
```

Fig. 15 A sample loop that can be unrolled

In Figure 16, the pipeline chart and stalls due to control hazards are shown. As long as the loop condition is true, the execution returns to instruction I1, and for the execution of each branch instruction, the pipeline receives a two-cycle penalty.

instructions \ cycles	1	2	3	4	5	6	7	8	9	10	11	12	13	14
I1	F	D	E	M	W									
I2		F	D	E	M	W								
I3			F	D	stall	E	M	W						
I4				F	stall	D	E	M	W					
I5					stall	F	D	E	M	W				
I6							F	D	E	M	W			
I1								stall	stall	F	D	E	M	W

Fig. 16 The schedule for a single iteration of the loop

After unrolling the loop once, it is transformed into a form as shown in Figure 17. In this version of the loop, two loop iterations become one, and the compiler generates a slightly modified version of the first iteration during the replication process. Note that the number of branch instructions and the number of ADDI instructions that are executed, which is the N term in Equation 1, are also halved. This is an additional performance benefit of the algorithm on top of removing half of the two-cycle stalls from the pipeline. Also note that a short prologue code, which is not shown in Figure 17 for the sake of code clarity, might be necessary when the total number of iterations in the original loop is not even.

```
I1:   L:   LOAD R1, R3, #100 /* R1 ← mem[R3+100]*/
I2:        LOAD R2, R3, #400 /* R2 ← mem[R3+400]*/
I3:        ADD R2, R2, R1    /* R2 ← R2 + R1 */
I4:        STORE R2, R3,#800 /* mem[R3+800] ← R2 */
I5:        LOAD R1, R3, #104 /* R1 ← mem[R3+104]*/
I6:        LOAD R2, R3, #404 /* R2 ← mem[R3+404]*/
I7:        ADD R2, R2, R1    /* R2 ← R2 + R1 */
I8:        STORE R2, R3,#804 /* mem[R3+804] ← R2 */
I9:        ADDI R3, R3, #8   /* R3 ← R3 + 8 */
I10:       BLT R3, R5, L     /* if R3 < R5 jump L */
```

Fig. 17 The sample loop of Figure 15 is unrolled once

Finally, Figure 18 shows the pipeline schedule for the transformed loop. While, in the original loop, n iterations finish in 11+9(n-1) cycles, in this new version of the loop, n iterations finish in 16+14(n/2-1) cycles. This indicates that approximately 1.3 times better performance is achieved when the loop is unrolled only once. This improvement is very close to the maximum performance that can be achieved (1.5 times better performance) when the loop is unrolled several times until no branch instructions is remained in the schedule. The drawback of applying this algorithm in its limits is that static code size may increase into an amount that is larger than the size of the instruction cache. In this case, the performance degradation due to instruction cache misses may be much higher than the performance gain obtained from the loop unrolling algorithm.

cycles / instructions	1	2	3	4	5	6	7	8	9	10	11	12	13	14	15	16	17	18	19
I1	F	D	E	M	W														
I2		F	D	E	M	W													
I3			F	D	stall	E	M	W											
I4				F	stall	D	E	M	W										
I5					stall	F	D	E	M	W									
I6							F	D	E	M	W								
I7								F	D	stall	E	M	W						
I8									F	stall	D	E	M	W					
I9											F	D	E	M	W				
I10												F	D	E	M	W			
I1													stall	stall	F	D	E	M	W

Fig. 18 The pipeline schedule for the loop given in Figure 17

4 Dynamic Instruction Scheduling and Out-of-Order Execution

While techniques such as instruction pipelining, branch prediction and data forwarding significantly improve the performance of microprocessors, the properties of typical instruction streams require additional mechanisms to push the performance envelope even further. For example, consider a long-latency instruction, such as a division operation or a *load* instruction that misses into the cache. In statically-scheduled in-order pipelines, if such an instruction is the oldest non-executed instruction in the processor, then "head-of-the-line" blocking effect occurs that inhibits any progress in the pipeline until the instruction in question finishes its execution and makes its result available to the dependent instructions. The problem here is that not only dependent instructions, but also the independent instruction are forced to wait, thus blocking the instruction flow and inhibiting fast reuse of pipeline resources.

To illustrate this problem, consider the following example.

```
I1:    LOAD R1, R2, #100 /* R1 ← mem[R2+100]*/
I2:    ADD R3, R1, R5    /* R3 ← R1 + R5   */
I3:    SUB R7, R8, R9    /* R7 ← R8 - R9   */
```

Fig. 19 Code example 1

In this case, if the instruction I1 misses into the cache and blocks, both instructions I2 and I3 (and all subsequent instructions in the pipeline) will also have to wait. Notice that the instruction I3 is independent of I1, and therefore could be executed while I1 and I2 were waiting, if appropriate support was added to the pipeline. This example motivates the idea *of dynamic instruction scheduling* with OoO execution.

In dynamically scheduled OoO processors, while the instruction I2 waits on the result of I1, the instruction I3 can be allowed to execute and even write back its generated result into the register file. This is accomplished by the following mechanism. All decoded instructions are placed into a queue called *Instruction Queue* (IQ)[4], from where they are eventually selected for the execution whenever all their data dependencies are resolved and the required physical execution unit is available. The instructions are thus placed into the IQ in program order, and are selected from of the IQ out of program order. The IQ can be quite sizable, typically several tens of entries; the larger size of the IQ allows for more tolerance to the

---

[4] The name, Instruction Queue, is due to early implementations of this structure for in-order SS processors, and in its recent implementations, IQ is no longer a queue structure. Although, instructions are placed into the IQ in program order, they may be placed into any unoccupied entries, and, therefore, IQ is also known as Instruction Dispatch Buffer (IDB).

long-latency operations. When the IQ becomes full, the instruction fetching and decoding process stalls until some IQ entries are made available (e.g., when instructions residing in these entries execute). Larger IQ also allows to reduce the number of cycles when the decode process is blocked, as the IQ becomes full less often.

The IQ-based scheduling mechanism effectively decouples the in-order pipeline front-end from the OoO execution back-end, allowing each to operate relatively independently: the front-end simply provides a supply of instructions to the back-end. To support dynamic scheduling, two key mechanisms are implemented within the IQ and its associated logic: instruction *wakeup* and instruction *selection*. Instruction wakeup refers to the process of matching the addresses of the source registers of the instructions waiting in the IQ against the destination registers of the instructions that are completing the execution. Specifically, whenever an instruction completes execution, its destination address (destination tag) is broadcasted across the IQ and all instructions currently waiting in the IQ associatively compare their source register addresses (source tags) against the destination tag being broadcasted. On a match, the corresponding source operand is marked as valid. When all source operands of an instruction become valid, then the instruction wakes up and becomes ready for execution. For example, referring to the code of Figure 19 above, when the *load* instruction finally completes the memory access stage, its destination register tag (1 in this case) is broadcasted and the first source of instruction I2 becomes valid, as it is a match against this destination tag.

The second component of the dynamic scheduling logic is the process of instruction selection. When the number of instructions that wake up in a cycle exceeds the number of available execution units, the arbitration has to be performed to select only some of the woken instructions for execution in the next cycle and delay the rest of the instructions until the following cycle. The most common examples of selection algorithms are as follows:

1) Position-Based Selection: The ready instructions are selected according to their physical order in IQ. Depending on a chosen start point for the search process, instructions that are closer to the start point become more favorable. As a result, some of the instructions that are placed into an IQ entry far from the starting search point may starve. This is not a fair policy, but it requires a very simple hardware.
2) The Oldest-First Selection: On the contrary, this is possibly the fairest policy among many, but its hardware implementation requires a very complex circuitry. The oldest and ready instruction in IQ is selected to be scheduled for execution, but that instruction may not be a critical instruction with a limited slack. The literature survey discusses *instruction criticality* and the term *slack*, in more detail.
3) Loads-First Selection: Memory instructions, especially *load* instructions that read from memory and consequently set a destination register, are instructions

with unpredictable latencies[5], and many data dependent instructions might be waiting their immediate execution. This policy gives priority to such instructions.

4) Longest-Latency-First Selection: In this selection policy, instead of giving priority to *load* instructions with unpredictable latencies, instructions with long latency values are favored. Examples to long latency instructions are multiplication, division and several floating-point instructions such as square root and logarithm.

Because of significant circuit delays involved in wakeup and selection activities, the scheduling logic is usually pipelined over two cycles. Such pipelining can make it impossible to execute dependent instructions in the back-to-back cycles in high-frequency implementations, but smart solutions have been proposed in the literature to address this problem [13][14].

While the conceptual idea of OoO execution is fairly simple, there are several new challenges that arise. We discuss these challenges and outline solutions to them below.

```
I1:    LOAD R1, R2, #100 /* R1 ← mem[R2+100] */
I2:    ADD R3, R1, R5    /* R3 ← R1 + R5   */
I3:    SUB R1, R8, R9    /* R1 ← R8 - R9   */
```

Fig. 20 Code Example 2

The first challenge is associated with *false data dependencies.* To understand the concept of false dependencies, consider a slight modification to the example of Figure 19, where the last instruction uses the same destination register as the first instruction. The new code is shown in Figure 20. In this case, which is quite common in typical programs as the registers are reused often, instruction I3 cannot simply write its result to register R1 before the instruction I1 does so; otherwise, the value of register R1 will be incorrect after the execution of this code. These "false" dependencies arise purely as an artifact of OoO execution and they have to be somehow satisfied to guarantee correctness.

A well-established solution to this problem in modern designs is a technique called *register renaming*, which is already discussed in Section 3.1. Renaming logic requires an additional pipeline stage at the front end, and the renamed instruction stream is inserted into the IQ, allowing previously discussed wakeup and selection mechanisms to be deployed without concerns for false dependencies. Of course, physical registers have to be added back to the free pool when they are no longer needed, and the microarchitecture provides mechanisms for doing so.

---

[5] A load instruction may be served by multiple levels of the memory hierarchy. When its address hits to the first level or the second level cache, its access latency may be a few or tens of cycles, respectively. Otherwise, if the main memory is accessed, the latency may be in the range of hundreds or thousands of cycles.

Another issue that has to be considered in these OoO designs is how to cope with branch mispredictions, exceptions and interrupts and ensure that a precise program state can always be reconstructed following these events. This is accomplished by maintaining all in-flight instructions in their program order in a new FIFO queue which is called the *Reorder Buffer* (ROB). The instructions are inserted into the ROB after they are renamed. Another stage (commit stage) is added at the back end of the pipeline to allow the instructions to commit their results to the architecturally visible program state strictly in program order. The order of instruction commitment is dictated by their positions in the ROB. Until the oldest instruction in the pipeline commits, the younger instructions have to wait in the ROB for their turn, even if their execution has already completed. On a branch misprediction, the wrong-path instructions can be determined by examining the ROB and removing instructions following the mispredicted branch from all pipeline structures. Thus, the presence of the ROB allows performing aggressive instruction reordering to maximize performance, and at the same time maintaining a safety net to guarantee correct execution in case of exceptional events.

Finally, additional considerations need to be taken into account for the processing of memory instructions in OoO microarchitectures. For example, for a *load* instruction to execute ahead of some earlier *store* instructions, it is necessary to ensure that the addresses targeted by these *stores* are not matching the address of the *load*. This is accomplished by maintaining another queue structure, called load-store queue (LSQ), which keeps all memory instructions in their program order. The LSQ is associatively checked for the address matches between the *loads* and the *stores*. On a match, the values are locally forwarded within the LSQ from a *store* to a matching *load*. On a mismatch, the cache access by the *load* is performed. This technique is known as *memory disambiguation*.

5 A Case Study: Intel Pentium 4 Processor

In Intel's "tick-tock" model, Intel promises continued innovations in both manufacturing process technology and processor architecture in alternating "tick" and "tock" cycles, respectively. Regarding this model, Intel's Pentium 4 processor architecture can be considered as a major "tock".

The microarchitectural design of the P4 processor consists of two major stages: in-order front-end and out-of-order execution core. In the front-end of the microprocessor, a decoder is responsible for decoding IA-32 instructions into basic operations called uops (micro-operations). Then, these uops are sent to a structure called Trace Cache, which is an advanced form of a Level 1 instruction cache. In case the IA-32 instruction must be replaced with more than four uops, the IA-32 instruction is sent to the microcode ROM, which sequences the uops necessary to complete the operation. The uops from the Trace Cache and microcode ROM are buffered in a simple in-order uop queue.

From there, uops are sent to the *Allocator*, which is responsible for allocating an entry in ROB, a register entry for the result, an entry in one of the uop queues in front of the instruction schedulers, and an entry for LSQ, if the uop is a memory operation. If such allocation is impossible, the Allocator stalls and prevents new uops from entering the pipeline. When the destination register entry is being allocated, register renaming is applied on the uop to eliminate false dependencies. The renaming logic keeps track of the most recent version of each register so that a new instruction coming down the pipeline can determine the correct register entry for its source operands. After register renaming, uops enter one of the two queues according to their type: load-store queue or non-memory instruction queue. These queues are in-order. However, they can be read out-of-order, which allows for the out-of-order execution. There are several types of uop schedulers that are used to schedule different types of uops for various execution units. These schedulers determine when uops are ready to be executed based on the readiness of uop's input operands and the availability of function units. After completing their execution, uops are retired from the ROB to maintain the original program order.

Pentium 4 processor has a very deep pipeline (more than 20 stages), and scheduling of instructions with long and unpredictable latency values becomes a major performance issue. The *replay* mechanism of the machine is used to speculatively execute operations that depend on a load instruction assuming that the address of the load exists in the L1 data cache. This is somewhat necessary to keep the deep pipeline full most of the time. However, when the load in question misses the L1 data cache and its destination register value is delayed, the dependent instructions get temporarily incorrect data. Then, the replay mechanism tracks down and re-executes those instructions that use incorrect data when the correct data value is served from the L2 cache or main memory.

6 Conclusion

In this paper, we reviewed the approaches to instruction scheduling in contemporary microprocessors. Scheduling can generally be implemented using either static or dynamic approaches. While static scheduling does not require complex hardware, it often results in sub-optimal performance and in some cases does not retain binary compatibility (i.e. when specific pipeline configuration is taken into account while creating an execution schedule in hardware). In contrast, dynamic instruction scheduling provides high performance through out-of-order execution of instructions and also retains binary compatibility, because the schedule is implemented transparently in hardware. The drawback is the increased complexity, delay and power consumption of the instruction scheduling logic.

References

[1] Bernstein, D., Cohen, D., Lavon, Y., Rainish, V.: Performance Evaluation of Instruction Scheduling on the IBM RISC System/6000. In: MICRO, vol. 25, pp. 226–235 (1992)

[2] Moon, S.M., Ebcioglu, K.: Parallelizing Nonnumerical Code With Selective Scheduling and Software Pipelining. TOPLAS 19(6), 853–898 (1997)

[3] Mahajan, A., Ali, M.S., Patil, M.: Instruction Scheduling Using Evolutionary Programming. In: ACC 2008, pp. 137–144 (2008)

[4] Weiss, S., Smith, J.E.: Instruction Issue Logic in Pipelined Supercomputers. In: ISCA 1984, pp. 110–118 (1984)

[5] Tomasulo, R.M.: An Efficient Algorithm for Exploiting Multiple Arithmetic Units. IBM J. Res. Development 11(1), 25–33 (1967)

[6] Thornton, J.E.: Parallel Operation in the Control Data 6600. In: AFIPS 1964, Part 2, pp. 33–40 (1964)

[7] Tune, E., Tullsen, D.M., Calder, B.: Quantifying Instruction Criticality. In: PACT 2002, p. 104 (2002)

[8] Lebeck, A.R., Koppanalil, J., Li, T., Patwardhan, J., Rotenberg, E.: A Large, Fast Instruction Window for Tolerating Cache Misses. In: ISCA 2002, pp. 59–70 (2002)

[9] Wang, H., Sangireddy, R.: Streamlining Long Latency Instructions For Seamlessly Combined Out-Of-Order and In-Order Execution. Microprocessors & Microsystems 32, 375–385 (2008)

[10] Sharkey, J.J., Ponomarev, D.V.: Non-Uniform Instruction Scheduling. In: Cunha, J.C., Medeiros, P.D. (eds.) Euro-Par 2005. LNCS, vol. 3648, pp. 540–549. Springer, Heidelberg (2005)

[11] Ernst, D., Hamel, A., Austin, T.: Cyclone: A Broadcast-Free Dynamic Instruction Scheduler with Selective Replay. In: ISCA 2003, pp. 253–263 (2003)

[12] Buyuktosunoglu, A., Albonesi, D.H., Bose, P., Cook, P.W., Schuster, S.E.: Tradeoffs in Power-Efficient Issue Queue Design. In: ISLPED 2002, pp. 184–189 (2002)

[13] Brown, M.D., Stark, J., Patt, Y.N.: Select-Free Instruction Scheduling Logic. In: MICRO, vol. 34, pp. 204–213 (2001)

[14] Stark, J., Brown, M.D., Patt, Y.N.: On Pipelining Dynamic Instruction Scheduling Logic. In: MICRO, vol. 33, pp. 57–66 (2000)

Sports Scheduling: Minimizing Travel for English Football Supporters

Graham Kendall and Stephan Westphal

Abstract. The football authorities in England are responsible for generating the fixtures for the entire football season but the fixtures that are played over the Christmas period are given special consideration as they represent the minimum distances that are traveled by supporters when compared with fixtures played at other times of the year. The distances are minimized at this time of the year to save supporters having to travel long distances during the holiday period, which often coincides with periods of bad weather. In addition, the public transport system has limited services on some of the days in question. At this time of the year every team is required to play, which is not always the case for the rest of the season. When every team is required to play, we refer to this as a *complete fixture*. Additionally, each team has to to play a home game and an away game. Therefore, over the Christmas period we are required to produce two complete fixtures, where each team has to have a Home/Away pattern of HA or AH. In some seasons four complete fixtures are generated where each team is required to have a Home/Away pattern of HAHA (or AHAH). Whether two or four fixtures are generated there are various other constraints that have to be respected. For example, the same teams cannot play each other and we have to avoid (as far as possible) having some teams play at home on the same day. This chapter has three main elements. i) An analysis of seven seasons to classify them as two or four fixture seasons. ii) The presentation of a single mathematical model that is able to generate both two and four fixture schedules which adheres to all the required constraints. Additionally, the model is parameterized so that we can conduct a series of experiments. iii) Demonstrating that the model is able to produce solutions which are superior to the solutions that were used in practise (the *published fixtures*)

Graham Kendall
University of Nottingham, Nottingham, UK and Malaysia
e-mail: graham.kendall@nottingham.ac.uk

Stephan Westphal
Institute for Numerical and Applied Mathematics,
Georg-August University, Germany
e-mail: s.westphal@math.uni-goettingen.de

A.Ş. Etaner-Uyar et al. (eds.), *Automated Scheduling and Planning*,
Studies in Computational Intelligence 505,
DOI: 10.1007/978-3-642-39304-4_3,

and which are also superior to our previous work. The solutions we generate are near optimal for the two fixture case. The four fixture case is more challenging and the solutions are about 16% of the lower bound. However, they are still a significant improvement on the fixtures that were actually used. We also show, through three experimental setups, that the problem owner might actually not want to accept the best solution with respect to the overall minimized distance but might want to take a slightly *worse* solution but which offers a guarantee as to the maximum distance that has to be traveled by the supporters within each division.

1 Introduction

In England, the football (soccer in the USA) league structure comprises four main divisions. These are generically called "FA Premiership" (20 teams), "FL Championship" (24 teams),"FL Championship" (24 teams) and "FL Championship 2" (24 teams). These names change with sponsorship arrangements for the given season. Within each division, a double round robin tournament is held, resulting in 2036 fixtures that have to be scheduled each season. Even though each division is an independent double round robin tournament, they cannot be scheduled in isolation from one another as there are a number constraints which operate across the divisions. For example, we should avoid, irrespective of which division they play in, certain teams playing at home on the same day (*pairing constraint*), only a certain number of FA Premiership teams based in London can play at home on the same day, only a total number of London based clubs (across all four divisions) can play at home on the same day and only a certain number of Manchester based clubs can play at home on the same day. These constraints are collective referred to as *geographical constraints*. All these constraints are captured in the model presented in Section 4.

When generating a schedule for the entire season, it is our belief that the football authorities initially schedule fixtures for the Christmas period. This means creating two or four sets of fixtures that will be used over two or four days. At this time of the year, every team is required to play (which is not always the case for other times in the season). We refer to such a schedule as a *complete fixture*. That is, a complete fixture ensures that all 92 teams play, representing 46 fixtures. Therefore, over the Christmas period we are required to generate either two or four complete fixtures. As well as respecting the pairing and geographical constraints, there are a number of further constraints that we have to respect over the Christmas period. For a two fixture schedule a team must play one game at home and one at an away venue (or away and then home); a so called home/away pattern of HA (resp. AH). For four complete fixtures the home away pattern must be HAHA (or AHAH). Furthermore, it is not permissible for teams to play each other twice over these two, or four, complete fixtures. For example, Chelsea cannot play Liverpool and later in the two or four sequence, Liverpool play Chelsea.

When generating these fixtures, the overall aim is to minimize the overall distance for all the clubs. Analyzing previous seasons (and personal correspondence with the football authorities) shows that this is indeed the primary objective of these fixtures. In this chapter, we are able to generate fixtures that are significant improvements over the published fixtures (i.e. those that were actually used) but we also present a number of experiments which indicates that the problem owner might prefer slightly worse solutions but which appear to be fairer to the clubs as it limits the maximum distance that a club would have to travel.

It is not clear why some seasons require two sets of fixtures to be generated, yet other seasons require four sets of fixtures to be generated. We thought that four fixtures were generated in order to complete the football season slightly earlier that usual to enable the national side more time to prepare for a Summer tournament (the FIFA World Cup or the UEFA European Championship). However, the data does not support this view (see Section 3). However, due to the methodology proposed in this chapter, the football authorities could easily generate both two or four completes fixtures and decide which one they prefer.

To assist other researchers we note that all the published fixtures were obtained from the Rothmans/Sky Sports Yearbooks Rollin and Rollin [2002, 2003, 2004, 2005, 2006, 2007, 2008]. The distance information was collated by ourselves using one of the UK motoring web sites where we entered the to/from postcodes of the football clubs to get the driving distance between the clubs. This, we believe is preferable to using the straight line distance. As the driving distances will change over time, we have made these distances available at (*for reviewers: we will make the data available on or our own web site*).

This chapter is organised as follows. In the next section we provide some background to sports scheduling. In Section 3 we analyze the previous season's fixtures to try and ascertain when it is required to generate two or four complete fixtures. The analysis is inconclusive but we believe that it is interesting to present this data for future researchers. In Section 4 we present our mathematical model, which is capable of generating two or four complete fixtures. In Section 5 we describe the various experiments that we conduct, followed by the results for each experiment. We discuss the results in Section 6 and conclude the chapter in Section 7.

2 Background

Various algorithms exist which produce double round robin tournaments, with the most well known probably being the polygon construction method (Dinitz et al [2006]). We are unable to use this method, in its raw form, as the generated fixtures would not be acceptable to all interested parties. That is, it would generate a valid double round robin tournament but the schedule would not adhere to other constraints imposed by the football clubs, football authorities, the supporters, the police etc. Nor would it minimize the distances, which is the prime objective.

Previous work has considered the minimization of travel distances for sports schedules. Costa Costa [1995], for example, investigated the scheduling requirements of the National Hockey League, where one of the factors was to minimize the distances traveled. Recent work Westphal [2011] has investigated reducing the distances that have to be driven on 2nd January 2012 for the German Basketball League. The fixtures were such that they form a minimum weight perfect matching (with respect to distances). This provides evidence that this area of sports scheduling is important even for relatively small leagues, and even when only one day is involved. The introduction of the Traveling Tournament Problem Easton et al [2001], using distances based on road trips that have to be undertaken by Major League Baseball teams in the United States, has helped promote research interest in this area. See, for example Crauwels and van Oudheusden [2002]; Ribeiro and Urrutia [2004]; Easton et al [2003]; Westphal and Noparlik [2010], with the best results being reported in Anagnostopoulos et al [2006]. An up to date list of the best known solutions, as well as details of all the instances, can be found at Trick [2009].

Urrutia and Ribeiro Urrutia and Ribeiro [2004] have shown that minimizing distance and maximizing breaks (two consecutive home or away games) is equivalent. This followed previous work de Werra [1981, 1988]; Elf et al [2003] showing how to construct schedules with the minimum number of breaks.

Overviews and surveys of sports scheduling can be found in Easton et al [2004]; Knust [2009]; Rasmussen and Trick [2008]; Kendall et al [2010a].

The problem that we consider in this chapter is the minimization of the distance traveled for two (or four) complete fixtures. These two (or four) complete fixtures can be used over the Christmas period when, for a variety of reasons, teams wish to limit the amount of traveling. Note, that this is a different problem to the Traveling Tournament Problem (Easton et al [2003]), which assumes that teams go on road trips, and so the total distance traveled over a season can be minimized. In English football, there is no concept of road trips, so the overall distance cannot be minimized. However, we are able to minimize the distance on certain days. Kendall Kendall [2008] adopted a two-phase approach to produce two complete fixtures with minimal distances. A depth first search was used to produce a complete fixture for one day, for each division. A further depth first search created another set of fixtures for another day. This process produced eight separate fixtures which adhered to some of the constraints (e.g. a team plays at home on one day and away on the other) but had not yet addressed the constraints with respect to pair clashes (where certain teams cannot play at home on the same day, see Appendix C and Table A2 in Kendall [2008]), the number of teams playing in London etc. (see Appendix D in Kendall [2008]). The fixture lists from the depth first searches were input to a local search procedure which aimed to satisfy the remaining constraints, whilst minimizing the overall distance traveled. The output of the local search, and a post-process operation to ensure feasibility, produced the results in Table 32.

3 Fixture Analysis

In Kendall [2008] an analysis was given of the four seasons considered in that paper. In this section we provide a more comprehensive analysis as we are now considering three additional seasons and we also extend the analysis to include four fixtures. For each season we consider the fixtures that were played around the Christmas period, seeking to find home and away patterns that we can use to classify it as a two or a four fixture. We also look at the distances and state whether the distances traveled for these fixtures are the minimum when compared to other complete fixtures in the season. We end up with a classification for each season.

3.1 Season 2002-2003

This season has four sets of complete fixtures (see Table 1) around the Christmas/New Year period. The fixtures played on 26th December and 1st January represent the lowest distances of any complete fixtures throughout the season. They also exhibit the property that if a team plays home on one day, they play away on the other (and vice versa) (i.e. HA or AH). The other complete fixtures (20/21/22/23 Dec and 28/29 Dec) are significantly higher with respect to distances, and there are no other complete fixtures in the season that have lower distances. In addition, the four complete fixtures do NOT have a HAHA (resp. AHAH) sequence for home and away patterns for each team. Therefore, this season is classified as a two fixture season, with a total of (3820+3964)=7784.

Table 1 Candidate complete fixtures for the 2002-2003 season. The selected fixtures are in bold and this season is classified as a *two* fixture season (see text for details).

Dates	# of fixtures	Distance
20th Dec 2002	4	484
21st Dec 2002	40	6016
22nd Dec 2002	1	1
23rd Dec 2002	1	199
	Total	6700
26th Dec 2002	46	3820
	Total	3820
28th Dec 2002	43	6871
29th Dec 2002	3	712
	Total	7583
1st Jan 2003	46	3964
	Total	3964

3.2 Season 2003-2004

This season has three sets of complete fixtures (see Table 2) around the Christmas/New Year period. The fixtures played on 26th and 28th December represent the lowest distances of any complete fixtures throughout the season. They also exhibit the property that if a team plays at home on one day, they play away on the other (and vice versa) (i.e. HA or AH). The other complete fixture (20th December) is higher with respect to distances, and there are other complete fixtures in the season that have lower distances. Therefore, this season is classified as a two fixture season, with a total of (3837+4342)=8179.

Table 2 Candidate complete fixtures for the 2003-2004 season. The selected fixtures are in bold and this season is classified as a *two* fixture season (see text for details).

Dates	# of fixtures	Distance
20th Dec 2003	46	6295
	Total	6295
26th Dec 2003	46	3837
	Total	3837
28th Dec 2003	46	4342
	Total	4342

3.3 *Season 2004-2005*

This season has five sets of complete fixtures (see Table 3) around the Christmas/New Year period. The fixtures played on 26th December are the lowest distances of any complete fixtures throughout the entire season. The fixtures on the 28th/29th are also amongst the minimal distances. There are some lower distances (e.g. 11th-13th September, 4985; 5th March, 5852; 23rd April, 5813) but we have to bear in mind that the fixtures on 26th December and 28th/29th December adhere to the HA (resp. AH) constraint. The fixtures on the 1st and 3rd Jan, although not being the lowest distances in the season for complete fixtures, do adhere to the HA (resp. AH) constraint. The fixtures on the 18th-20th Dec can be ignored as they do not have a HA (resp. AH) relationship with any of the other fixtures. Therefore, this season is classified as a four fixture season, with four sets

Table 3 Candidate complete fixtures for the 2004-2005 season. The selected fixtures are in bold and this season is classified as a *four* fixture season (see text for details).

Dates	# of fixtures	Distance
18th Dec 2004	44	5758
19th Dec 2004	1	79
20th Dec 2004	1	15
	Total	5852
26th Dec 2004	46	4563
	Total	4563
28th Dec 2004	45	6164
29th Dec 2004	1	285
	Total	6449
1st Jan 2005	46	5122
	Total	5122
3rd Jan 2005	46	7139
	Total	7139

of complete fixtures (4563+6449=11,012 and 5122+7139=12,261), giving a total of (11,012+12,261)=23,273). When we later analyze this season (see Section 6.3.2) as a two fixture schedule we use 26th December (4563) and 28th/29th December (6449) (total of 11012) as the comparator as these follow a HA (resp. AH) pattern and these are the lowest distances from the two sets of complete fixtures.

3.4 Season 2005-2006

This season has four sets of complete fixtures (see Table 4) around the Christmas/New Year period. The fixtures are amongst the lowest across the entire season. There are some equally low distances, however, those on the 17th April and 1st April are a reverse of those on 26th December and 31st December resp., and so could not be used over Christmas as it would violate the no reverse constraint. The four sets of fixture adhere to the HAHA (resp. AHAH) constraint. Therefore, this season is classified as a four fixture season, with two sets of complete fixtures (4295+6331=10,626 and 4488+6645=11,333), giving a total of (10,626+11,333)=21,959 for the four complete fixtures. When we analyze the two fixture case (see Section 6.2), we use the 26th/28th December as these are the minimum of the two sets of complete fixtures.

Table 4 Candidate complete fixtures for the 2005-2006 season. The selected fixtures are in bold and this season is classified as a *four* fixture season (see text for details).

Dates	# of fixtures	Distance
26th Dec 2005	46	4295
	Total	4295
28th Dec 2005	46	6331
	Total	6331
31st Dec 2006	46	4488
	Total	4488
2nd Jan 2006	45	6648
3rd Jan 2006	1	197
	Total	6845

3.5 Season 2006-2007

This season has four sets of complete fixtures (see Table 5) around the Christmas/New Year period. Although each team plays four complete fixtures, the home/away patterns are HAAH (resp. AHHA), rather than the more usual HAHA (resp. AHAH). However, we have still classified this season as a four fixture season, with two sets of complete fixtures (7904+3857=11,761 and 7324+4582=11,906), giving a total of (11,761+11,906)=23,667 for the four complete fixtures. When we later analyze this season as a two fixture schedule (see Section 6.3.3) we will 26th/27th

Table 5 Candidate complete fixtures for the 2006-2007 season. The selected fixtures are in bold and this season is classified as a *four* fixture season (see text for details).

Dates	# of fixtures	Distance
23rd Dec 2006	46	7904
	Total	7904
26th Dec 2006	45	3843
27th Dec 2006	1	14
	Total	3857
30th Dec 2006	46	7324
	Total	7324
1st Jan 2007	46	4582
	Total	4582

December 2006 (3857) and 1st January 2007 (4582) (total of 8439) as the comparator as these follow a HA (resp. AH) pattern and these are the two lowest distances, so it is a fairer comparison.

3.6 Season 2007-2008

This season has four sets of complete fixtures (see Table 6) around the Christmas/New Year period. Like 2006/2007 the home away patterns follow HAAH (resp. AHHA), rather than HAHA (resp. AHAH). However, we still classify this season as a four fixture season, with two sets of complete fixtures (6943+4459=11,402 and 7226+4085=11,311), giving a total of (11,402+11,311)=22,713 for the four complete fixtures. When we later analyze this season as a two fixture schedule (see Section 6.3.4) we will 26th December 2007 (4459) and 1st/2nd January 2008 (4085) (total of 8544) as the comparator as these follow a HA (resp. AH) pattern and these are the two lowest distances, so it is a fairer comparison.

Table 6 Candidate complete fixtures for the 2007-2008 season. The selected fixtures are in bold and this season is classified as a *four* fixture season (see text for details).

Dates	# of fixtures	Distance
21st Dec 2007	4	276
22nd Dec 2007	42	6667
	Total	6943
26th Dec 2007	46	4459
	Total	4459
29th Dec 2007	46	7226
	Total	7226
1st Jan 2008	45	3991
2nd Jan 2008	1	94
	Total	4085

3.7 *Season 2008-2009*

This season has three sets of complete fixtures (see Table 7) around the Christmas/New Year period. The three sets of fixture follow a HAH (resp. AHA) pattern. However, the fixtures on the 26th and 28th December are the lowest distances and we use those fixtures and classify the season as a two fixture season, with two complete fixtures (4548+4764=9,312).

Table 7 Candidate complete fixtures for the 2008-2009 season. The selected fixtures are in bold and this season is classified as a *two* fixture season (see text for details).

Dates	# of fixtures	Distance
20th Dec 2008	46	7709
	Total	7709
26th Dec 2008	46	4548
	Total	4548
28th Dec 2008	46	4764
	Total	4764

3.8 *Discussion*

Of the seven seasons that we study in this chapter, three of them are classified as *two* fixture seasons, with the other four being classified as *four* fixture seasons (see Table 8). We initially believed that the reason a season was classified as a four fixture season was because the football authorities wanted the season to end slightly early to enable the national team to train together in preparation for the tournament. However, this appears not to be the case as we would have expected seasons 2003-2004, 2005-2006 and 2007-2008 to be classified as four fixture seasons and to finish earlier than the other seasons (at least with respect to the Premier division).

Table 8 This table shows whether the football authorities generated a two or four fixture schedule over the holiday period. In all cases, these fixtures represent the minimum distances between clubs when compared against fixtures that are used at other times in the season. We also show whether the season was a World Cup or European Championship year.

Season	Two or Four	End Date (Prem)	End Date (Others)	World or Euro?
2002-2003	Two	11th May 2003	4th May 2003	
2003-2004	Two	15th May 2004	9th May 2004	Euro
2004-2005	Four	14th May 2005	8th May 2005	
2005-2006	Four	7th May 2006	6th May 2006	World
2006-2007	Four	13th May 2007	6th May 2007	
2007-2008	Four	11th May 2008	4th May 2008	Euro
2008-2009	Two	28th May 2009	3rd May 2009	

The data does not support this assumption and we are unsure why some seasons have four complete fixtures at Christmas, and others have two.

When we carry out our experiments, we treat each season as both two and four fixture season so that other researchers have the data for comparative purposes and also to demonstrate that we are able to generate both type of fixtures for the seven seasons that we study.

4 Mathematical Model

In earlier work we presented a naive approach Kendall [2008], and a slightly more sophisticated approach Kendall et al [2010b], in order to tackle the problem addressed in this chapter. These previous works had shortcomings, which are addressed here. Firstly, we only generated two complete fixtures, with the generation of four fixtures being left as future work. Secondly, both previous approaches used a two phase methodology. In the first phase fixtures were generated for individual divisions, without taking into account any constraints that operated across division boundaries. In the second phase, a local search was utilized that removed any hard constraints that were present and also minimized the soft constraint violations. The previous approaches could be time consuming. In particular, Kendall [2008], took upwards of 20 hours for the depth first search phase. Finally, the previous approaches utilized meta-heuristics and so the solutions were not provably optimal. Indeed, the results presented in this chapter are superior to our previous work which had already improved on the published fixtures. For reference, our previous results are summarized in Appendix A.

In this chapter we address these issues by presenting a mathematical formulation that attempts to solve the model in a single phase. That is, we consider all four divisions, eliminating the need for a local search phase to resolve hard constraint violations as the minimization of soft constraint violations.

The model is as follows, with explanations after:

Indices

L	the set of leagues
T	the set of teams
T_l	the set of teams belonging to league l
H	the set of days $\{1,2,\ldots,k\}$
P	the set of paired teams
R	the set of divisions

Decision Variables

$x_{i,j,d}$	1 if team i is playing team j on day d at i's site
$h_{i,d}$	1 if i is playing at home on day d
$y_{i,j,d}$	1 if the paired teams i and j play both at home on day d

Parameters

$D_{i,j}$ the distance (in miles) between team i and team j
L_i 1 if team i is a London-based club
M_i 1 if team i is a Greater Manchester-based club
Q_i 1 if team i is a Premier club
β_l The maximum number of clubs based in London which can play at home on the same day. $\beta_l = 6$.
β_m The maximum number of clubs based in Greater Manchester which can play at home on the same day. $\beta_m = 4$.
β_q The maximum number of Premier Division clubs based in London which can play at home on the same day. $\beta_q = 3$.
δ_r The maximum allowed travel distance for teams in division r.
γ The maximum number of allowed pair clashes.

Objective Function

The objective function minimizes the total distance by all the teams and furthermore helps to adjust the y-variables correctly.

$$\min \sum_{h \in H, l \in L, i,j \in T_l} D_{i,j} \cdot x_{i,j,h} + \sum_{h \in H, \{i,j\} \in P} 0.01 \cdot y_{i,j,h} \tag{1}$$

Subject to

Every team plays exactly one match per day.

$$\sum_{j \in T_l \setminus \{i\}} (x_{i,j,d} + x_{j,i,d}) = 1 \qquad \forall\, l \in L,\ i \in T_l,\ d \in H \tag{2}$$

Every pair of teams meets each other at most once.

$$\sum_{d \in H} \left(x_{i,j,d} + x_{j,i,d}\right) \leq 1 \qquad \forall\, l \in L,\ i,j \in T_l \tag{3}$$

Paired teams are not allowed to play against each other.

$$x_{i,j,d} = 0 \qquad \forall\, d \in H, l \in L, \{i,j\} \in (T_l \times T_l) \cap P \tag{4}$$

Teams in division r are not allowed to travel a distance greater than δ_r miles.

$$x_{i,j,d} = 0 \qquad \forall\, d \in H, l \in L, i,j \in T_l : D_{i,j} > \delta_r \tag{5}$$

The following constraints couple fixture variables x to home-variables h.

$$\sum_{j \in T_l \setminus \{i\}} x_{i,j,d} = h_{i,d} \qquad \forall\, l \in L,\ i \in T_l,\ d \in H \tag{6}$$

Every team plays exactly one home game in two successive days (which implies exactly one away every in those two days)

$$h_{i,d} + h_{i,d+1} = 1 \qquad \forall\, l \in L,\ i \in T_l,\ d \in H \setminus \{k\} \tag{7}$$

Together with the objective function the following inequality ensures that $y_{i,j} = 1$ if and only if the paired teams i and j play both at home on day d.

$$h_{i,d} + h_{j,d} \leq 1 + y_{i,j,d} \qquad \forall\, d \in H, \{i,j\} \in P \tag{8}$$

There are not more than γ pair clashes.

$$\sum_{d \in H, \{i,j\} \in P} y_{i,j,d} \leq \gamma \tag{9}$$

The maximum number of Greater Manchester-based clubs playing at home on any of the holidays must not exceed a certain threshold.

$$\sum_{i \in T} L_i \cdot h_{i,d} \leq \beta_l \qquad \forall\, d \in H \tag{10}$$

The maximum number of London-based clubs playing at home on any of the holidays must not exceed a certain threshold.

$$\sum_{i \in T} Q_i \cdot h_{i,d} \leq \beta_q \qquad \forall\, d \in H \tag{11}$$

The maximum number of London-based Premier Division clubs playing at home on any of the holidays must not exceed a certain threshold.

$$\sum_{i \in T} M_i \cdot h_{i,d} \leq \beta_m \qquad \forall\, d \in H \tag{12}$$

Notes

1. $L = \{$*FA Premeirship, FL Championship, FL Championship 1, FL Championship* 2$\}$. These are the four main divisions in the English league. The names of the divisions change in line with sponsorship agreements.
2. $|T|$ = 92, these being made up from 20 teams in the Premier division and 24 teams in each of the other three divisions.
3. H is the set of days, which will either be {1,2} when generating a two fixture schedule or {1,2,3,4} when generating a four fixture schedule.
4. P = a set of teams that are paired. If two teams are *paired* they, ideally, should not play at home on the same day. However, it is impossible to have zero pairing violations so we allow the same number that were present in the published fixtures (see Table 33 for the number of pairing violations that we allow). Details of the actual paired teams are given in Kendall [2008]. They are not reproduced here for reasons of space.

5. γ defines the number of pair clashes that we allow. In the model, γ takes different values for each season (see Table 33). In previous work we defined separate values for Boxing Day and News Years Day (see Kendall [2008]) but this is no longer valid as we are producing schedules for both these days and also for four days. However the values used in Kendall [2008] are still used but are added together.
6. δ_r defines the maximum distance that can be traveled by any single team in division r. In our experiments, we try different values for δ_r, to test its effect.
7. Equation 1 minimizes the overall distance. The second term ensures that $y_{i,j,h} = 0$ if the paired teams i and j do not both play at home on day d.

5 Experimental Setup and Results

Using the model from Section 4 we used CPLEX 12.2 in order to solve various instances of the model so that we could explore several scenarios. All experiments were run on an Acer Ferrari 1100 laptop, with a 2.29 GHz processor (AMD Turion 64X2 Mobile, Technology TL-66), with 2.29GB of RAM and running Windows XP Professional (Version 2002, SP3)). We allowed CPLEX to run for 300 seconds (five minutes) and 7200 seconds (2 hours) for each experiment. Each scenario is presented below, along with the results. For each scenario we run two experiments $H = \{1,2\}$ (to capture the two day case) and $H = \{1,2,3,4\}$ (to capture the four day case).

We note that for all the solutions we present, they are an improvement on the published fixtures, as well as being an improvement on our previous work (see Table 32 in Appendix A).

In presenting the results, if the *Gap* is less than 0.01% (the default termination criteria for CPLEX), this indicates that a near optimal solution was found before the time expired. If the value in the *Seconds* column is less than 7200 (and the gap is greater than 0.01%), it indicates that CPLEX ran out of memory at the time shown and the result reported is the incumbent solution at that time. We do not report the time for the 300 second experiments as CPLEX never reported an out of memory condition and, unless a value of 0.01% is reported, it ran for the full 300 seconds.

5.1 Experiment 1: $\delta_r = \infty$

This experiment sets no limit on the distance that teams are allowed to travel. This, provides the most flexibility, as any team can play against any other team. The potential drawback is that some teams may travel far greater distances than others, although the total overall distance might be suitably minimized. In these experiments $\delta_0=\delta_1=\delta_2=\delta_3 = \infty$.

5.2 Results for Experiment 1: $\delta_r = \infty$

Allowing the algorithm to run for 300 seconds for the two day case (Table 9) we are able to find near optimal solutions. Allowing the algorithm to run for 7200 seconds (2 hours) we are able to further improve on the solutions generated. Season 2002-2003 was the only one where we could not get within 1% of the optimal solution.

For the four day case (Table 10) we are able to find solutions which are typically around 15% of the lower bound, although it seems appropriate to allow the algorithm to run for 7200 seconds, as this does provide better solutions that just allowing 300 seconds. There is further work to be done to decrease the optimality gap even further. However we note that for the four seasons that are classified as four fixture seasons (see Table 8) that we are able to produce much better solutions than the published fixtures (compare with see Table 32).

Table 9 Results: Experiment 1a: $\delta_r = \infty$ (two day case: $H = \{1,2\}$)

	300 Seconds			7200 Seconds			
Season	**LB**	**Found**	**% Gap**	**LB**	**Found**	**% Gap**	**Seconds**
2002-2003	4556.49	4905.09	7.11	4692.07	4801.09	2.27	5640
2003-2004	5172.79	5225.11	1.00	5185.91	5209.11	0.45	
2004-2005	5107.88	5182.10	1.43	5134.64	5161.10	0.51	4065
2005-2006	5037.63	5038.13	0.01	5037.63	5038.13	0.01	
2006-2007	5271.32	5373.11	1.89	5294.53	5308.11	0.26	
2007-2008	5002.78	5043.12	0.80	5019.79	5034.12	0.28	
2008-2009	5212.14	5245.10	0.63	5243.42	5244.10	0.01	3624

Table 10 Results: Experiment 1b: $\delta_r = \infty$ (four day case: $H = \{1,2,3,4\}$)

	300 Seconds			7200 Seconds			
Season	**LB**	**Found**	**% Gap**	**LB**	**Found**	**% Gap**	**Seconds**
2002-2003	11368.16	13995.18	18.77	11376.85	13813.18	17.64	
2003-2004	11890.66	14288.22	16.78	11896.10	13966.22	14.82	
2004-2005	12036.40	14255.20	15.56	12039.56	13605.20	11.51	
2005-2006	12221.16	14075.26	13.17	12221.19	13785.26	11.35	
2006-2007	12388.06	14706.22	15.76	12408.73	14262.22	13.00	
2007-2008	11982.16	14971.22	19.97	11984.65	14089.24	14.94	
2008-2009	12264.46	19015.14	35.50	12282.52	14671.20	16.28	

5.3 Experiment 2: δ_r = Maximum

A potential problem with experiment 1 is that some teams may have to travel large distances so that others can travel shorter instances. In this experiment we set a *global* maximum distance such that no team can exceed that distance. If this value is too restrictive there will not be any feasible solutions. To give an example. In the 2003-2004 season, Plymouth's distances from the other teams in its division are (in ascending order) {119, 134, 162, 210, 214, 216, 219, 231, 246, 248, 254, 276, 277, 282, 284, 290, 292, 296, 300, 303, 321, 347, 389}. By simple inspection we can see that if set the maximum travel distance too low (in this case below 210) then it is impossible to generate a four fixture schedule as Plymouth will not be able to play four fixtures. Plymouth is often the team that will define the maximum travel distance, but it may not always be the case. By inspecting each season we can set the maximum distance, both for two season fixtures and for four season fixtures. These are presented in Table 11. It should be noted that there is just one value for each season. That is, $\delta_0=\delta_1=\delta_2=\delta_3$. To continue the example from above, Plymouth must travel 210 miles, in the 2003-2004 season, and this is the value that is applied to every team, *in every division*. This means that teams in the other leagues can also travel up to this maximum distance.

Table 11 Maximum distances for each season that will enable a feasible schedule to be generated. These values define the maximum distance that at least one team has to travel and we set this as a maximum distance that all teams are able to travel. With reference to the model we set δ_r to the value in each cell depending on the season and whether we are generating a two or a four fixture schedule. For each experiment $\delta_0=\delta_1=\delta_2=\delta_3$.

Season	Two	Four
2002-2003	153	165
2003-2004	134	210
2004-2005	199	214
2005-2006	160	209
2006-2007	154	212
2007-2008	153	202
2008-2009	153	190

One potential drawback with this approach is that we are effectively dictating the fixtures for certain teams. Plymouth (in 2003-2004) will be forced to play Bournemouth (134 miles), Brentford (210 miles), Bristol City (119 miles), Swindon Town (162 miles). For the two fixture season (with a maximum distance of 134), Plymouth will be forced to play Bournemouth (134 miles), Bristol City (119 miles).

Another potential drawback is that we are giving too much scope to other divisions, as they are allowed to use the same maximum traveling distance as the division which has imposed the upper limit. We consider an extension of the model in Section 5.5 by setting δ_r for each division.

5.4 Results for Experiment 2: δ_r = Maximum

Table 12 presents the results for the two day case. Like experiment 1 we are able to produce results to within 1% of optimality, with the exception of season 2002-2003.

For the four day case (Table 13), the 2002-2003 season proved to be intractable in that we never found an incumbent solution even after 7200 seconds. The other seasons produced similar results to the first experiment (i.e. around 15% of the lower bound).

Table 12 Results: Experiment 2a: δ_r = *maximum* (two day case: $H = \{1,2\}$)

	300 Seconds			7200 Seconds			
Season	**LB**	**Found**	**% Gap**	**LB**	**Found**	**% Gap**	**Seconds**
2002-2003	4742.30	4862.09	2.46	4767.61	4862.09	2.54	1800
2003-2004	5284.99	5311.11	0.49	5295.95	5311.11	0.29	
2004-2005	5176.87	5212.10	0.68	5199.59	5212.10	0.24	
2005-2006	5039.63	5040.13	0.01	5039.63	5040.13	0.01	
2006-2007	5305.31	5358.11	0.99	5323.76	5358.11	0.64	
2007-2008	5084.33	5096.12	0.23	5095.62	5096.12	0.01	
2008-2009	5337.95	5365.10	0.51	5364.56	5365.10	0.01	

Table 13 Results: Experiment 2b: δ_r = *maximum* (four day case: $H = \{1,2,3,4\}$)

	300 Seconds			7200 Seconds			
Season	**LB**	**Found**	**% Gap**	**LB**	**Found**	**% Gap**	**Seconds**
2002-2003	11394.16	-	-	11404.53	-	-	
2003-2004	11984.64	14825.22	19.16	11990.19	13805.22	13.15	
2004-2005	12088.16	14614.20	17.28	12089.37	13782.20	12.28	
2005-2006	12237.16	18128.22	32.50	12237.80	13884.26	11.86	
2006-2007	12439.16	15004.22	17.10	12445.41	14619.22	14.87	
2007-2008	12007.80	14938.24	19.62	12010.51	14011.24	14.28	
2008-2009	12427.14	-	-	12469.91	14962.20	16.66	

5.5 *Experiment 3: δ_r = Maximum for Each Division*

In experiment 2, a global δ_r (i.e. $\delta_0=\delta_1=\delta_2=\delta_3$) was used across all divisions. In this experiment we explore if having a δ_r for each division is beneficial. We did plan to derive δ_r in the same way as experiment 2. However, it cannot easily be done by inspection, An example will explain why. Consider the 2005-2006 season, Premier Division. If we look for the team that has to travel the furthest, we find that Portsmouth has the following travel distances (in ascending order); $\{70, 71, 79, 79, 82, 82, 153, 161, 165, 243, 245, 247, 255, 256, 257, 268, 311, 336, 341\}$. Portsmouth has to travel at least 71 miles to complete a two fixture schedule (and 79 miles for a four fixture schedule). Therefore we can set δ_0=71. However, if we analyze this, we can see that this will mean that Portsmouth will play Fulham (70 miles away) and Chelsea (71 miles away). Looking at other teams, we note that Newcastle also only has two potential fixtures (Sunderland 15 miles away and Middlesbrough 45 miles away). The other fixtures for Newcastle are all greater than 71 miles. The problem is, Newcastle cannot play Sunderland, as they are paired. Therefore, if we set δ_0=71, there is no feasible solution. However, this is a simple case and it is not always obvious what values should be used, especially when we look at the four fixture case. We could use something such as constraint programming to determine suitable values but as we already have a model we decided to use that. Therefore, in order to derive the value for each division we proceeded as follows:

1. For each team in the Premiership, obtain their distance vectors (as we did for Portsmouth above) and sort each one in ascending order.
2. Find the team for the Premiership that has to travel the largest distance in the second position of the sorted distance vectors. In the example above, this will be Portsmouth whose distance vector is $\{70, 71, 79, 79, 82, 82, 153, 161, 165, 243, 245, 247, 255, 256, 257, 268, 311, 336, 341\}$. Element two is 71, which is the largest value for all teams.
3. Find the team that has the largest distance in the third position of the sorted distance vectors. In our 2005-2006 example, this is Newcastle $\{15, 45, 128, 144, 152, 153, 155, 170, 171, 200, 203, 212, 279, 280, 280, 285, 286, 287, 341\}$, so we take the value of 128.
4. We continue this process, taking the maximum values from the fourth, fifth, sixth values etc. from the sorted distance vectors. We do not need to carry out a complete analysis (although it is not time consuming, we simply used the SMALL function in Excel) as we do not need all the values.
5. This leads to a vector of $\{71, 128, 144, 152, ...\}$.
6. We now solve the model using $\delta_0 = 71$ and δ_r=∞ for all the other divisions (i.e. δ_1=δ_2=δ_3=∞). If CPLEX reports an infeasible solution, or has not generated an incumbent solution in 1800 seconds (30 minutes), we set δ_0 to the next value and try to solve again. Eventually, we will solve the model, or at least have an incumbent solution, so that we know that there is a feasible solution.
7. We now fix that δ_0 value and move onto the next division and repeat the process.
8. After carrying out this process for each division, we will have four δ_r values that we can use to solve the model.
9. A similar process is repeated for the four fixture case but the initial index into the distance vectors is element four, rather than element two.

We believe that this process has the benefit that as we consider the Premiership first, this will establish the lowest maximum distance for that division. This seems the right thing to do as more fans are affected by the Premierships teams (as they have larger fan bases, larger stadiums, attract more media interest etc.) so minimizing their distances first seems worthwhile.

The δ_r values we derived are shown in Table 14.

Table 14 Maximum distances for each division using the process presented in Section 5.5

	Prem		Champ		Div 2		Div 3	
Season	Two (δ_0)	Four (δ_0)	Two (δ_1)	Four (δ_1)	Two (δ_2)	Four (δ_2)	Two (δ_3)	Four (δ_3)
2002-2003	128	144	117	127	153	162	124	168
2003-2004	104	153	124	148	134	210	116	202
2004-2005	128	170	199	214	109	183	106	147
2005-2006	128	153	160	209	111	166	134	199
2006-2007	135	150	154	212	109	176	143	183
2007-2008	142	152	153	202	124	150	86	141
2008-2009	143	161	153	199	124	156	145	180

5.6 Results for Experiment 3: δ_r = Maximum for Each Division

For the two day case (see Table 15), this experiment manages to produce similar solutions to the other experiments, in that solutions within 1% of optimality are obtained, with the exception of the 2002-2003 season. The four fixture case is more challenging (see Table 16). Only two seasons could generate a solution within 300

Table 15 Results: Experiment 3a: δ_r = *maximum for each division* (two day case: $H = \{1,2\}$)

	300 Seconds			7200 Seconds			
Season	LB	Found	% Gap	LB	Found	% Gap	Seconds
2002-2003	4613.07	4965.09	7.09	4829.12	4958.09	2.62	2617
2003-2004	5339.89	5398.11	1.08	5360.32	5377.00	0.31	
2004-2005	5305.41	5345.10	0.74	5331.64	5345.10	0.25	
2005-2006	5081.13	5082.13	0.02	5081.13	5082.13	0.02	41
2006-2007	5325.86	5393.11	1.25	5365.08	5376.11	0.21	
2007-2008	5107.56	5153.12	0.88	5131.62	5132.12	0.01	362
2008-2009	5368.23	5385.10	0.31	5384.56	5385.10	0.01	1241

Table 16 Results: Experiment 3a: δ_r = *maximum for each division* (four day case: $H = \{1,2,3,4\}$)

	300 Seconds			7200 Seconds			
Season	LB	Found	% Gap	LB	Found	% Gap	Seconds
2002-2003	-	-	-	11525.04	14246.18	19.10	
2003-2004	-	-	-	12040.16	14465.22	16.76	
2004-2005	-	-	-	12110.68	14107.20	14.15	
2005-2006	12314.16	16909.26	27.18	12323.17	14284.26	13.73	
2006-2007	12367.57	16435.22	24.75	12511.40	14659.22	14.65	
2007-2008	-	-	-	12071.43	14471.24	16.58	
2008-2009	-	-	-	12525.00	14946.20	16.20	

seconds. If we allow 7200 seconds, a solution was always returned and, similar to the other experiments, the solutions were about 15% of the lower bound. The solutions, with regard to the overall distance, are slightly higher than the other experiments but as we will discuss in the next section, this is not necessarily a problem.

6 Discussion

The results we reported in Section 5 are difficult to interpret, just by looking at the tables. In this section, we analyze the results for two seasons but they are representative of the underlying themes throughout the seven seasons (we summarize the other seasons in Section 6.3).

6.1 *Season 2002-2003*

We choose this season to analyze as it appears to be the most *difficult* season given that the gap is consistently over 1% whereas all other results (for the two day case) are under 1%. In Table 17 we present a summary of the various experiments. The table shows the total distance for the generated schedule, the maximum distance traveled by a team, the number of times that a team has to travel 180 miles or more (we chose this figure as 180 miles represents about three hours of driving time which seems a reasonable time limit for travel at this time of the year) and the number of Derby Clashes (i.e. when paired teams play each other). Our model actually treats Derby Clashes as a hard constraint (eq. 4), so for our experiments this value is always zero, but the published fixtures sometimes allow them.

Table 17 also shows the published fixture for the 2002-2003 season. The total distance was 7884 miles and the maximum distance for any one fixture was 171 miles (Newcastle vs Liverpool). No team had to travel over 180 miles but Rotherham and Sheffield Wednesday (which are paired) played each other. Having paired teams play each other is often beneficial, as far as minimizing the distance is concerned, as they are often local derbies and, by definition, the teams are close to each other (the distance between Rotherham and Sheffield Wednesday is 7 miles).

In all cases our model (we would suggest) is a significant improvement over the published fixtures (distances of under 5000 miles compared to 7884 miles). For our experiments, the maximum distance traveled by a single team is also an improvement over the published fixture (153 or 157 miles compared to 171 miles).

Choosing which experimental setup a user should choose would initially suggest 2a (as it has the lowest overall distance of 4862 miles) but we would urge caution. Experiment 3a sets a limit at the division level whereas both experiment 1a and 2a could allow greater distances, especially 1a, which allows infinite (of course the maximum distance is actually capped) travel distances. If we compare experiment 2a and 3a, we find that for the Premiership the total distance traveled is 908 miles (resp. 825 miles) for experiment 3a (resp. 2a). Therefore, it might appear that it would be more sensible to select experiment 2a as the methodology of choice. However, for the 3a experiment, as we set the maximum distance at the division level, no team had to travel more than 115 miles (Tottenham vs Aston Villa). For experiment 2a Southampton had to travel to Aston Villa (143 miles). Looking at the other divisions for experiment 2a, the maximum distances for each division are (we give all four divisions) Exp-2a={143, 116, 153, 145}. The maximum distances for experiment Exp-3a={115, 117, 153, 123}. For experiment 1a the values are Exp-1a={143, 156, 157, 124}. Apart from a single mile (116 vs 117), experiment 3a produces the same, or lower, maximum distances than the other two experiments. Therefore, there would be a decision to be made. Does the problem owner want to minimize the total distance or take a more local view and ensure that no one club has to travel over a certain distance? There is no definitive answer as to which experiment returns the best result but as each experiment only takes five minutes, there is no reason why we cannot simply provide the user with all the solutions and let them decide which one is best.

Running the experiments for longer (7200 seconds) makes little difference to the overall results (see Table 18). The maximum distances for each division, for each experiment is as follows; Exp-1a={135, 117, 157, 124}, Exp-2a={143, 117, 153, 124}, Exp-3a={115, 117, 153, 124}. Exp-3a has the lowest (or equal) maximums across all divisions.

Table 17 Analysis: Season 2002-2003 (two day case, 300 seconds)

Experiment	Total Distance	Maximum Distance	# > 180	# of paired teams playing each other
Published	7884	171	0	1
1a (table 9)	4905	157	0	0
2a (table 12)	4862	153	0	0
3a (table 15)	4965	153	0	0

Table 18 Analysis: Season 2002-2003 (two day case, 7200 seconds)

Experiment	Total Distance	Max Distance	# > 180	# of paired teams playing each other
Published	7884	171	0	1
1a (table 9)	4801	157	0	0
2a (table 12)	4862	153	0	0
3a (table 15)	4958	153	0	0

For the four day case, we only consider the 7200 second experiment as we are likely to run the experiment for this amount of time if we were planning to use the results, as the 300 second experiment does not always return a solution. In fact, experiment 2b did not return a solution for the 7200 experiment so we cannot analyze it here.

The summary is presented in Table 19. We do not show the published fixtures as this season is classified as a two fixture season, so no data is available. Similar to the two days case, experiment 3b has a larger overall distance (14246 cf 13813) but has no fixtures that require a team to travel 180 miles or more. By comparison, experiment 1b has four fixture that requires teams to travel 180 miles or more. In fact, for experiment 3b, the maximum distance is only 168 miles. Experiment 1b has five fixtures greater than this, the four above 180 miles and another of 169 miles.

We note that the maximum distances for each division are as follows; Exp-1b={152, 166, 210, 202}; Exp-3b={144, 124, 162, 168}. Experiment 3b returns the lowest maximums across all four divisions.

Table 19 Analysis: Season 2002-2003 (four day case, 7200 seconds)

Experiment	Total Distance	Max Distance	# > 180	# of paired teams playing each other
1b (table 9)	13813	210	4	0
2b (table 12)	-	-	-	-
3b (table 15)	14246	168	0	0

6.2 *Season 2005-2006*

We chose to analyze the 2005-2006 season as this appears to be the *easiest* season as the gap in Table 9 is the lowest (0.01%) of all the seasons. However, it would appear that the football authorities had problems scheduling these fixtures as for the two day case there were 17 teams (see Table 20) that had to travel 180 miles or more and for the four day case (see Table 21 and also Appendix B) there were 37 teams that had to travel 180 miles or more.

At first sight, the fixtures that we have generated for the two fixture case seem to be a lot better than the published fixtures. However, we need to bear in mind that the 2005-2006 season is classified as a four fixture season (see section 3.4) so it is not really a fair comparison. However, we give our results to enable others to compare against our results. We also note that of the two figures available (see Table 32) we take the lowest as a comparison (i.e. of 10,626 and 11,333, we report 10,626 in this analysis).

All the experiments produced similar results with the maximum travel distance being either 160 or 161 miles. Experiment 3a produced a slighter higher overall distance (5082 miles) but, like 2002-2003 we are guaranteed to have a maximum distance traveled for each division. We have not shown the results for the 7200 second experiment, for the two day case, as the results are identical.

The maximum distances for each division are as follows; Exp-1a={161, 160, 112, 134}; Exp-2a={153, 160, 112, 134}; Exp-3a={128, 160, 111, 134}. Experiment 3a returns the lowest (or equal) maximums across all for divisions.

The more interesting analysis is for the four fixture schedule. These results are summarized in Table 21. We only provide the results for the 7200 second experiments as the gap tends to be quite large when we only allow 300 seconds, if a solution is found at all.

Table 20 Analysis: Season 2005-2006 (Two day case, 300 seconds)

Experiment	Total Distance	Max Distance	# > 180	# of paired teams playing each other
Published	10626	304	17	3
1a (table 9)	5038	161	0	0
2a (table 12)	5040	160	0	0
3a (table 15)	5082	160	0	0

All the experiments give superior results to the published fixtures. Experiment 3b only has four fixtures where teams have to travel more 180 miles or more, whereas all the other solutions have at least eight teams traveling 180 miles or more and the published fixtures has 37 teams. This does come at the expense of a slightly higher overall total. If we look at the best solution, with regard to overall distance, (13785) the result from experiment 3b is 499 miles higher. This represents an average of just under three extra miles across the 184 fixtures. It would be up to the problem owner to make the final judgement which solution they prefer.

We note that the maximum distances for each division are as follows; Exp-1b={200, 220, 233, 219}; Exp-2b={203, 209, 182, 207}; Exp-3b={153, 209, 165, 199}. Experiment 3b returns the lowest (or equal) maximums across all for divisions.

Table 21 Analysis: Season 2005-2006 (Four day case, 7200 seconds)

Experiment	Total Distance	Max Distance	# > 180	# of paired teams playing each other
Published	21959	352	37	4
1b (table 10)	13785	233	8	0
2b (table 13)	13884	209	8	0
3b (table 16)	14284	209	4	0

6.3 *Other Seasons*

For completeness, we provide the summary tables, for the seasons not analyzed above. As above, if the season was not classified as a four season fixture we cannot provide the published fixture figures, although we still calculate our values for that season. Also, similar to above, when a season is classified as a four fixture season, we use the two fixture schedule that we indicated in the relevant section. We note again that this is not really a fair comparison with the published fixtures, but we are using the minimum distances in order to be as fair as possible.

6.3.1 2003-2004

As noted above, for the two day case, we are using an overall distance total of 8179 (see Section 3.2). As this season has been classified as a two fixture season, we cannot provide the statistics for the published fixture for the four day case, but we still generate our own set of fixtures.

Tables 22 and 23 summarizes the results for the 2003-2004 season.

The maximum distances for each division, for the two day case, are as follows; Exp-1a={188, 156, 134, 165}; Exp-2a=128, 128, 134, 134; Exp-3a={104, 124, 134, 116}. Experiment 3a returns the lowest (or equal) maximums across all four divisions

The maximum distances for each division, for the four day case, are as follows; Exp-1b={158, 157, 248, 243}; Exp-2b=170, 159, 210, 202; Exp-3b={153, 148, 210, 202}. Experiment 3b returns the lowest (or equal) maximums across all for divisions.

Table 22 Analysis: Season 2003-2004 (Two day case, 7200 seconds)

Experiment	Total Distance	Max Distance	# > 180	# of paired teams playing each other
Published	8179	210	3	0
1a (table 9)	5209	188	1	0
2a (table 12)	5311	134	0	0
3a (table 15)	5377	134	0	0

Table 23 Analysis: Season 2003-2004 (Four day case, 7200 seconds)

Experiment	Total Distance	Max Distance	# > 180	# of paired teams playing each other
1b (table 10)	13966	248	4	0
2b (table 13)	13805	210	3	0
3b (table 16)	14465	210	3	0

6.3.2 2004-2005

Tables 24 and 25 summarizes the results for the 2004-2005 season.

The maximum distances for each division, for the two day case, are as follows; Exp-1a={162, 219, 109, 124}; Exp-2a=161, 199, 109, 118; Exp-3a={128, 199, 209, 106}. Experiment 3a returns the lowest (or equal) maximums across all for divisions

The maximum distances for each division, for the four day case, are as follows; Exp-1b={200, 242, 208, 182}; Exp-2b=165, 214, 197, 144; Exp-3b={165, 214, 183, 147}. Experiment 3b returns the lowest (or equal) maximums across all for divisions

Table 24 Analysis: Season 2004-2005 (Two day case, 7200 seconds)

Experiment	Total Distance	Max Distance	# > 180	# of paired teams playing each other
Published	11012	257	9	0
1a (table 9)	5161	219	1	0
2a (table 12)	5212	199	1	0
3a (table 15)	5345	199	1	0

Table 25 Analysis: Season 2004-2005 (Four day case, 7200 seconds)

Experiment	Total Distance	Max Distance	# > 180	# of paired teams playing each other
Published	21959	???	??	?
1b (table 10)	13605	242	8	0
2b (table 13)	13782	214	8	0
3b (table 16)	14107	214	6	0

6.3.3 2006-2007

As noted above, for the two day case, we are using an overall distance total of 8439 (see Section 3.5). For the four day case, we are using a distance total of 23667

Tables 26 and 27 summarizes the results for the 2006-2007 season.

Table 26 Analysis: Season 2006-2007 (Two day case, 7200 seconds)

Experiment	Total Distance	Max Distance	# > 180	# of paired teams playing each other
Published	8439	213	4	1
1a (table 9)	5308	157	0	0
2a (table 12)	5358	154	0	0
3a (table 15)	5376	154	0	0

Table 27 Analysis: Season 2006-2007 (Four day case, 7200 seconds)

Experiment	Total Distance	Max Distance	# > 180	# of paired teams playing each other
Published	23667	364	42	2
1b (table 10)	14262	291	7	0
2b (table 13)	14619	212	8	0
3b (table 16)	14659	212	5	0

The maximum distances for each division, for the two day case, are as follows; Exp-1a={135, 157, 109, 150}; Exp-2a=150, 154, 137, 150; Exp-3a={135, 154, 109, 143}. Experiment 3a returns the lowest (or equal) maximums across all for divisions

The maximum distances for each division, for the four day case, are as follows; Exp-1b={174, 291, 202, 207}; Exp-2b=175, 212, 194, 199; Exp-3b={150, 212, 176, 183}. Experiment 3b returns the lowest (or equal) maximums across all for divisions

6.3.4 2007-2008

As noted above, for the two day case, we are using an overall distance total of 8644 (see Section 3.6). For the four day case, we are using a distance total of 22713

Tables 28 and 29 summarizes the results for the 2007-2008 season.

The maximum distances for each division, for the two day case, are as follows; Exp-1a={150, 154, 140, 86}; Exp-2a=148, 153, 140, 86; Exp-3a={142, 153, 124, 86}. Experiment 3a returns the lowest (or equal) maximums across all for divisions

The maximum distances for each division, for the four day case, are as follows; Exp-1b={205, 226, 176, 148}; Exp-2b=176, 202, 156, 141; Exp-3b={152, 202, 150, 141}. Experiment 3b returns the lowest (or equal) maximums across all for divisions

Table 28 Analysis: Season 2007-2008 (Two day case, 7200 seconds)

Experiment	Total Distance	Max Distance	# > 180	# of paired teams playing each other
Published	8644	213	1	1
1a (table 9)	5034	154	0	0
2a (table 12)	5096	153	0	0
3a (table 15)	5132	153	0	0

Table 29 Analysis: Season 2007-2008 (Four day case, 7200 seconds)

Experiment	Total Distance	Max Distance	# > 180	# of paired teams playing each other
Published	22713	311	17	2
1b (table 10)	14089	226	6	0
2b (table 13)	14011	202	2	0
3b (table 16)	14471	202	2	0

6.3.5 2008-2009

As noted above, for the two day case, we are using an overall distance total of 9312 (see Section 3.7). As this season has been classified as a two fixture season, we cannot provide the statistics for the published fixture for the four day case, but we still generate our own set of fixtures.

Tables 30 and 31 summarizes the results for the 2008-2009 season.

The maximum distances for each division, for the two day case, are as follows; Exp-1a=$\{143, 190, 152, 199\}$; Exp-2a=143, 153, 152, 145; Exp-3a=$\{143, 153, 124, 145\}$. Experiment 3a returns the lowest (or equal) maximums across all for divisions

The maximum distances for each division, for the four day case, are as follows; Exp-1b=$\{192, 307, 208, 213\}$; Exp-2b=171, 190, 184, 188; Exp-3b=$\{161, 199, 156, 180\}$. Experiment 3b returns the lowest (or equal) maximums for divisions, except for one, where it returns a maximum of 199 in experiment 3b, whereas experiment 2b returned a maximum of 190. This season's results are also different in that experiment 3b returns the lowest overall total.

Table 30 Analysis: Season 2008-2009 (Two day case, 7200 seconds)

Experiment	**Total Distance**	**Max Distance**	**# > 180**	**# of paired teams playing each other**
Published	9312	189	2	1
1a (table 9)	5244	199	2	0
2a (table 12)	5365	153	0	0
3a (table 15)	5385	153	0	0

Table 31 Analysis: Season 2008-2009 (Four day case, 7200 seconds)

Experiment	**Total Distance**	**Max Distance**	**# > 180**	**# of paired teams playing each other**
1b (table 10)	14671	307	7	0
2b (table 13)	14962	190	4	0
3b (table 16)	14946	199	4	0

7 Conclusion

We have presented a single model that minimizes the distance traveled by football supporters over the holiday season. The model is able to produce two or four complete fixtures, depending on the requirements of the football authorities. Several experiments were conducted, varying the parameters of the model. The model is able to produce solutions which are superior to those that are currently used. Previous discussions with the football league have suggested that we meet all the requirements, but it would be useful to hold further discussions with the authorities, as well as the police, in order to establish whether the model needs further refinement.

Of the three experiments that we conducted we would suggest that the option to limit each division to a maximum travel distance would probably be the most suitable to be used in practise as, although it usually produces slightly higher total distances, the solutions produced would probably be seen as being fairer when viewed by the supporters.

For our future work, the model presented in this chapter opens up the possibility to carry out more in-depth and *what-if* analysis on the fixtures for the holiday period.

For example, are we able to reduce pair clashes whilst still minimizing the distance. If this is possible it could make significant savings for the police as they will not have to devote the same amount of resources to police the fixtures. We would also like to investigate weighting each pair clash, and including that in the objective function. This would be interesting as, at the moment, a pair clash between Liverpool and Everton, for example, is the same as a pair clash between Liverpool and Tranmere. However, the police would rather have Liverpool and Tranmere playing at home on the same day as this will be easier to police that the *Merseyside Derby*.

We will also turn our attention to generating schedules for the entire season. Given the experiences reported in this chapter, we do not believe that we will be able to produce optimal solutions and we feel that a (meta-)heuristic approach will be required.

Appendix

Summary of Published Fixtures and Previous Results

The 2005-2006 Fixtures Where Supporters Had to Travel 180 Miles or More

This appendix lists the 37 fixtures from the published fixture for the 2005-2006 season where a team is required to travel 180 miles or more. The figure in brackets is the distance in miles.

Table 32 This table summaries the distances traveled for the published fixtures (i.e. those that were actually played) and also the two fixture schedules that were generated in two previous papers (Kendall [2008]; Kendall et al [2010b]) in order to provide a comparison with the results reported here. Note that there are slight differences from the figures shown in Kendall [2008] for 2002-2003 (7791 cf 7784), 2003-2004 (8168 cf 8179) and 2005-2006 (10631 cf 10626) due to minor errors found in the input data. Where N/A is specified, this indicates that this season did not produce a four fixture schedule. Only two fixture distances are shown for Kendall [2008]; Kendall et al [2010b] as these papers did not generate four fixture schedules, with ** indicating that that paper did not generate schedules for those seasons.

	Published		Kendall [2008]	Kendall et al [2010b]
Season	**Two Day**	**Four Day**	**Two Day**	**Two Day**
2002-2003	7784	N/A	6040	5243
2003-2004	8179	N/A	6359	5464
2004-2005	11,012/12,261	23,273	6784	5365
2005-2006	10,626/11,333	21,959	6917	5234
2006-2007	11,761/11,906	23,667	**	5713
2007-2008	11,402/11,311	22,713	**	5366
2008-2009	9312	N/A	**	5564

1. Darlington vs Torquay (352)
2. Plymouth Argyle vs Leeds (321)
3. Plymouth Argyle vs Preston (304)
4. Newcastle vs Charlton (287)
5. Fulham vs Sunderland (281)
6. Tottenham vs Newcastle (280)
7. Hartlepool vs Southend (276)
8. Blackpool vs Southend (271)
9. Blackburn vs Portsmouth (268)
10. Hartlepool vs Swindon (267)
11. Bournemouth vs Scunthorpe (258)
12. Stockport vs Torquay (257)
13. Bournemouth vs Barnsley (242)
14. Norwich vs Preston (237)
15. Huddersfield vs Gillingham (235)
16. Swansea vs Gillingham (233)
17. QPR vs Burnley (232)
18. Darlington vs Barnet (230)
19. Norwich vs Burnley (229)
20. Tranmere vs Yeovil (229)
21. Doncaster vs Yeovil (228)
22. Everton vs Charlton (227)
23. Torquay vs Rushden & D'monds (224)
24. West Ham United vs Wigan (214)
25. Boston vs Carlisle (212)
26. Manchester City vs Chelsea (210)
27. Wolverhampton vs Plymouth Argyle (209)
28. Bradford vs Brentford (205)
29. Southampton vs Sheffield United (200)
30. Torquay vs Wycombe (199)
31. Manchester City vs Tottenham (199)
32. Arsenal vs Man Utd (197)
33. Stoke vs Ipswich (193)
34. Hull vs Ipswich (192)
35. Grimsby vs Carlisle (187)
36. Colchester vs Scunthorpe (185)
37. Brentford vs Swansea (182)

Number of Allowed Pairing Violations

Table 33 shows the number of pairing violations that were present in the published fixtures. We allow ourselves the same number of violations in our solutions. Note that if (for example) Manchester United and Manchester are both playing at home, this counts as one pairing violation. In Kendall [2008] the same counts were used

Table 33 Number of Allowed Pairing Violations

Year	γ (Two fixtures)	γ (Four fixtures)
2002-2003	9	18
2003-2004	11	22
2004-2005	10	20
2005-2006	13	26
2006-2007	11	22
2007-2008	12	24
2008-2009	10	20

but each violation was counted as two. Further note that the number of pairing violations allowed for the four fixture schedule is simply double that of the two fixture schedule.

References

Anagnostopoulos, A., Michel, L., van Hentenryck, P., Vergados, Y.: A simulated annealing approach to the traveling tournament problem. Journal of Scheduling 9, 177–193 (2006), doi:10.1007/s10951-006-7187-8

Costa, D.: An evolutionary tabu search algorithm and the nhl scheduling problem. In: INFOR, vol. 33, pp. 161–178 (1995)

Crauwels, H., van Oudheusden, D.: A generate-and-test heuristic inspired by ant colony optimization for the traveling tournament problem. In: Burke, E.K., Causmaecker, P.D. (eds.) Proceedings of the 4th International Conference on the Practice and Theory of Automated Timetabling, PATAT 2002, pp. 314–315 (2002)

Dinitz, J., Lamken, E., Wallis, W.: Scheduling a Tournament. In: CRC Handbook of Combinatorial Designs. CRC Press (2006) (a previous edition of this book was published in 1995)

Easton, K., Nemhauser, G.L., Trick, M.A.: The traveling tournament problem description and benchmarks. In: Walsh, T. (ed.) CP 2001. LNCS, vol. 2239, pp. 580–3349. Springer, Heidelberg (2001)

Easton, K., Nemhauser, G.L., Trick, M.A.: Solving the travelling tournament problem: A combined integer programming and constraint programming approach. In: Burke, E.K., De Causmaecker, P. (eds.) PATAT 2002. LNCS, vol. 2740, pp. 100–109. Springer, Heidelberg (2003); a previous version of this paper was published in the PATAT conference

Easton, K., Nemhauser, G., Trick, M.: Sports Scheduling, ch. 52, pp. 52-1–52-19. Chapman & Hall (2004)

Elf, M., Jünger, M., Rinaldi, G.: Minimizing breaks by maximizing cuts. Operations Research Letters 31, 343–349 (2003)

Kendall, G.: Scheduling English football fixtures over holiday periods. Journal of the Operational Research Society 59(6), 743–755 (2008)

Kendall, G., Knust, S., Ribeiro, C., Urrutia, S.: Scheduling in sports: An annotated bibliography. Computers & Operations Research 37, 1–19 (2010a)

Kendall, G., McCollum, B., Cruz, F., McMullan, P.: Scheduling English football fixtures: Consideration of two conflicting objectives. In: McCollum, B., Burke, E. (eds.) Proceedings of the 8th International Conference on the Practice and Theory of Automated Timetabling (PATAT 2010), pp. 1–15 (2010b)

Knust, S.: Classification of literature on sports scheduling (2009), `http://www.inf.uos.de/knust/sportssched/sportlit_class/` (last accessed on October 15, 2009)

Rasmussen, R., Trick, M.A.: Round robin scheduling a survey. European Journal of Operational Research 188, 617–636 (2008)

Ribeiro, C., Urrutia, S.: Heuristics for the mirrored traveling tournament problem. In: Proceedings of the 5th International Conference on the Practice and Theory of Automated Timetabling (PATAT 2004), pp. 323–342 (2004)

Rollin, R., Rollin, J. (eds.): Rothmans Football Yearbook 2002-2003. Headline Book Publishing, London (2002) ISBN: 0-7553-1100-0

Rollin, R., Rollin, J. (eds.): Sky Sports Football Yearbook 2003-2004. Headline Book Publishing, London (2003) ISBN: 0-7553-1228-7

Rollin, R., Rollin, J. (eds.): Sky Sports Football Yearbook 2004-2005. Headline Book Publishing, London (2004) ISBN: 0-7553-1311-9

Rollin, R., Rollin, J. (eds.): Sky Sports Football Yearbook 2005-2006. Headline Book Publishing, London (2005) ISBN: 0-7553-1385-2

Rollin, R., Rollin, J. (eds.): Sky Sports Football Yearbook 2006-2007. Headline Book Publishing, London (2006) ISBN: 0-7553-1526-X

Rollin, R., Rollin, J. (eds.): Sky Sports Football Yearbook 2007-2008. Headline Book Publishing, London (2007) ISBN: 0-7553-1664-9

Rollin, R., Rollin, J. (eds.): Sky Sports Football Yearbook 2008-2009. Headline Book Publishing, London (2008) ISBN: 0-7553-1820-9

Trick, M.: Traveling tournament instances (2009), `http://mat.gsia.cmu.edu/TOURN` (last accessed on October 16, 2009)

Urrutia, S., Ribeiro, C.: Minimizing travels by maximizing breaks in round robin tournament schedules. Electronic Notes in Discrete Mathematics 18-C, 227–233 (2004)

de Werra, D.: Scheduling in sports. In: Studies on Graphs and Discrete Programming, pp. 381–395. North-Holland (1981)

de Werra, D.: Some models of graphs for scheduling sports competitions. Discrete Applied Mathematics 21, 47–65 (1988)

Westphal, S.: Scheduling the german basketball league. Under review (2011)

Westphal, S., Noparlik, K.: A 5.875-approximation for the traveling tournament problem. In: Proceedings of the Practice and Theory of Automated Timetabling (PATAT 2010), pp. 417–426 (2010)

Educational Timetabling

Jeffrey H. Kingston

Abstract. This chapter is an introduction to the problems of timetabling educational institutions such as high schools and universities. These are large problems with multiple sources of NP-completeness, for which robust solvers do not yet exist, although steady progress is being made. This chapter presents the three main problems found in the literature: high school timetabling, university examination timetabling, and university course timetabling. It also examines some major subproblems of these problems: student sectioning, single student timetabling, and room assignment. This chapter also shows how real-world instances of these problems, with their many constraints, can be modelled in full detail, using a case study in high school timetabling as an example.

1 Introduction

Educational timetabling is not a single problem. For each kind of timetable needed by each kind of institution there is a separate problem with a separate literature. These literatures are too large to survey comprehensively within the limits of a book chapter, so only a selection, including recent survey papers, is referenced here. Schaerf (1999) is a good general survey.

Solving a real instance of an educational timetabling problem (an instance taken without simplification from an actual institution) by hand can take weeks of tedious and error-prone work by an expert. Hand-generated timetables are still common, although automated or semi-automated methods are making inroads. For example, the traditional way to timetable students in North American universities, which is to publish lists of course sections and times and expect the students to create their own timetables and sign up for the sections they want, is giving way to a semi-automated method, in which the number of sections of each course and the times they run may still be decided by hand, but the students' timetables are generated automatically.

Jeffrey H. Kingston
School of Information Technologies, University of Sydney, Australia
e-mail: jeff@it.usyd.edu.au

A.Ş. Etaner-Uyar et al. (eds.), *Automated Scheduling and Planning*,
Studies in Computational Intelligence 505,
DOI: 10.1007/978-3-642-39304-4_4,

The lead in educational timetabling has always been given by researchers who are trying to solve real problems from real institutions. This practical orientation has informed the selection of topics for this chapter.

2 Educational Timetabling Problems

The educational timetabling literature mainly studies problems found in *high schools* (schools for older children) and universities. High schools need to timetable their normal activities once per year, or sometimes more often. This is the *high school timetabling problem* (Sect. 4). Universities need to timetable their normal activities once per semester. This is called the *university course timetabling problem* (Sect. 6), to distinguish it from the other main university problem, the *(university) examination timetabling problem* (Sect. 5): the timetabling of examinations after the end of semester. There are other kinds of institutions and problems, but educational timetabling, as it appears in the literature, is essentially about these three problems.

When the number of students enrolled in a high school or university course is large, the course may need to be broken into *sections*: copies of the course, each with its own time, room, and teacher. Each student enrolled in the course must then be assigned to one of its sections. If these assignments are made early in the timetabling process, the result is the *student sectioning* problem (Sect. 7), which aims to assign students to sections so as to facilitate the assignment of times to sections later, by minimizing the number of pairs of sections that share at least one student.

A *phase* is one part of a solver's work, carried out more or less independently of its other phases. Student sectioning is one example of a phase. Other commonly encountered examples are *single student timetabling* (Sect. 8), the creation of a timetable for one student after courses are broken into sections and the sections are assigned times, and *room assignment* (Sect. 9), the assignment of suitable rooms to events after the events' times are fixed. It is often necessary to divide a solve process into phases when faced with large, real instances of timetabling problems, even though it usually rules out all hope of finding a globally optimal solution. This makes individual phases worthy subjects of study in their own right.

All these problems are concerned with assigning times and *resources* (students, teachers, rooms, and so on) to events so as to avoid *clashes* (cases where a resource attends two events at the same time) and violations of various other constraints, such as unavailable times for resources, or restrictions on when events may occur. Some times and resources may be preassigned; others are left open to the solver to assign. Informally, a *timetabling problem* is any problem that fits this description.

Several problems outside the scope of this chapter are timetabling problems by this definition. Nurse rostering is one example. The events are the shifts. The events' times are preassigned, and resources (nurses in this case) must be assigned to them. Another example is sport competition timetabling. The events are the matches, to which times must be assigned. Their resources, largely preassigned, are the teams and venues. These problems are always treated separately, however, and rightly so, because their constraints, other than the avoid-clashes constraint, are very different.

Many timetabling problems can be proved to be NP-complete using a reduction from graph colouring due to Welsh and Powell (1967). For each node of the graph to be coloured, create one event of duration 1. For each edge, create one resource preassigned to the two events corresponding to the edge's endpoints. Then assigning a minimal number of colours to the nodes so that no two adjacent nodes have the same colour is equivalent to assigning a minimal number of times to the events so that no resources have clashes.

The inverse of this construction is the *clash graph*, a widely used conceptual aid. It has one node for each event. An edge joins each pair of events, weighted by the number of resources the two events have in common (Figure 1). Colouring

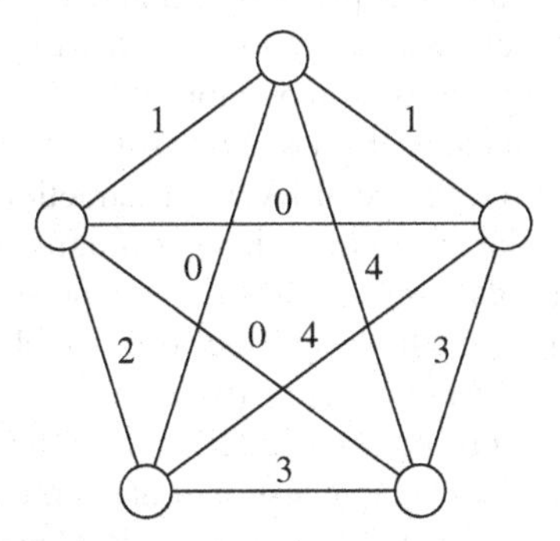

Fig. 1 A clash graph. Each node represents one event; each edge is weighted by the number of resources the events at its endpoints have in common.

this graph to minimize the total weight of edges that join nodes of the same colour is equivalent to assigning a time to each event which minimizes the number of clashes.

It would be a mistake to consider educational timetabling as a branch of graph colouring, however. Timetabling problems have many kinds of constraints. Each may be a source of NP-completeness in its own right, and much of the difficulty lies in handling all of them together. For example, ensuring that students have breaks between examinations is a travelling salesman problem (Sect. 5), balancing teacher workloads is bin packing, and so on (Cooper and Kingston, 1996).

3 Educational Timetabling Models

For most of its history, educational timetabling research has been very fragmented. Each research group has used its own definitions of the problems, and its own data in its own format. There has been little exchange of data, except for one set of instances of the examination timetabling problem.

What is primarily needed to break down these barriers is for researchers to reach consensus on a model, or format, in which instances and solutions can be expressed, including unambiguous rules for calculating the cost of solutions. The terms 'model' and 'format' are roughly interchangeable; 'model' emphasizes the ideas, 'format' emphasizes the concrete syntax that realizes those ideas.

Pure algorithmic problems, such as graph colouring and the travelling salesman problem, are easy to model. Real timetabling problems have many details which vary from institution to institution, and modelling them is a daunting problem—so much so, that it is the largest obstacle to progress in many cases.

Collaborative work on modelling began with a discussion session at the first Practice and Theory of Automated Timetabling (PATAT) conference (Cumming and Paechter, 1995). The work has been carried on continuously since then, primarily within the PATAT conferences. There have been many arguments and some wrong turnings, but the fog is lifting, and data exchange is now becoming common.

One issue has been whether the problems should be modelled in full detail, or simplified to highlight their essence and reduce the implementation burden for solvers. Researchers strive to improve on previous work, and one way to improve is to work with more realistic data, so this issue is resolving itself as time passes.

A second issue has been the choice of level of abstraction. An example of a very abstract format is the input language of an integer programming package. Input in this format allows many kinds of constraints to be expressed, but it is too general to permit the use of solution methods specific to timetabling. Successful formats instead offer a long but finite list of concrete (timetabling-specific) constraint types. Modelling each timetabling problem separately, rather than using a single model for all of them, also makes for concreteness, and has turned out to be best, if only because the work needed to reach a consensus is less, and researchers are more likely to take an interest when their own problem is discussed specifically.

Some generalizations are natural and desirable, however. For example, a model which treats teachers, students, and rooms separately must define a 'no clashes' constraint for each kind of entity. Generalizing to *resources*, which are entities that attend events and may represent individual students, groups of students, teachers, rooms, or anything else (but not times), simplifies the model. Allowing arbitrary sets of times to be defined and named, rather than, for example, just the days and weeks (and similarly for resources), is another useful generalization, as is recognizing each source of cost in a solution as a violation of some kind of constraint. The value of a good generalization is often underestimated: it simplifies solvers as well as models.

The oldest well-known educational timetabling model, and the most successful as measured by the amount of data sharing effected, is the one used by the Toronto data set, which contains 13 real instances of the examination timetabling problem collected by Carter et al. (1996). In recent years it has been criticised for being too simple: it does not model rooms, and its constraints are implicit and so have fixed weights. It is discussed in more detail in Sect. 5.

Another landmark is a format created for nurse rostering in 2005. It models many more constraints than the Toronto data set does, and it models them explicitly, which allows individual instances to choose to include them or not, and to vary their weights. It also uses XML, which is verbose but has the great advantage of being clear and definite. At the time of writing, 20 instances were available in the current version of the format, collected from researchers in 13 countries (Curtois, 2012).

As a case study in modelling educational timetabling problems, the remainder of this section presents a model of the high school timetabling problem called XHSTT (Post, 2012a), developed recently by a group of high school timetabling researchers. The model, which was influenced by the nurse rostering model just described, was refined over several years and tested against real instances. It offers 15 types of constraints, has been used to model about 30 real instances from 10 countries so far, and has achieved widespread acceptance within the high school timetabling research community. The following description, written by this author, is from Post (2012b). Syntactic details are omitted; they may be found online (Kingston, 2009).

An XHSTT file is an XML file containing one *archive*, which consists of a set of instances of the high school timetabling problem, plus any number of *solution groups*. A solution group is a set of solutions to some or all of the archive's instances, typically produced by one solver. There may be several solutions to one instance in one solution group, for example solutions produced using different random seeds.

Each instance has four parts. The first part defines the instance's *times*, that is, the individual intervals of time, of unknown duration, during which events run. Taken in chronological order these times form a sequence called the instance's *cycle*, which is usually one week. Arbitrary sets of times, called *time groups*, may be defined, such as the Monday times or the afternoon times. A *day* is a time group holding the times of one day, and a *week* is a time group holding the times of one week. To assist display software, some time groups may be labelled as days or weeks.

The second part defines the instance's *resources*: the entities that attend events. The resources are partitioned into *resource types*. The usual resource types are a *Teachers* type whose resources represent teachers, a *Rooms* type of rooms, a *Classes* type of *classes* (sets of students who attend the same events), and a *Students* type of individual students. However, an instance may define any number of resource types. Arbitrary sets of resources of the same type, called *resource groups*, may be defined, such as the set of Science laboratories, the set of senior classes, and so on.

The third part defines the instance's *events*: meetings between resources. An event contains a *duration* (a positive integer), a *time*, and any number of *resources* (sometimes called *event resources*). The meaning is that the resources are occupied attending the event for *duration* consecutive times starting at *time*. The duration is a fixed constant. The time may be preassigned or left open to the solver to assign. Each resource may also be preassigned or left open to the solver to assign, although the type of resource to assign is fixed. Arbitrary sets of events, called *event groups*, may be defined. A *course* is an event group representing the events in which a particular class studies a particular subject. Some event groups may be labelled as courses.

For example, suppose class *7A* meets teacher *Smith* in a Science laboratory for two consecutive times. This is represented by one event with duration 2, an open time, and three resources: one preassigned *Classes* resource *7A*, one preassigned *Teachers* resource *Smith*, and one open *Rooms* resource. Later, a constraint will specify that this room should be selected from the *ScienceLaboratories* resource group, and define the penalty imposed on solutions that do not satisfy that constraint.

If class *7A* meets for Science several times each week, several events would be created and placed in an event group labelled as a course. However, it is common in high school timetabling for the total duration of the events of a course to be fixed, but for the way in which that duration is broken into events to be flexible. For example, class *7A* might need to meet for Science for a total duration of 6 times per week, in events of duration 1 or 2, with at least one event of duration 2 during which the students carry out experiments. One acceptable outcome would be five *sub-events*, as these fragments are called, of durations 2, 1, 1, 1, and 1. Another would be three sub-events, of durations 2, 2, and 2. This is modelled in XHSTT by giving a single event of duration 6. Later, constraints specify the ways in which this event may be split into sub-events, and define the penalty imposed on solutions that do not satisfy those constraints.

The last part of an instance contains any number of *constraints*, representing conditions that an ideal solution would satisfy. At present there are 15 types of constraints, stating that events should be assigned times, prohibiting clashes, and so on. The full list appears in Table 1.

Table 1 The 15 types of XHSTT constraints, with informal explanations of their meaning

Name	Meaning
Assign Resource constraint	Event resource should be assigned a resource
Assign Time constraint	Event should be assigned a time
Split Events constraint	Event should split into a constrained number of sub-events
Distribute Split Events constraint	Event should split into sub-events of constrained durations
Prefer Resources constraint	Event resource assignment should come from resource group
Prefer Times constraint	Event time assignment should come from time group
Avoid Split Assignments constraint	Set of event resources should be assigned the same resource
Spread Events constraint	Set of events should be spread evenly through the cycle
Link Events constraint	Set of events should be assigned the same time
Avoid Clashes constraint	Resource's timetable should not have clashes
Avoid Unavailable Times constraint	Resource should not be busy at unavailable times
Limit Idle Times constraint	Resource's timetable should not have idle times
Cluster Busy Times constraint	Resource should be busy on a limited number of days
Limit Busy Times constraint	Resource should be busy a limited number of times each day
Limit Workload constraint	Resource's total workload should be limited

Each type of constraint has its own specific attributes. For example, a Prefer Times constraint lists the events whose time it constrains, and the preferred times for those events, while a Link Events constraint lists sets of events which should be assigned the same time. Each constraint also has attributes common to all constraints, including a Boolean value saying whether the constraint is hard or soft, and an integer weight.

Traditionally, a *hard constraint* is one that must be satisfied if the timetable is to be used at all, although in practice a few violations of hard constraints are often acceptable, since the school can overcome them by undocumented means (moving a class to after school hours, assigning the deputy principal to a class with no teacher, and so on). A *soft constaint*, on the other hand, is a constraint that is violated quite routinely, although the total cost of those violations should be minimized.

As stated above, solutions are stored separately from instances, in solution groups within the archive file. A solution is a set of sub-events, each containing a duration, a time assignment, and some resource assignments. A solution's *infeasibility value* is the sum over the hard constraints of the number of violations of the constraint times its weight. Its *objective value* is similar, but using the soft constraints. One solution is better than another if it has a smaller infeasibility value, or an equal infeasibility value and a smaller objective value. A web site (Kingston, 2009) has been created which calculates the infeasibility and objective values of the solutions of an archive, and displays comparative tables, lists of violations, and so on.

It used to be said that real instances of timetabling problems required too many types of constraints for complete modelling to be possible. The nurse rostering and high school timetabling models disprove this; they show that careful elucidation of constraints, aided by suitable generalizations, can lead to complete models of real instances which are small enough to be usable. As instances appear that require other types of constraints, those constraints can be added gradually.

4 High School Timetabling

In the *high school timetabling problem*, a set of events of arbitrary integer *duration* is given, each of which contains a *time* and some *resources*: students, classes (sets of students who attend the same events, at least for the most part), teachers, and rooms. The time and resources may be preassigned to specific values, or left open to the solver to assign. The meaning is that the resources assigned to the event are occupied for *duration* consecutive times starting at *time*. The problem is to assign the unpreassigned values so as to avoid clashes and satisfy a variety of other constraints, such as those listed in Table 1 (Sect. 3).

When student sectioning (Sect. 7) is needed, it is carried out as a separate initial phase. Accordingly, it is usually considered not to be part of high school timetabling proper. This is one way in which high school timetabling differs from university course timetabling. Another is that the high school problem timetables groups of students (classes) which are usually occupied together for all, or almost all, of the times of the cycle, whereas the university problem timetables individual students whose timetables contain a significant amount of free time. Some clashes are probably inevitable among the thousands of individual university students' timetables, whereas clashes are not acceptable for classes.

A major division exists within high school timetabling instances with respect to teachers. On one side lie schools whose teachers are mostly part-time and tend to be preassigned to specific courses, and the emphasis is on providing timetables for the teachers which require their attendance for a minimal number of days per week and give them few idle times (free times in between busy times) on those days. On the other side lie schools whose teachers are mostly full-time, making it impractical to preassign teachers to specific courses except in the senior years, and the emphasis is on finding a timetable with teacher assignments that assign a qualified teacher to every lesson of every course.

Schmidt et al. (1980) comprehensively surveys the early history of high school timetabling research, including a description of a very basic version of the problem called *class-teacher timetabling*, which can be solved in polynomial time by edge colouring. Appleby et al. (1960), Gotlieb (1962), and De Werra (1971) are examples of papers that were influential in their day. Carter et al. (1997) is a good snapshot of a more recent era. The PATAT 2012 conference (Kjenstad, 2012) contains several high school timetabling papers, most of them stimulated by the Third International Timetabling Competition (Post, 2012b).

The only recent survey is Pillay (2010). It classifies about 40 papers which solve high school timetabling problems. Their methods include (in decreasing order of popularity) evolutionary algorithms, tabu search, integer programming, simulated annealing, and constraint programming. Many papers hybridize several methods. All of these papers pre-date the creation of standard benchmarks (for which see Sect. 3), so any attempt to rank them would be futile. At the time of writing the only paper with any objective claim to eminence is Fonseca et al. (2012), which describes the work that won the Third International Timetabling Competition (Post, 2012b). A more recent on-line version of this survey (Pillay, 2012) lists many more papers.

5 Examination Timetabling

The (university) examination timetabling problem has one event for each course, representing the course's final examination. The durations of the events may differ, although that is often a minor consideration. Each event's resources are the students who attend the course, and possibly a room. The aim is to assign a time to each event, avoiding clashes in the students' examination timetables. Room assignment (Sect. 9) may be required, and it is characteristic to include *proximity constraints*, expressing in some way the undesirability of attending two examinations close together in time. Proximity constraints might prohibit two examinations for one student on one day, for example, or two consecutive examinations ignoring day boundaries.

Insight into proximity constraints can be gained from the unrealistic special case where the number of examinations equals the number of times, and examinations may not occur simultaneously, ruling out all possibility of clashes. (This could arise in practice if the examinations are clustered before they are timetabled.) If there are two examination sessions per day, then minimizing the number of cases where a student attends two examinations in one day is equivalent to finding a maximum matching of minimum weight in the clash graph, for which there is a polynomial time algorithm. Alternatively, ignoring day boundaries and simply minimizing the number of cases where a student attends examinations at two consecutive times is a travelling salesman problem in the clash graph and so is NP-hard (Figure 2).

Despite involving many of the same resources, examination timetabling is much easier to model than university course timetabling. Coming as it does later in the semester, after enrolments have settled, it has not the dynamic character of course timetabling. Student sectioning is not required, even if different examinations are given to different sections, because it has already been done; and room requirements

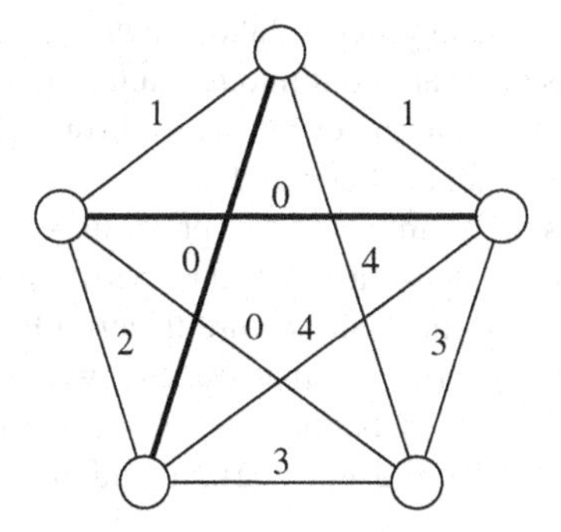

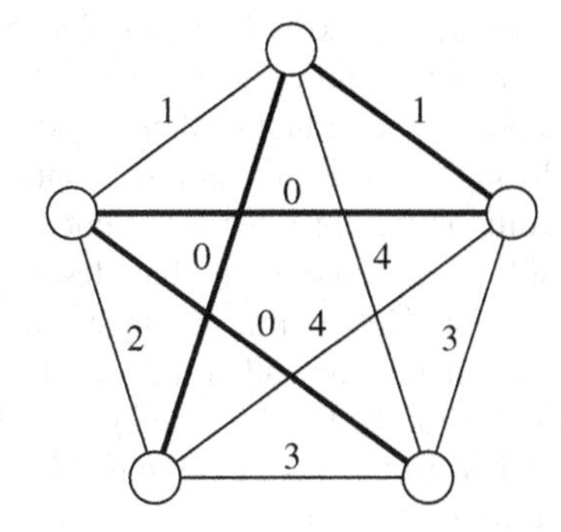

Fig. 2 A clash graph, showing (left) a minimum matching, which defines a pairing of the events which minimizes the number of cases where a resource attends both events of a pair; and (right) a travelling salesman path, which defines an ordering of the events which minimizes the number of cases where a resource attends two consecutive events. Example taken from Kingston (2010).

are usually more uniform, except perhaps for a few practical examinations which take place in laboratories.

Examination timetabling has a relatively long history of use of benchmark data, specifically the 13 real instances of the Toronto data set (Carter et al., 1996), still available and in use. These have quite elaborate proximity constraints, assigning a high penalty for two consecutive examinations, a lower penalty for examinations separated by one time, and so on; but these and the student no-clashes constraints are the only constraints. A more recent data set used by the Second International Timetabling Competition follows a more realistic model (McCollum et al., 2012). A guide to examination timetabling data sets is available online (Qu, 2012).

The leading recent survey is Qu et al. (2009), online at (Qu, 2012), a major work which describes other surveys, models, methods, and data sets, and cites 160 papers and 15 PhD theses. Carter (1986) and Carter et al. (1996) are still worth reading.

6 University Course Timetabling

University course timetabling aims to break university courses into sections, and assign times, students, rooms, and possibly instructors to the sections, subject to constraints like those of high school timetabling, only applied to individual students (each of whom has a distinct timetable) rather than to groups of students.

A *static* timetabling problem is one that is set up once, solved, and used. There may be some exploration of alternative scenarios, but once an acceptable timetable has been found, the work is finished. In contrast, a *dynamic* problem is one whose requirements and solution evolve over time. Most timetabling problems are static, and they are also small enough to be set up by a single person, the local expert.

University course timetabling is different. It is so large that no single person ever understands it all; local experts are scattered across the faculties and departments of the university, working largely independently of each other. It takes months to set up the problem, and even after semester begins there are continuous changes to the timetables of individual students, and some changes to the events as well (opening and closing sections in response to late changes in student demand).

Traditionally, the tiger was tamed by re-using as much of the previous year's timetable as possible (more than deserved to be, in many cases), and by partitioning the problem. The central administration controlled the main lecture theatre block, the departments controlled the rest. Each faculty made sure that students working entirely within the faculty could get workable timetables, but liaison across faculties was limited to the bare essentials. And if something did not work, a student was simply advised not to do it.

Vestiges of this approach can still be found, but its deficiencies are so glaring (poor room utilization and unhappy students, who are often paying customers these days) that it cannot survive for much longer. At the same time, the presence of a web browser on every desk has resolved the dilemma of a non-partitionable problem whose data are distributed: the departments are required to send their data via the web to the centre, which owns all the resources and does all the timetabling.

So the problem as it stands today is to timetable the entire university, not one department or faculty, including delivering an individual timetable to every student. Most changes after the start of semester can be handled by single student timetabling (Sect. 8), so the focus is on finding a good timetable before semester begins.

Two general approaches to university course timetabling may be distinguished. Emphasis may be placed on ensuring that certain sets of courses can be taken in combination, because students need them to satisfy the degree rules. Such sets of courses are called *curricula*, and this approach to the problem is called *curriculum-based university course timetabling*. Although ultimately each student must receive an individual timetable, in its pure form the curriculum-based approach does not utilize enrolment data for individual students.

In universities where students have large-group lectures and small-group tutorials and laboratories, there is an important sub-problem: assigning times and rooms to the lectures, given some basic information about what combinations of courses the students are likely to choose. Curriculum-based timetabling addresses this kind of problem, and serves as a model of what it may be worthwhile to do before the dynamic phase of university course timetabling begins.

Declaring a set of events to be a curriculum amounts to saying that a number of students will be taking those events in combination. In high school timetabling the same declaration is made by placing a preassigned class resource into the events. This relationship between curriculum-based timetabling and high school timetabling has been exploited in a few papers, such as Nurmi and Kyngäs (2008).

The alternative to curriculum-based timetabling is *enrolment-based university course timetabling*, in which the enrolment data for individual students are used to determine which courses should be able to be taken in combination.

The two approaches are not mutually exclusive. Student enrolment data often becomes available fairly late, in which case curriculum-based timetabling may be used early to lay out the basic structure of the timetable (such as the times of large-group lectures), while enrolment-based timetabling is used later to fine-tune it.

When constructing a timetable with sections based on student enrolment data, a basic dilemma emerges: whether to assign students to sections before or after assigning times to sections. Neither alternative is fully satisfactory, so, in practice, implementations of both approaches always include some way of reconsidering the first phase after completing the second.

Assigning times first has the advantage that students can then be assigned to sections using single student timetabling (Sect. 8), which is known to work well. If the result is poor, the time assignment can be adjusted.

A semi-automatic version of this method is used at the author's university. Initial values for the number of sections of each course, and their times, are chosen manually, based on curricula, history, and incomplete student enrolment data. Then a dynamic process of refinement begins. As student enrolment information improves, dummy runs of single student timetabling applied to each student (but not published to the students) are carried out at the centre. Results are distributed to departmental coordinators, who respond by opening and closing sections as enrolment numbers become clearer, and moving sections left underfilled by single student timetabling to other and hopefully better times, subject to room and teacher availability.

It is possible that a fully automated timetabling system could be built by this process of repeated time assignment then testing by single student timetabling. Even if single student timetabling is highly optimized and virtually instantaneous for one student, it will still take several seconds to timetable every student, which is too slow to support an extensive search through the space of time assignments. So there would probably be time to re-timetable only those students directly affected by each time adjustment. But this method does not seem to have ever been tried, except by Aubin et al. (1989), who used it to timetable 'a large high school in Montreal'.

Assigning students first leads naturally to a three-phase method (Carter, 2001). First, assign students to sections before times are assigned, aiming to minimize the number of pairs of sections with students in common. This is the *student sectioning* problem, discussed in detail in Sect. 7. Second, assign times to the sections. This is a graph colouring problem similar to examination timetabling without proximity constraints. Finally, make one pass over the entire student list, re-timetabling each student using single student timetabling. (Experience at the author's university has shown that two or even three passes help to even out section numbers.)

Only two systems which solve the full university course timetabling problem have been published in detail. The first is the system described in Carter (2001), which has been in use since 1985. It is specific to one university, although of course the ideas are portable. It performs enrolment-based timetabling using the three-phase method just described.

The second system, UniTime (2012), is free, open-source, and not specific to one university—a combination of features apparently unavailable elsewhere. It offers both curriculum-based and enrolment-based timetabling, following Carter (2001) for the latter, and is in use at several universities, although development continues. A long list of papers is given on its web site (UniTime, 2012); only a selection can be cited here. Murray et al. (2002) is the original work. Müller et al. (2004) considers the problem of finding minimally perturbed timetables when circumstances change after the timetable has been published, an important problem whose study has barely begun. Murray et al. (2007) and Murray et al. (2010) present the mature system.

The dynamic nature of the university course timetabling problem is an obstacle to designing a realistic data model for it. No data sets are available for the full problem, although partial data are available. The UniTime web site offers data sets for several sub-problems: departmental problems and so on. Several timetabling competitions have targeted university course timetabling problems. The most recent of these is the Second International Timetabling Competition (McCollum, 2007), within which Track 2 is enrolment-based, and Track 3 is curriculum-based. Both tracks offer only drastically simplified instances: they are static, they model one faculty rather than the whole university, their events have equal duration, and every course has just one section. The data are still available and are the focus of many papers. No comprehensive survey of these papers is known to the author.

7 Student Sectioning

As explained earlier, a course may break into *sections*: copies of it, each with its own time, room, and teacher. Each student enrolled in the course must be assigned to one of its sections. The *student sectioning* problem asks for an assignment of students to sections which is likely to work well when times are assigned to the sections later, typically by minimizing the number of pairs of sections that have at least one student in common.

Some formulations of the problem also ask for a clustering of the sections, such that if the sections in each cluster run at the same time, but different clusters run at different times, then no students have clashes. This is a natural extension, since a good clustering proves that a student sectioning is successful.

Student sectioning arises in university course timetabling (Sect. 6), when the choice is made to assign students to sections before assigning times to sections.

Essentially the same situation arises in high schools. High school instances often have complex events called *electives*: sets of courses that the school decides to run simultaneously. Each student chooses one course from each elective. Electives are usually determined by surveying the students to find out which courses they intend to take, and ensuring that there are enough sections of each course to accommodate the students who wish to take it, and that the sections of popular combinations of courses lie in different electives. This problem of defining the electives is a student sectioning problem, including the clustering extension.

The student sectioning literature is very fragmentary. There is no survey, and one must search for discussions of student sectioning in papers on university course timetabling and high school timetabling. To add to the confusion, the term 'student sectioning' is sometimes used for the full university course timetabling problem, for which it is clearly a misnomer.

Carter (2001) is the seminal paper for student sectioning. It has several pages of practical discussion of the problem. For each course, each enrolled student is made into a node of a graph. An edge joins each pair of nodes, with a weight between 0 and 1 determined by how similar the two students' selection of courses is: 0 means identical, 1 means disjoint. The students are then grouped into sections by a standard graph clustering algorithm. This heuristic method produces relatively few pairs of sections with students in common. Murray et al. (2007) follow Carter (2001) in the student sectioning phase of their university course timetabling algorithm.

The student sectioning problem is smaller in high schools than in universities, which may explain why the few high school sectioning papers known to the author use more ambitious methods: de Haan et al. (2007) use branch and bound, while Kristiansen and Stidsen (2012) use adaptive large scale neighbourhood search.

8 Single Student Timetabling

The *single student timetabling* problem, usually encountered as a phase of university course timetabling (Sect. 6), asks for a timetable for a single student after courses are broken into sections and the sections are assigned times. Its first priority is to find a clash-free timetable for the student; its second is to assign the student to less full sections (say, by minimizing the total enrolment of the sections chosen), so that as it is run for many students, the sections are kept approximately equally full.

If all sections have duration 1, this is a weighted bipartite matching problem. One set of nodes represents the courses, the other represents the times of the week. An edge is drawn from a course to a time whenever a section of that course occurs at that time, weighted by the current enrolment of that section.

In practice, sections of different courses may have different durations, and there may be additional constraints, such as that the assignment of sections of two courses be correlated in some way (ordered in time, for example). Even so, real instances can be solved to optimality very quickly using a tree search. Each node of the tree represents one course, and each downward edge out of that node represents the assignment of the student to a section of that course. Edges that produce clashes with higher edges are not followed.

In the senior years, where course enrolments are lower and the number of sections is correspondingly fewer, the tree search just described may be adequate as it stands. But junior courses may have many sections. A course with 500 students that breaks into sections of 20 students each will have about 25 sections, and if a student takes several such courses the search will need to be optimized.

One obvious optimization is to assign the courses with the fewest sections first. Then courses with only one section are (in effect) preassigned, and there are more choices at the lower levels of the tree.

A second optimization, sometimes called *intelligent backtracking*, is as follows. Suppose the search is at some node of the tree, and that every section of that node's course has been tried and has failed owing to a clash with sections assigned higher in the tree. A simple tree search would return to the parent of that node and continue with its next alternative. But if the parent was not involved in any of the clashes, that is futile: the same clashes will recur. Instead of backtracking to the parent, intelligent backtracking backtracks to the closest ancestor involved in a clash.

A third optimization focuses on minimizing the total enrolment of the sections. First, the cost of assigning a section is changed from the current enrolment of that section to the amount by which that current enrolment exceeds the current enrolment of the least full section of its course. For example, enrolling a student in a least full section costs 0. Then branch and bound is used to terminate a search path when its cost equals or exceeds the cost of the best solution found so far; and if a complete solution is found whose cost is 0, the entire search is terminated early. Combined with sorting the sections so that the least full ones are tried first, this optimization is effective at reducing the size of the search tree when there are many solutions.

It seems likely that many universities would have such solvers, given their need to timetable thousands of students, and to re-timetable them when their enrolments change. Laporte et al. (1986) describes one, only without intelligent backtracking. Another, employing all these optimizations, has been used routinely at the author's university for many years (unpublished). It timetables a single student on demand virtually instantaneously, producing virtually equal section enrolment numbers. So single student timetabling is a solved problem.

9 Room Assignment

The *room assignment problem* asks for an assignment of rooms to events after the events' times are fixed. Each event has its own room requirements, such as for a specialist room (a Science laboratory, a lecture theatre, and so on), or for a room capable of holding at least a certain number of students. This problem occurs in all kinds of educational timetabling.

Carter et al. (1992) is a fascinating compendium of results on room assignment. It observes that room assignment is exactly list colouring of interval graphs, a well-known NP-complete problem, and shows that it remains NP-complete even when the cycle contains only two times, a remarkable result.

Each room may be tested against each event's room requirements, and in this way any combination of room requirements for an event may be reduced to a set of suitable rooms for that event before solving begins. Among suitable rooms some may be more suitable than others, in which case the outcome of the testing is an integer rating of each room's suitability for each event.

When all events have duration 1, and all rooms lie within easy walking distance of each other, the instance of the room assignment problem for the events assigned a given time t is independent of the instances at other times. It can be solved to optimality by finding a maximum matching in the bipartite graph whose nodes are the events assigned time t and the rooms available at time t, with an edge joining an event node to a room node whenever the room is suitable for the event (Figure 3). If rooms have integer ratings, the edges are weighted by the ratings.

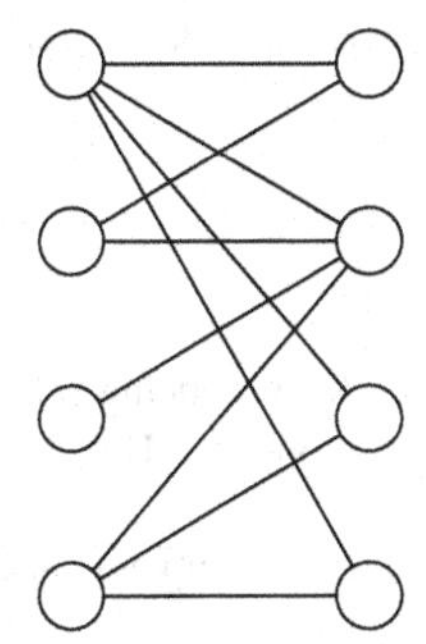

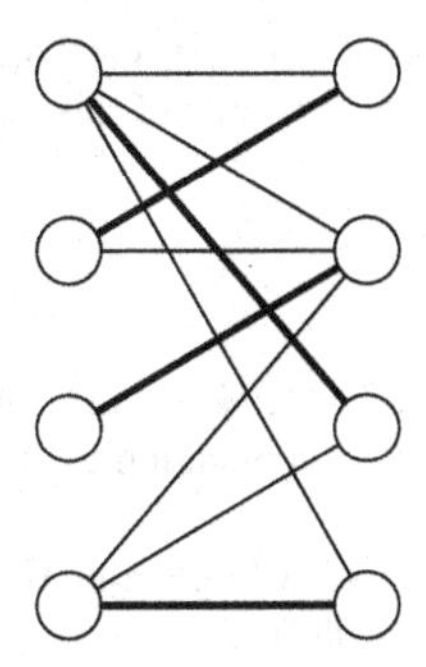

Fig. 3 A bipartite graph (left), and the same graph showing a maximum matching (right). Each left-hand node represents an event assigned a particular time *t* and demanding one room, each right-hand node represents a room, and edges indicate which rooms are suited to which events.

Kingston (2012) reports perfect results from a room assignment algorithm which assigns rooms using a constructive heuristic followed by adjustment using ejection chains, while maintaining the existence of an unweighted maximum matching of optimal size at each time as an invariant throughout the solve. The algorithm runs in polynomial time and takes less than one second on real instances. In practice, then, room assignment, like single student timetabling, is a solved problem.

Room assignment is usually a phase of a larger problem. Kingston (2012) gives techniques for efficiently maintaining the invariant while times are being assigned to events. In this way, a nearly exact guarantee can be given that room assignment will succeed, without actually assigning any rooms.

Can other resources be assigned in the same way as rooms? Students and classes are usually preassigned, leaving nothing to do. Teachers, too, are often preassigned. When they are not, there is usually a requirement, called *teacher stability* or the *avoid split assignments constraint*, that the teacher assigned to the events of one course be the same. Kingston (2012) investigates this variant with some success, although it is significantly more difficult in practice than room assignment.

10 Conclusion

Educational timetabling has been an active area of research for over 50 years. While there is no sign that provably optimal solutions to large, real instances will be found any time soon, the more practical goal of quickly and reliably finding timetables that are preferred to manually produced ones is in sight. More realistic models and data sets are needed in some areas, and more powerful solving techniques are needed in others, but steady progress continues to be made on both fronts, the transfer of research methods into commercial software is growing, and there is every reason to believe that success is not far off.

Acknowledgements. The author thanks Keith Murray, Ender Özcan, and Gerhard Post for their comments on earlier drafts of this chapter.

References

Appleby, J.S., Blake, D.V., Newman, E.A.: Techniques for producing school timetables on a computer and their application to other scheduling problems. The Computer Journal 3, 237–245 (1960)

Aubin, J., Ferland, J.A.: A large scale timetabling problem. Computers and Operations Research 16, 67–77 (1989)

Carter, M.W.: A survey of practical applications of examination timetabling algorithms. Operations Research 34, 193–202 (1986)

Carter, M.W., Tovey, C.A.: When is the classroom assignment problem hard? Operations Research 40, S28–S39 (1992)

Carter, M.W., Laporte, G.: Recent developments in practical examination timetabling. In: Burke, E.K., Ross, P. (eds.) PATAT 1995. LNCS, vol. 1153, pp. 3–21. Springer, Heidelberg (1996)

Carter, M.W., Laporte, G., Lee, S.Y.: Examination timetabling: algorithmic strategies and applications. Journal of Operational Research Society 47, 373–383 (1996)

Carter, M.W., Laporte, G.: Recent developments in practical course timetabling. In: Burke, E.K., Carter, M. (eds.) PATAT 1997. LNCS, vol. 1408, pp. 3–19. Springer, Heidelberg (1998)

Carter, M.W.: A comprehensive course timetabling and student scheduling system at the University of Waterloo. In: Burke, E., Erben, W. (eds.) PATAT 2000. LNCS, vol. 2079, pp. 64–81. Springer, Heidelberg (2001)

Cooper, T.B., Kingston, J.H.: The complexity of timetable construction problems. In: Burke, E.K., Ross, P. (eds.) PATAT 1995. LNCS, vol. 1153, pp. 283–295. Springer, Heidelberg (1996)

Cumming, A., Paechter, B.: Standard formats for timetabling data. In: Unpublished Discussion Session at the First International Conference on the Practice and Theory of Automated Timetabling, PATAT 1995, Edinburgh (August 1995)

Curtois, T.: Employee scheduling benchmark data sets, `http://www.cs.nott.ac.uk/~tec/NRP/` (Cited September 15, 2012)

De Cesco, F., Di Gaspero, L., Schaerf, A.: Benchmarking curriculum-based course timetabling: formulations, data formats, instances, validation, and results. In: Proceedings, 7th International Conference on the Practice and Theory of Automated Timetabling, PATAT 2008, Montreal (August 2008)

de Haan, P., Landman, R., Post, G., Ruizenaar, H.: A case study for timetabling in a Dutch secondary school. In: Burke, E.K., Rudová, H. (eds.) PATAT 2007. LNCS, vol. 3867, pp. 267–279. Springer, Heidelberg (2007)

De Werra, D.: Construction of school timetables by flow methods. INFOR—Canadian Journal of Operational Research and Information Processing 9, 12–22 (1971)

Fonseca, G.H.G., Santos, H.G., Toffolo, T.A.M., Brito, S.S., Souza, M.J.F.: A SA-ILS approach for the high school timetabling problem. In: Proceedings of the Ninth International Conference on the Practice and Theory of Automated Timetabling (PATAT 2012), Son, Norway (August 2012)

Gotlieb, C.C.: The construction of class-teacher timetables. In: Popplewell, C.M. (ed.) Information Processing 1962 (Proceedings of the 1962 IFIP Congress), pp. 73–77 (1962)

Kingston, J.H.: The HSEval high school timetable evaluator (2009), `http://www.it.usyd.edu/au/~jeff/hseval.cgi` (Cited September 15, 2012)

Kingston, J.H.: Timetable construction: the algorithms and complexity perspective. In: Proceedings of the Eighth International Conference on the Practice and Theory of Automated Timetabling (PATAT 2010), Belfast, UK (August 2010)

Kingston, J.H.: Resource assignment in high school timetabling. Annals of Operations Research 194, 241 (2012)

Kjenstad, D., Riise, A., Nordlander, T.E., McCollum, B., Burke, E.: In: Proceedings, of the 9th International Conference on the Practice and Theory of Automated Timetabling, PATAT 2012, Son, Norway (August 2012)

Kristiansen, S., Stidsen, T.R.: Adaptive large neighborhood search for student sectioning at Danish high schools. In: Proceedings of the Ninth International Conference on the Practice and Theory of Automated Timetabling (PATAT 2012), Son, Norway (August 2012)

Laporte, G., Desroches, S.: The problem of assigning students to course sections in a large engineering school. Computers and Operations Research 13, 387–394 (1986)

McCollum, B.: The Second International Timetabling Competition (ITC 2007), Track 3 (2007), `http://www.cs.qub.ac.uk/itc2007` (Cited September 17, 2012)

McCollum, B., McMullan, P., Parkes, A.J., Burke, E.K., Qu, R.: A new model for automated examination timetabling. Annals of Operations Research 194, 291–315 (2012)

Müller, T., Rudová, H., Barták, R.: Minimal perturbation problem in course timetabling. In: Burke, E.K., Trick, M.A. (eds.) PATAT 2004. LNCS, vol. 3616, pp. 126–146. Springer, Heidelberg (2005)

Müller, T., Rudová, H.: Real-life curriculum-based timetabling. In: Proceedings of the 9th International Conference on the Practice and Theory of Automated Timetabling, PATAT 2012, Son, Norway (August 2012)

Murray, K., Rudová, H.: University course timetabling with soft constraints. In: Burke, E.K., De Causmaecker, P. (eds.) PATAT 2002. LNCS, vol. 2740, pp. 310–328. Springer, Heidelberg (2003)

Murray, K., Müller, T., Rudová, H.: Modeling and solution of a complex university course timetabling problem. In: Burke, E.K., Rudová, H. (eds.) PATAT 2007. LNCS, vol. 3867, pp. 189–209. Springer, Heidelberg (2007)

Murray, K., Müller, T.: Comprehensive approach to student sectioning. Annals of Operations Research 181, 249–269 (2007)

Nurmi, K., Kyngäs, J.: A conversion scheme for turning a curriculum-based timetabling problem into a school timetabling problem. In: Proceedings, of the 7th International Conference on the Practice and Theory of Automated Timetabling, PATAT 2008, Montreal (August 2008)

Pillay, N.: An overview of school timetabling. In: Proceedings, of the 8th International Conference on the Practice and Theory of Automated Timetabling, PATAT 2010, Belfast, UK, pp. 321–335 (August 2010)

Pillay, N.: Classification of school timetabling research, `http://titan.cs.unp.ac.za/~nelishiap/st/classification.htm` (Cited September 15, 2012)

Post, G.: Benchmarking project for (high) school timetabling, `http://www.utwente.nl/ctit/hstt/` (Cited September 15, 2012)

Post, G., Di Gaspero, L., Kingston, J.H., McCollum, B., Schaerf, A.: The third international timetabling competition. In: Proceedings, of the 9th International Conference on the Practice and Theory of Automated Timetabling, PATAT 2012, Son, Norway (August 2012)

Qu, R., Burke, E.K., McCollum, B., Merlot, L.T.G., Lee, S.Y.: A survey of search methodologies and automated system development for examination timetabling. Journal of Scheduling 12, 55–89 (2009)

Qu, R.: Benchmark data sets in exam timetabling, `http://www.cs.nott.ac.uk/~rxq/data.htm` (Cited September 15, 2012)

Schaerf, S.: A survey of automated timetabling. Artificial Intelligence Review 13, 87–127 (1999)

Schmidt, G., Ströhlein, T.: Timetable construction–an annotated bibliography. The Computer Journal 23, 307–316 (1980)

UniTime: a comprehensive university timetabling system, `http://www.unitime.org/` (Cited September 18, 2012)

Welsh, D.J.A., Powell, M.B.: An upper bound for the chromatic number of a graph and its application to a timetabling problem. The Computer Journal 10, 85–86 (1967)

Automated Shift Design and Break Scheduling

Luca Di Gaspero, Johannes Gärtner, Nysret Musliu,
Andrea Schaerf, Werner Schafhauser, and Wolfgang Slany

Abstract. Shift design and break scheduling are important employee scheduling problems that arise in many contexts, especially at airports, call centers, and service industries. The aim is to find a minimum number of legal shifts, the number of workers assigned to them, and a suitable number of breaks so that the deviation from predetermined workforce requirements is minimized. Such problems have been extensively investigated in Operations Research and recently have been also tackled with Artificial Intelligence techniques. In this chapter we outline major characteristics of these problems and provide a literature survey over solution techniques to solve them. We then describe in detail two state-of-the-art approaches based on local search techniques. Finally, we discuss our experiences with the application of one of these techniques in a real life case study.

1 Introduction

Designing shifts and breaks is one of the most important phases in the general employee scheduling problem. The typical employee scheduling process in an organization consists of several stages. Given a planning period (often called the temporal

Luca Di Gaspero · Andrea Schaerf
DIEGM, Università degli Studi di Udine, Italy
e-mail: {l.digaspero, schaerf}@uniud.it

Johannes Gärtner · Werner Schafhauser
XIMES XIMES GmbH, Austria
e-mail: {gaertner, schafhauser}@ximes.com

Nysret Musliu
DBAI, Technische Universität Wien, Austria
e-mail: musliu@dbai.tuwien.ac.at

Wolfgang Slany
IST, Technische Universität Graz, Austria
e-mail: wolfgang.slany@tugraz.at

A.Ş. Etaner-Uyar et al. (eds.), *Automated Scheduling and Planning*,
Studies in Computational Intelligence 505,
DOI: 10.1007/978-3-642-39304-4_5,

horizon), in a first phase the temporal requirements are collected which then determine the number of employees needed for each day and time interval during the planning period. After these requirements have been established, the shifts can be designed. At this stage the number of employees for each shift on each day in the planning period has to be determined. In some cases, moreover, for each employee assigned to a shift, a set of breaks must also be scheduled.

Shift design and break assignment problems are known also under other names, for example as *shift design* [19, 10], *shift scheduling* [3, 2, 25, 28], and *break scheduling* [4, 5, 20, 31]. Such problems arise at airports, call centers, and in the service industry in general, and have been extensively investigated in Operations Research and Artificial Intelligence.

On the one hand, the quality of solutions for shift design and break scheduling is particularly relevant because of the legal issues concerning the working time of employees, the well-being of employees, and the importance of reducing costs while guaranteeing a high level of service. In some working environments the quality of solutions is even critical: For example, an inadequate break assignment to air traffic controllers can be the cause of safety problems due to diminished concentration. On the other hand, the search space of these problems is big and they are subject to conflicting constraints which makes it practically impossible to tackle them by means of exact methods.

In this chapter, after the formal introduction, we will describe solving methods for these problems. In particular two state-of-the-art approaches based on local search techniques will be described in detail. Moreover, we will provide a real life case study that includes shift design. We will describe the solving process and present our experiences with the application of these techniques in a real life case study. Furthermore, we provide a literature survey over solution techniques for these problems.

2 The Shift Design Problem

The shift design problem is a variant of the shift scheduling problem and has been first introduced in [12], [21], and [19]. The definition of this problem is given below (this definition is taken almost verbatim from [21] and [19]).

The input of the problem:

- n consecutive *time slots* $[a_1, a_2)$, $[a_2, a_3)$, ..., $[a_n, a_{n+1})$, all with the same length *slotlength* in minutes. Time point a_1 represents the begin of the planning period and time point a_{n+1} represents the end of the planning period. A value w_i, related with each slot $[a_i, a_{i+1})$, indicates the optimal number of employees that should be present during that slot.
- y *shift types* $v_1, \ldots, v_y$. Each shift type v_j has the related parameters v_j.min-start and v_j.max-start, which represent the earliest and latest start of the shift, and v_j.min-length and v_j.max-length, which represent the minimum and maximum length of the shift. An example of shift types is given in Table 1.

- Two scalar real-valued quantities, necessary to define the distance from the average number of duties.

  AS: the upper limit for the average number of working shifts per week per employee

  AH: the average number of working hours per week per employee

Table 1 Possible shift types

Shift type	min-start	max-start	min-length	max-length
M	06:00	08:00	07:00	09:00
D	10:00	11:00	07:00	09:00
A	13:00	15:00	07:00	09:00
N	21:00	23:00	07:00	09:00

The aim is to generate a set of k shifts $s_1, \ldots, s_k$. Each shift s_l is completely determined by the parameters s_l.start and s_l.length and must belong to one of the shift types. Additionally, each shift s_p has a set of parameters $s_p.w_i, \forall i \in \{1, \ldots, C\}$ (C is the number of days in the planning period) indicating the number of employees assigned to shift s_p during day i.

The objective of the problem is to minimize the following four components:

F_1 : sum of the *excesses* of workers in each time slot during the planning period.
F_2 : sum of the *shortages* of workers in each time slot during the planning period.
F_3 : *number of shifts*.
F_4 : *distance* to the average number of duties per week in case it is above a certain threshold. This component is meant to avoid an excessive fragmentation of workload in too many short shifts.

The problem is a multi criteria optimization problem, in which each criterion has different importance depending on the situation. The objective function is the weighted sum of the four components.

Usually the designing shifts for a week is considered (less days are also possible) and the schedule is cyclic (the element following the last time slot of a planning period is equal to the first time slot of the planning period).

Formal Definitions

The generated shifts belong to at least one of the shift types if:

$$\forall l \in \{1, \ldots, k\}\, \exists j \in \{1, \ldots, y\} \,/$$

$$v_j.\text{min-start} \leq s_l.\text{start} \leq v_j.\text{max-start}$$

$$v_j.\text{min-length} \leq s_l.\text{length} \leq v_j.\text{max-length}$$

In order to define the shortage and excess of workers, we first define the load l_d for a time slot d as

$$l_d = \sum_{p=1}^{k} x_{p,d}$$

where

$$x_{p,d} := \begin{cases} s_p.w_i & \text{if the time slot } d \text{ belongs to the interval of shift } s_p \text{ on day } i \\ 0 & \text{otherwise.} \end{cases}$$

The total excess F_1 and total shortage F_2 (in minutes) of workers in all time slots during the planning period is then defined as

$$F_1 = \sum_{d=1}^{n} (\max(l_d - w_d, 0) * slotlength)$$

$$F_2 = \sum_{d=1}^{n} (\max(w_d - l_d, 0) * slotlength)$$

Regarding F_3, this is simply equal to the number of shifts k.

The average number of working shifts per week per employee (*AvD*) is defined below:

$$AvD = \frac{(\sum_{i=1}^{k} \sum_{j=1}^{C} s_i.w_j) * AH}{\sum_{i=1}^{k} \sum_{j=1}^{C} s_i.w_j * s_i.length}$$

The penalty associated with the average number of duties is defined as.

$$F_4 = \max(AvD - AS, 0)$$

It is worth noticing that the criterion F_4 is not always taken into consideration in the literature.

A problem similar to this problem, called the *shift scheduling problem*, has been considered by several authors previously in the literature. We will describe later in the literature section the main differences between the shift design and shift scheduling problem.

3 Break Scheduling

In this section we present a break scheduling problem which has been first mentioned in [5] and [26]. This problem appeared in the area of supervisory personnel, whereas a slightly different variant of the problem, in the case of call centers, has been solved in [4].

The definition of the break scheduling problem for supervisory personnel is as follows (this definition is taken almost verbatim from [5] and [26]):

In the same setting of the shift design problem, we are additionally given:

- Fixed shifts $(s_1, s_2, ..., s_n)$ representing employees working within the planning period. Each shift, s_i, has an adjoined parameter, $s_i.breaktime$, that specifies the required amount of break time for s_i in time slots.
- An employee is considered to be working during time slot $[a_t, a_{t+1})$ if that employee neither has a break during time slot $[a_t, a_{t+1})$ nor he/she has stopped working at time point. After a break, an employee needs a full time slot, usually 5 minutes, to become reacquainted with the altered situation. Thus, during the first time slot following a break, an employee is not considered to be working.

Similarly to shifts, also breaks b_j are characterized by two parameters, b_j.start and b_j.end, representing the time slots in which a shift or break starts and ends. Subtracting the value for start from the value for end gives the duration of shifts and breaks in time slots. The durations of shifts and breaks are stored in an additional parameter, duration. Moreover, each break is associated with a certain shift in which it is scheduled. We distinguish between two different types of breaks: lunch breaks and monitor breaks.

Formal Definitions

Given a planning period, a set of shifts, the associated total break times, and the staffing requirements, a feasible solution to the break scheduling problem is a set of breaks with the following characteristics:

- Each break, b_j, lies entirely within its associated shift, s_i. That is,

$$s_i.\text{start} \leq b_j.\text{start} \leq b_j.\text{end} = s_i.\text{end}$$

- Two distinct breaks (b_j, b_k) associated with the same shift, s_i, do not overlap in time:

$$b_j.\text{start} \leq b_j.\text{end} \leq b_k.\text{start} \leq b_k.\text{end} \vee b_k.\text{start} \leq b_k.\text{end} \leq b_j.\text{start} \leq b_j.\text{end}$$

- In each shift, s_i, the sum of durations of its associated breaks equals the required amount of break time:

$$\sum_{b_j \in s_i} b_j.\text{duration} = s_i.\text{breaktime}$$

Criteria for Finding an Optimal Solution

Among all feasible solutions for the break scheduling problem, we try to find an optimal one according to seven criteria, which we model as soft constraints:

C_1: Break Positions. Each break, b_j, may start, at the earliest, a certain number of time slots after the beginning of its associated shift s_i, and may end, at the latest, a given number of time slots before the end of its shift:

$$b_j.\text{starts} \geq s_i.\text{start} + \text{distance to shift start}$$
$$b_j.\text{end} \leq s_i.\text{end} - \text{distance to shift end}$$

C_2: Lunch Breaks. A shift s_i can have several lunch breaks, each required to last a specified number of time slots (min lunch break duration), and should be located within a certain time window after the shift start. Let b_{lb} be a lunch break. Then,

$$b_{lb}.\text{start} \geq s_i.\text{start} + \text{distance to shift start}\, lb$$
$$b_{lb}.\text{end} \leq s_i.\text{end} - \text{distance to shift end}\, lb$$

C_3: Duration of Work Periods. Breaks divide a shift into work and rest periods. The duration of work periods within a shift must range between a required minimum and maximum duration:

$$\text{min work duration} \leq b_1.\text{start} - s_i.\text{start} \leq \text{max work duration}$$
$$\text{min work duration} \leq b_{j+1}.\text{start} - b_j.\text{end} \leq \text{max work duration}$$
$$\text{min work duration} \leq s_i.\text{end} - b_m.\text{end} \leq \text{max work duration}$$

where $(b_1,...,b_j,b_{j+1},...,b_m)$ are the breaks of s_i in temporal order.

C_4: Minimum Break Times after Work Periods. If the duration of a work period exceeds a certain limit, the break following that period must last a given minimum number of time slots (min ts count):

$$b_1.\text{start} - s_i.\text{start} \geq \text{work limit} \Rightarrow b_1.\text{duration} \geq \text{min ts count}$$
$$b_{j+1}.\text{start} - b_j.\text{end} \geq \text{work limit} \Rightarrow b_{j+1}.\text{duration} \geq \text{min ts count}$$

where, once again, $(b_1,...,b_j,b_{j+1},...,b_m)$ are the breaks of s_i in temporal order.

C_5: Break Durations. The duration of each break, b_j, must lie within a specified minimum and maximum value:

$$\text{min duration} \leq b_j.\text{duration} \leq \text{max duration}$$

C_6: Shortage of Employees. In each time slot, $[a_t, a_{t+1})$, at least r_t employees should be working.

C_7: Excess of Employees. In each time slot, $[a_t, a_{t+1})$, a_t most r_t employees should be working.

Objective Function

For each constraint, we define a violation degree, $violation(C_k)$, specifying the deviation (in time slots or employees) from the requirements stated by the respective constraint. The importance of each criterion and its corresponding constraint varies from task to task. Consequently, the objective function for the break scheduling problem is the weighted sum of the violation degrees of all the constraints:

$$F(solution) = \Sigma_{k=1}^{7} W_k \times violation(C_k)$$

where W_k is a weight indicating the importance assigned to constraint C_k. Given an instance of the break scheduling problem, our goal is to find a feasible solution that minimizes this objective function.

4 Literature Review

The shift design problem described in this chapter has been first introduced in [12], [21], and [19]. In these works a tabu search based algorithm to solve this problem is proposed. Additionally, their method orders moves and applies these first to regions with larger conflicts (larger over/under-staffing). The proposed solution methods have been used since several years in the commercial software package OPA of XIMES Inc. Di Gaspero et al. [14, 10] proposed the application of hybrid approaches based on local search and min-cost max-flow techniques. The hybrid algorithm improved results reported in [19] on benchmark examples.

A similar problem called shift scheduling problem has been extensively investigated in the literature. Although shift design problem and shift scheduling show some similarities there are some further specifics that characterize the shift design problem. In the shift design problem the number of employees for each shift over a whole week should be determined and the problem is cyclic (the shift starting on the last day that will last overnight will finish in the first day of the schedule). Furthermore, a different objective function is used for the shift design problem that also allows undercover and tries to minimize the number of used shifts.

Regarding the shift scheduling problem the first approaches have been proposed many years ago. Dantzig developed the original set-covering formulation [8] for the shift scheduling problem, in which feasible shifts are enumerated based on possible shift starts, shift durations, breaks, and time windows for breaks. Integer programming formulations for shift scheduling include [3], [29], and [1]. Aykin [1] introduced an implicit integer programming model for the shift scheduling problem in 1996 and later he compared an extended version [2] of his previous model with a similarly extended formulation introduced by Bechtold and Jacobs [3]. He observed that Bechtold and Jacobs' approach needed fewer variables whereas his approach needed fewer constraints. Several problems were solved using both models with the integer programming software LINDO. The model proposed by Aykin was shown to be superior.

Rekik et al. [25] developed two other implicit models and managed to improve upon previous approaches among them Aykin's original model. Tellier and White [28] developed a tabu search algorithm to solve a shift scheduling problem originating in contact centers which is integrated into the workforce scheduling system Contact Center Scheduling 5.5 from PrairieFyre Soft Inc. (`http://www.prairiefyre.com`). The solution of a shift scheduling problem with a planning period of one day, and at most three breaks (two 15 minutes breaks and a lunch break of 1 hour) has been considered in [24] and [6]. In [6] the authors make use of automata and context-free grammars to formulate constraints on sequences of decision variables. Quimper and Rousseau [24] investigate modeling of the regulations for the shift scheduling problem by using regular and context-free languages and solved the overall problem with Large Neighborhood Search. In addition to the previous model, the authors applied their methods in single and multiple activity shift scheduling problems. A new implicit formulation for multi-activity shift scheduling problems using context-free grammars has been proposed recently by Côté et al. [7].

Previously, break scheduling has been addressed mainly as part of the shift scheduling problem. Several approaches have been proposed for problem formulations that include a small number of breaks. These approaches schedule the breaks within a shift scheduling process. Such approaches include the previous mentioned methods for shift scheduling [8], [3], [29], [1], [2], [25], [28], [24], [6], [7]. Generation of three to four breaks per shift after the design of shifts with a greedy approach has been investigated in [13].

Some important break scheduling problems arising in call centers, airports, and other areas include a much higher number of breaks compared to the problem formulations in previous works on shift scheduling. Also, additional requirements like time windows for lunch breaks or restrictions on the length of breaks and work time emerged. These new constraints significantly enlarge the search space. Therefore, researchers recently started to consider a new approach which regards shift scheduling and break scheduling as two different problems and tries to solve them in separate phases.

The break scheduling problem that imposes same constraints for breaks defined in this chapter has been investigated in [5, 26, 20, 31, 30]. Beer et al. [5] applied a min-conflict based heuristic to solve this problem. This method has been applied in a real-life application for the supervision personnel. Beer et al. [5] also introduce real life benchmark instances containing shifts that include more than 10 breaks. The results presented in [5] have been further improved by memetic algorithms proposed in [20, 31, 30]. A similar break scheduling problem, which origins in call centers and includes also meetings and some slightly changed constraints, has been described in [4] and [26]. Note that a simplified break scheduling problem can be formulated as temporal constraint satisfaction problem (STP) [9] therefore it can be solved in polynomial time. This algorithm can be applied to find legal position of breaks, but without taking into consideration over-staffing and under-staffing that are very important criteria when solving shift design and break scheduling problems.

Recently, an algorithm based on constraint programming and local search for solving of this break scheduling problem together with shift design has been investigated in [15]. This approach could not improve the results obtained by solving the break scheduling problem separately (after generating shifts).

5 Local Search for Shift Design

Local search is a search paradigm which has evidenced to be very effective for a large number of AI problems [17]. This paradigm is based on the idea of navigating through the search space by iteratively stepping from one state to one of its "neighbors", which are obtained by applying a simple local change to the current solution.

The first local search algorithm for the shift design problem has been proposed in [12], [21], and [19]. These initial works proposed different move types that were used for the exploration of the neighborhood. Regarding the local search techniques, the basic principles of tabu search [16] were used to escape from local minima. In order to make the search more effective, a new method was introduced to explore only parts of the neighborhood. The search is focused on days in which a shortage or excess is present. Moreover, some of the moves are applied only to specific shift types in the region in which the shortage or excess appears. Additionally, an approach for generating a good initial solution, which is based on the idea of starting (ending) a shift whenever the requirements increase (decrease), was proposed. These techniques were implemented in a commercial product called Operating Hours Assistant, which has been used in different areas to solve shift design problems.

A hybrid local search approach for shift design has been presented in [10]. This solver comprises two stages, namely a greedy construction for the initial solution followed by a tabu search procedure [16] that iteratively improves it. The greedy construction method relies on the equivalence of the (non-cyclic) shift design problem to a variant of the Min-Cost Max-Flow network problem [23]. The greedy heuristic employs a polynomial subroutine which can easily compute the optimal staffing with minimum (weighted) deviation, but it is not able to simultaneously minimize the number of shifts used.

The second stage of the proposed heuristic is based on the local search paradigm and relies on multiple neighborhood relations. The local search model is defined by specifying three entities, namely the *search space*, the *neighborhood relation*, and the *cost function*. In details, the local search entities are described in the following.

5.1 Search Space and Initial Solution

We consider as a state for shift design the set of shifts $Q = \{s_1, s_2, \ldots\}$ together with their associated parameters. The shifts of a state are split into two categories:

- *Active* shifts: at least one employee is assigned to a shift of this type on at least one day.

- *Inactive* shifts: no employees are assigned to a shift of this type on any day. These shifts do not contribute to the solution and to the objective function. Their role is explained later.

More formally, we say that a shift $s_i \in Q$ is active (resp. inactive) if and only if $\sum_{j=1}^{C} s_i.w_j \neq 0\ (= 0)$.

5.2 *Neighborhood Relations*

For tackling the shift design problem we consider three different neighborhood relations that are combined in the spirit of the multi-neighborhood search [11], and are a subset of those applied in [19] with some modifications.

Given a state Q of the search space the types of moves considered are the following:

ChangeStaff (CS): The staff of a shift is increased ($\uparrow$) or decreased ($\downarrow$) by one employee. This kind of move is described as a triple $\langle s_i, j, a\rangle$, where $s_i \in Q$ is a shift, $j \in \{1,\ldots,C\}$ is a day, $a \in \{\uparrow,\downarrow\}$.

ExchangeStaff (ES): One employee in a given day is moved from one shift to another one of the same type. The move is described by the triple $\langle s_{i_1}, s_{i_2}, j\rangle$, where $s_{i_1}, s_{i_2} \in Q$, and $j \in \{1,\ldots,C\}$.

ResizeShift (RS): The length of the shift is increased or decreased by 1 time-slot, either on the left-hand side or on the right-hand side. The move is described by the triple $\langle s_i, l, p\rangle$, where $s_i \in Q$, $l \in \{\uparrow,\downarrow\}$, and $p \in \{\leftarrow,\rightarrow\}$. In order to apply this move, the shift s_i', obtained from s_i must be feasible with respect to its original type.

When the effect of the move will transform an inactive shift in an active one (e.g., by increasing the staff of an inactive shift at a given day), a new inactive shift of the same type is randomly created.

Inactive shifts allow us to insert new shifts and to move staff between shifts in a uniform way. This approach limits the creation of new shifts only to the current inactive ones, rather than considering all possible shifts belonging to the shift types (which are many more). The possibility of creating any legal shift is rescued if we insert as many (distinct) inactive shifts as compatible with the shift type. Experimental results, though, show that there is a trade-off between computational cost and search quality which seems to have its best compromise in having 2 inactive shifts per type.

5.3 *Search Strategies*

Our local search solver is driven by tabu search. A full description of this technique is out of the scope of this chapter and we refer to the book of Glover and Laguna [16] to for a general introduction. We describe the specialization of this technique to our problem.

Differently from [19], that uses tabu search as well, we employ the three neighborhood relations selectively in various phases of the search, rather than exploring the overall neighborhood at each iteration.

Our strategy is to combine the neighborhood relations CS, ES, and RS, according to the following scheme made of compositions and interleaving. In detail, our algorithm interleaves three different tabu search *runners* using the following neighborhoods:

- the ES alone
- the RS alone
- the set-union of the two neighborhoods CS and RS

The runners are invoked sequentially and each one starts from the best state obtained from the previous one. The overall process stops when a full round of all of them does not find an improvement. Each single runner stops when it does not improve the current best solution for a given number of iterations (called *idle iterations*).

This composite solver is further improved by performing a few changes on the final state of each runner, before handing it over as the initial state of the following runner. In details, the modifications at the performed are the merge of identical shifts and the recreation of a suitable number of inactive shifts.

For all three runners, the size of the tabu list is dynamic and each move remains in this list for a random number of iterations selected within a given range. Moreover, the tabu status of a move is dropped if it leads to a state better than the current best, in accordance with the standard *aspiration* criterion of tabu search.

6 Local Search for Break Scheduling

In this section we present in details a solution method based on local search for the break scheduling problem. The algorithm that will be described is based on the min-conflicts heuristic [18] and has been first proposed for break scheduling problem in [5], [4], and [26]. The following description is based on these references.

The solution of the break scheduling problem is represented as a set of breaks. For each shift, s_i, the breaks to be scheduled are instantiated at the beginning of a local search algorithm. At first, lunch breaks are generated, and then the remaining break time is distributed among monitor breaks. Hence, the duration of each lunch break is set to the exact number of time slots required by constraint C_2 (lunch breaks), and the duration of each monitor break lies within the specified minimum and maximum limits imposed by constraint C_5 (break durations).

In the min-conflicts-based algorithm, the start of a break, b_j.start, is an integer variable that can be altered during the search process. In contrast, we require that the duration of a break, b_j.duration, remains unchanged and keeps its initially assigned value. However, we allow multiple breaks to be scheduled consecutively so that breaks of longer duration can be created.

6.1 Initial Solution

Once the breaks are created, they must be placed in the given shift plan. We implemented two methods to schedule breaks within their associated shifts. The first simply schedules breaks randomly so that they do not overlap. The second schedules breaks so that the resulting break pattern completely satisfies constraints C_1 through C_5. This task is accomplished by formulating the problem as a simple temporal problem (STP) [9] and solving it by means of a randomized version of the Floyd-Warshall shortest-path algorithm [23].

A STP consists of a set of variables $X = X_1, ..., X_n$ and a set of constraints on those variables. The variables of an STP represent time points having continuous domains. Each constraint is represented as an interval that either restricts the domain values for a single variable X_i or restricts the difference $(X_j - X_i)$ of two distinct variables (X_i, X_j).

To schedule breaks correctly with respect to constraints C_1 through C_5, the start and end parameters of shifts and breaks are modeled as variables of an STP. The STP constraints will take care of the various limits imposed on break positions and on the duration of breaks and work periods. In detail the STP constraints are the following:

$$\begin{aligned} b_j.\text{start} &\in [(s_i.\text{start} + \text{distance to shift start}), s_i.\text{end}] \\ b_j.\text{end} &\in [s_i.\text{start}, (s_i.\text{end} - \text{distance to shift end})] \end{aligned} \qquad (C_1)$$

$$\begin{aligned} b_{lb}.\text{start} &\in [(s_i.\text{start} + \text{distance to shift start}\, lb), s_i.\text{end}] \\ b_{lb}.\text{end} &\in [s_i.\text{start}, (s_i.\text{end} - \text{distance to shift end}\, lb)] \end{aligned} \qquad (C_2)$$

$$\begin{aligned} b_1.\text{start} - s_i.\text{start} &\in [\text{min work duration}, \text{max work duration}] \\ b_{j+1}.\text{start} - b_j.\text{end} &\in [\text{min work duration}, \text{max work duration}] \\ s_i.\text{end} - b_m.\text{end} &\in [\text{min work duration}, \text{max work duration}] \end{aligned} \qquad (C_3)$$

$$\begin{aligned} b_1.\text{duration} \le \text{min ts count} &\iff b_1.\text{start} - s_i.\text{start} \in [\text{min duration}, \text{work limit}) \\ b_{j+1}.\text{duration} \le \text{min ts count} &\iff b_{j+1}.\text{start} - b_j.\text{end} \in [\text{min duration}, \text{work limit}) \end{aligned} \qquad (C_4)$$

The two temporal constraints for C_4 are inserted if and only if (b_1.duration $\le$ min length) and (b_{j+1}.duration $\le$ min length), respectively. Constraint C_5 is satisfied by construction, since the solution creation process creates only breaks whose durations range between the required time limits.

6.2 Neighborhood Relations

We developed two types of moves for the break scheduling problem. The first move (called assignment) assigns to a break a new start within its respective shift. The second move (denoted swap) exchanges the start times of two breaks associated with the same shift, meaning those breaks are actually swapped. Figure 1 illustrates these two moves. Given a feasible solution S to the break scheduling problem, the neighborhood $N(S)$ is the set of all solutions obtained by applying an assignment to a single break in S or by swapping two breaks within the same shift in S.

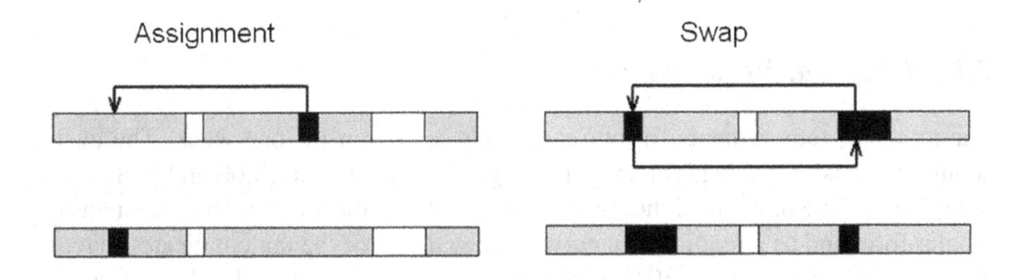

Fig. 1 The two moves used for the break scheduling problem. The assignment move assigns to a break a new start within its respective shift. The swap move exchanges the start times of two breaks associated with the same shift.

6.3 Min-conflicts Heuristics

The minimum-conflicts heuristics aims at improving the current solution by concentrating only on the parts that cause constraint violations. During an iteration, the minimum-conflicts heuristic selects a break that violates a constraint and determines a move that decreases (or leave unchanged) the violation degree of the current solution. If such a move exists, it is applied to the current solution, and the search continues until some stopping condition is satisfied.

The minimum-conflicts search method applies only moves that do not decrease the current solution quality. Thus, if the search reaches a local optimum solution, the algorithm will not proceed any further, since it will not find solutions of better quality than the local optimum. To avoid this undesirable behavior, we apply an additional strategy known as random walk, which has been used successfully in algorithms for satisfiability problems [27], rotating workforce scheduling [22], etc.

The random walk strategy also selects a break that violates a constraint. However, unlike the minimum-conflicts heuristic, it applies an arbitrary move to that break. On the one hand, the violation degree of the resulting solution could be worse than the previous one. However, on the other hand, performing such moves can help the algorithm to escape from local optima. We call the combination of both strategies min-conflicts random walk. The random walk strategy is carried out with a small probability p, whereas the ordinary minimum-conflicts search is carried out with a higher probability of $1-p$. The specific value of p is determined experimentally.

7 A Case Study

In this section we consider a real life problem arising from the shift design domain. This example represents a particular sub-problem that was solved in an European airport. We selected this domain, because shift design problems in airports are of high practical relevance. To solve the particular shift design problem we apply the local search techniques proposed in [21, 19]. This method is included in a commercial scheduling system called Operating Hours Assistant(OPA) [12] owned by the XIMES Corp. OPA has been successfully applied by the consultants of the XIMES GmbH in many companies and institutions since more than ten years.

7.1 Temporal Requirements

The temporal requirements for our case study are given for one week. The cycle should be considered (a night shift that begins on Sunday at 23:00 and is 8 hours long, impacts the first day of the week). Figure 2 shows the temporal requirements in tabular form and as a requirement curve. The first row of the table indicates that on Monday between 05:00-05:30 9 employees are required, on Tuesday 10 employees are needed, etc. As we can see the number of employees varies during the planning period. The maximal number of employees in the planning period is 38, whereas the minimal number of employees is 5.

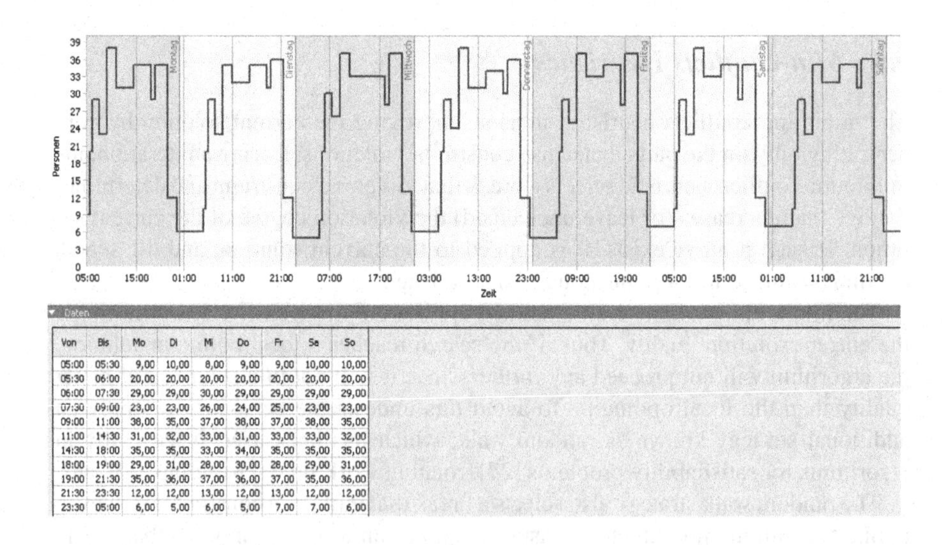

Von	Bis	Mo	Di	Mi	Do	Fr	Sa	So
05:00	05:30	9,00	10,00	8,00	9,00	9,00	10,00	10,00
05:30	06:00	20,00	20,00	20,00	20,00	20,00	20,00	20,00
06:00	07:30	29,00	29,00	30,00	29,00	29,00	29,00	29,00
07:30	09:00	23,00	23,00	26,00	24,00	25,00	23,00	23,00
09:00	11:00	38,00	35,00	37,00	38,00	37,00	38,00	35,00
11:00	14:30	31,00	32,00	33,00	31,00	33,00	33,00	32,00
14:30	18:00	35,00	35,00	33,00	34,00	35,00	35,00	35,00
18:00	19:00	29,00	31,00	28,00	30,00	28,00	29,00	31,00
19:00	21:30	35,00	36,00	37,00	36,00	37,00	35,00	36,00
21:30	23:30	12,00	12,00	13,00	12,00	13,00	12,00	12,00
23:30	05:00	6,00	5,00	6,00	5,00	7,00	7,00	6,00

Fig. 2 Temporal requirements for the airport case study

7.2 *Shift Types*

In the next step we have to define the shift types that determine the possible start and length of the real shifts. Table 1 represents the four shift types that are used for this case study: M-morning shift, D-day shift, A-afternoon shift, and N-night shift. Shifts generated by the algorithm should fulfill the criteria given by the shift types. For example, morning shifts can start between 6:00 and 8:00 and their length should be between 7 to 9 hours.

7.3 *Weights of the Criteria*

As explained in the problem definition the solution to the shift design problem is evaluated with an objective function, which combines four weighted criteria: excess in minutes, shortage in minutes, number of shifts, and distance from average number of duties per week. The weights of these criteria usually depend on the particular situation. In some cases it is very important to avoid completely the shortage of employees and in other cases the minimization of employees is more important. To find a solution for our case study we will illustrate further two scenarios regarding the importance of criteria.

7.4 *Generation of Shifts*

The local-search based algorithm for generating shifts iteratively improves the initial solution. OPA is an interactive software system that includes a graphical user interface that shows the improvements of the current solution and the relevant information regarding the overall shortage and excess, and the number of shifts. The decision maker can stop the algorithm at any time. If the decision maker is not satisfied with the current solution, he/she can change the weights and try to further improve the current solution.

Firstly, we assign weights to optimization criteria as follows: 1 - shortage, 1 - excess, 60 - number of shifts, 0 - distance from average number of duties per week. For the given requirements, constraints and the weights above, the local search algorithm generates the solution shown in Figure 3. This solution contains 8 shifts, the overall shortage is 4,2%, and the overall excess of employees is 2,32%. As we can see Figure 3 the shortage of employees appears in several time slots.

Note that the solution obtained by the Min-Cost Max-Flow algorithm has a total shortage of 3,63%, and a total excess of 2,45%. This polynomial algorithm computes the optimal staffing with minimum (weighted) deviation, however it is not able to simultaneously minimize the number of shifts used and the cycle is not taken into consideration. The number of shifts generated by this method is 24. Therefore, the solution obtained by the local search procedure is much better regarding the number of shifts, and regarding the covering of requirements it is only slightly worse than the solution obtained by the Min-Cost Max-Flow method.

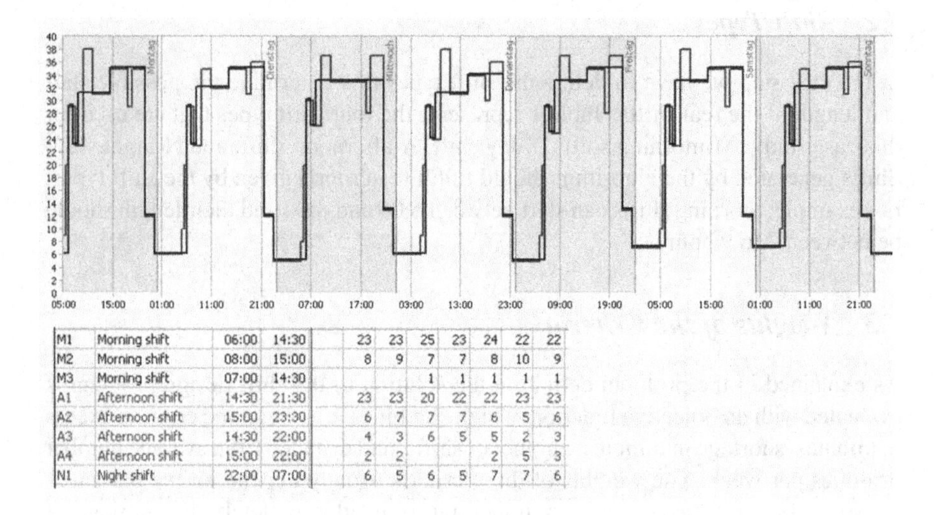

M1	Morning shift	06:00	14:30	23	23	25	23	24	22	22
M2	Morning shift	08:00	15:00	8	9	7	7	8	10	9
M3	Morning shift	07:00	14:30			1	1	1	1	1
A1	Afternoon shift	14:30	21:30	23	23	20	22	22	23	23
A2	Afternoon shift	15:00	23:30	6	7	7	7	6	5	6
A3	Afternoon shift	14:30	22:00	4	3	6	5	5	2	3
A4	Afternoon shift	15:00	22:00	2	2			2	5	3
N1	Night shift	22:00	07:00	6	5	6	5	7	7	6

Fig. 3 The shift design solution generated using local search

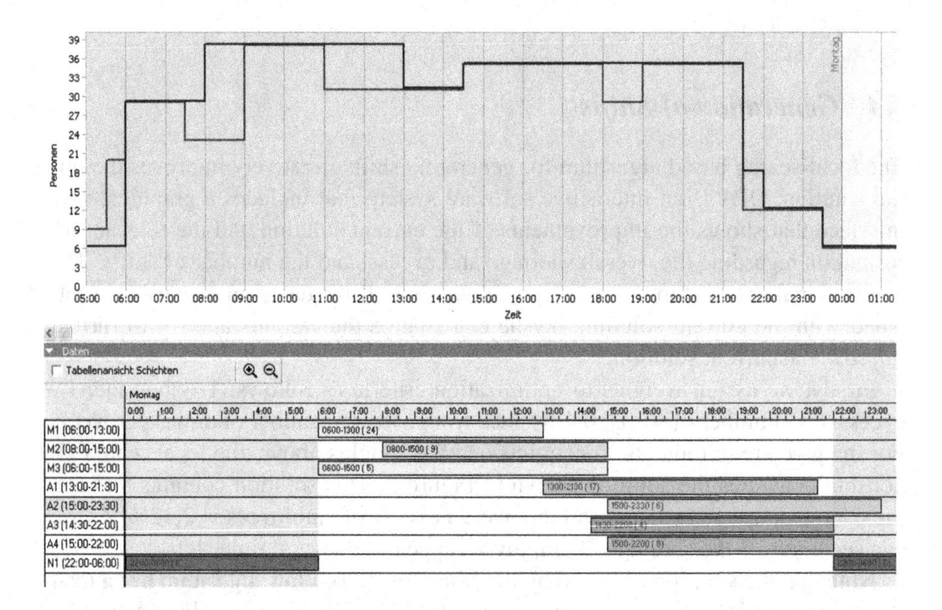

Fig. 4 The shift design solution generated by the local search. The weight of shortage is increased to 5.

In the second scenario we increased the weight of the shortage to 5. Other weights are not changed. Using these weights we can run the local search algorithm that uses the current solution as an initial solution. This algorithm finds a solution with 8 shifts that produce 1,49% shortage and 6,45% excess. The overall sum of the shortage and the excess is increased. However, the shortage which is now the most important criterion is significantly improved. The new generated solution is presented in Figure 4 (only the first day is shown). By increasing of the weight of the shortage, we could also find a solution with no shortage.

Such a result is quite typical for real life projects. It is not overall the minimization of shortage and excess but a multidimensional optimization problem with many factors to be considered. Typically users like to get an idea of consequences of weight changes (either by recalculation from scratch or by stepwise refinement of the solution at hand). This helps users to assess the potential and limits for further improvements and sometimes even causes a change in the requirements stated (...if there is no solution with ... we cannot ...). A quick optimization is crucial to facilitate such assessment process.

8 Conclusion

In this chapter we presented two real-life problems from the area of personnel scheduling. We gave a precise problem definition for shift design and break scheduling and then surveyed the existing literature for these two tasks and other related problems. The solution techniques based on local search that have been used successfully to solve different large real-life problems in these domains have been further introduced. Finally, we presented a case study and described the solution process in a commercial personnel scheduling system that exploits the local search techniques to solve the shift design problem.

The local search techniques presented in this chapter have been shown to be powerful tools for such problems. In our previous studies we have shown that the results of these techniques can be still improved by hybridization with other solution methods like network flow algorithms. Therefore, considering simpler problems that can be efficiently solved exactly can be very useful to get robust methods that combine the advantages of different solution paradigms.

Solving shift design problem and break scheduling (with a large number of breaks) simultaneously is still a challenging task. Although some work has been done in this direction, the results obtained by the approach that considers these two phases separately could still not be improved. Furthermore, it would be interesting to additionally consider the assignment of the shifts to the employees, include employee qualifications, and consider also the task assignment simultaneously.

Acknowledgments. This work was supported by the Austrian Science Fund (FWF): P24814-N23. Moreover, the research herein is partially conducted within the competence network Softnet Austria II (www.soft-net.at, COMET K-Projekt) and funded by the Austrian Federal Ministry of Economy, Family and Youth (bmwfj), the province of Styria, the Steirische Wirtschaftsförderungsgesellschaft mbH. (SFG), and the city of Vienna in terms of the center for innovation and technology (ZIT).

References

1. Aykin, T.: Optimal shift scheduling with multiple break windows. Management Science 42, 591–603 (1996)
2. Aykin, T.: A comparative evaluation of modelling approaches to the labour shift scheduling problem. European Journal of Operational Research 125, 381–397 (2000)
3. Bechtold, S.E., Jacobs, L.W.: Implicit modelling of flexible break assignments in optimal shift scheduling. Management Science 36(11), 1339–1351 (1990)
4. Beer, A., Gaertner, J., Musliu, N., Schafhauser, W., Slany, W.: Scheduling breaks in shift plans for call centers. In: Proceedings of the 7th International Conference on the Practice and Theory of Automated Timetabling, Montreal, Canada (2008)
5. Beer, A., Gärtner, J., Musliu, N., Schafhauser, W., Slany, W.: An AI-based break-scheduling system for supervisory personnel. IEEE Intelligent Systems 25(2), 60–73 (2010)
6. Côté, M.-C., Gendron, B., Quimper, C.-G., Rousseau, L.-M.: Formal languages for integer programming modeling of shift scheduling problems. Constraints 16(1), 55–76 (2011)
7. Côté, M.-C., Gendron, B., Rousseau, L.-M.: Grammar-based integer programming models for multiactivity shift scheduling. Management Science 57(1), 151–163 (2011)
8. Dantzig, G.B.: A comment on Eddie's traffic delays at toll booths. Operations Research 2, 339–341 (1954)
9. Dechter, R., Meiri, I., Pearl, J.: Temporal constraint networks. Artificial Intelligence 49(1-3), 61–95 (1991)
10. Di Gaspero, L., Gärtner, J., Kortsarz, G., Musliu, N., Schaerf, A., Slany, W.: The minimum shift design problem. Annals of Operations Research 155, 79–105 (2007)
11. Di Gaspero, L., Schaerf, A.: Multi-neighbourhood local search with application to course timetabling. In: Burke, E.K., De Causmaecker, P. (eds.) PATAT 2002. LNCS, vol. 2740, pp. 262–275. Springer, Heidelberg (2003)
12. Gärtner, J., Musliu, N., Slany, W.: Rota: a research project on algorithms for workforce scheduling and shift design optimization. AI Commun. 14(2), 83–92 (2001)
13. Gärtner, J., Musliu, N., Slany, W.: A heuristic based system for generation of shifts with breaks. In: Proceedings of the 24th SGAI International Conference on Innovative Techniques and Applications of Artificial Intelligence, Cambridge (2004)
14. Di Gaspero, L., Gärtner, J., Kortsarz, G., Musliu, N., Schaerf, A., Slany, W.: The minimum shift design problem: Theory and practice. In: Di Battista, G., Zwick, U. (eds.) ESA 2003. LNCS, vol. 2832, pp. 593–604. Springer, Heidelberg (2003)
15. Di Gaspero, L., Gärtner, J., Musliu, N., Schaerf, A., Schafhauser, W., Slany, W.: A hybrid LS-CP solver for the shifts and breaks design problem. In: Blesa, M.J., Blum, C., Raidl, G., Roli, A., Sampels, M. (eds.) HM 2010. LNCS, vol. 6373, pp. 46–61. Springer, Heidelberg (2010)
16. Glover, F., Laguna, M.: Tabu search. Kluwer Academic Publishers, Dordrecht (1997)
17. Hoos, H.H., Stützle, T.: Stochastic Local Search: Foundations & Applications. Elsevier / Morgan Kaufmann (2004)
18. Minton, S., Johnston, M.D., Philips, A.B., Laird, P.: Minimizing conflicts: A heuristic repair method for constraint satisfaction and scheduling problems. Artif. Intell. 58(1-3), 161–205 (1992)
19. Musliu, N., Schaerf, A., Slany, W.: Local search for shift design. European Journal of Operational Research 153(1), 51–64 (2004)
20. Musliu, N., Schafhauser, W., Widl, M.: A memetic algorithm for a break scheduling problem. In: 8th Metaheuristic International Conference, Hamburg, Germany (2009)

21. Musliu, N.: Intelligent Search Methods for Workforce Scheduling: New Ideas and Practical Applications. PhD thesis, Vienna University of Technology (2001)
22. Musliu, N.: Heuristic methods for automatic rotating workforce scheduling. International Journal of Computational Intelligence Research 2(4), 309–326 (2006)
23. Papadimitriou, C.H., Steiglitz, K.: Combinatorial Optimization: Algorithms and Complexity. Prentice Hall (1982)
24. Quimper, C.-G., Rousseau, L.-M.: A large neighbourhood search approach to the multi-activity shift scheduling problem. Journal of Heuristics 16(3), 373–391 (2010)
25. Rekik, M., Cordeau, J.F., Soumis, F.: Implicit shift scheduling with multiple breaks and work stretch duration restrictions. Journal of Scheduling 13, 49–75 (2010)
26. Schafhauser, W.: TEMPLE - A Domain Specific Language for Modeling and Solving Real-Life Staff Scheduling Problems. PhD thesis, Vienna University of Technology (2010)
27. Selman, B., Kautz, H.A., Cohen, B.: Local search strategies for satisfiability testing. In: Proceedings of the Second DIMACS Challange on Cliques, Coloring, and Satisfiability (1993)
28. Tellier, P., White, G.: Generating personnel schedules in an industrial setting using a tabu search algorithm. In: Burke, E.K., Rudova, H. (eds.) The 5th International Conference on the Practice and Theory of Automated Timetabling, pp. 293–302 (2006)
29. Thompson, G.: Improved implicit modeling of the labor shift scheduling problem. Management Science 41(4), 595–607 (1995)
30. Widl, M.: Memetic algorithms for break scheduling. Master's thesis, Vienna University of Technology, Vienna, Austria (2010), http://www.kr.tuwien.ac.at/staff/widl/thesis.pdf
31. Widl, M., Musliu, N.: An improved memetic algorithm for break scheduling. In: Blesa, M.J., Blum, C., Raidl, G., Roli, A., Sampels, M. (eds.) HM 2010. LNCS, vol. 6373, pp. 133–147. Springer, Heidelberg (2010)

Nurse Rostering: A Complex Example of Personnel Scheduling with Perspectives

Pieter Smet, Patrick De Causmaecker,
Burak Bilgin, and Greet Vanden Berghe

Abstract. Nurse rostering is an attractive research domain due to its societal relevance, while academics are intrigued by its combinatorial complexity. Descriptions of nurse rostering problems vary largely across the literature, which makes it almost impossible to track down scientific advances of models and corresponding approaches. The present chapter introduces a mathematical formulation of a generic nurse rostering model. It provides common elements present in most nurse rostering research as well as important hospital constraints that are usually omitted from academic models. The new mathematical model satisfies all the basic requirements for future nurse rostering research and practical developments. Finally, the importance of public datasets is discussed, together with the characteristics of the various benchmark instances and research results obtained working on these instances.

1 Introduction

The wellsprings of automated nurse rostering research go back to the '70s of the former century. Over the years, this particular subfield of automated scheduling has continued to draw attention in academia, reaching a climax in the last decade. As is often the case with many domains of operations research, the present evolution takes place in lockstep with the exponential increase in computer power, and with a growing understanding and sophistication in the field of models and algorithms.

Pieter Smet · Burak Bilgin · Greet Vanden Berghe
CODeS, KAHO Sint-Lieven, Gebr. De Smetstraat 1, 9000 Gent, Belgium
e-mail: {pieter.smet,burak.bilgin,greet.vandenberghe}@kahosl.be

Patrick De Causmaecker · Greet Vanden Berghe
CODeS & iMinds-ITEC-KU Leuven,
Department Computer Science, KULAK, KU Leuven
E. Sabbelaan 53, 8500 Kortrijk, Belgium
e-mail: Patrick.DeCausmaecker@kuleuven-kulak.be

A.Ş. Etaner-Uyar et al. (eds.), *Automated Scheduling and Planning*,
Studies in Computational Intelligence 505,
DOI: 10.1007/978-3-642-39304-4_6,

Early papers strongly focussed on practical problems while displaying a thorough and systematic analysis [Warner, 1976]. In the later decades, an evolution took place towards a comprehensive research domain with the models from nurse rostering acting as test cases for many new developments in optimisation and metaheuristics ([Brucker et al., 2010, Maenhout and Vanhoucke, 2007, Valouxis et al., 2012] are a few examples).

The purpose of the recent 'Cross-domain Heuristic Search Challenge' (CHeSC 2011) was to develop search algorithms that work well across different problem domains. Hyperheuristics [Burke et al., 2003] are such general search methods capable of addressing different problems without domain specific knowledge. The challenge provided domain specific heuristics together with a number of instances for a set of problem domains. A selection hyperheuristic [Burke et al., 2010b] is composed of a mechanism that selects one of the low level domain specific heuristics and applies it to a candidate solution. Whether or not the new solution is accepted, is determined by the hyperheuristics' acceptance criterion. The challenge revealed that for hyperheuristics, it was harder to solve the nurse rostering problem than the other optimisation problems partaking in the competition [Misir et al., 2012]. Typical heuristics for nurse rostering consume a considerable amount of time, therefore limiting the adaptive capabilities of general search algorithms when the computation time is limited. For real-world nurse rostering problems, hyperheuristics offer a suitable solution approach [Smet et al., to appear]. Typically, in these cases computation time is not critical, however, instances can display a large diversity and therefore a general solution method is required. Furthermore, in practice, hyperheuristics offer the advantage that one can focus on the problem model, rather than on the solution approach. Once implemented, hyperheuristics can cope with increasingly complex models without requiring changes to the algorithmic design.

Ever stronger results are being obtained in laboratory conditions, whereas only a few of those results find their way into practice. The present chapter attempts to find the reasons why this is the case, investigating to what extent this situation is problematic, while suggesting possible remedies. Simultaneously, the power of the academic rostering solutions allows for another level of analysis to support decisions at higher organisational levels. Without capturing the full complexity of real nurse rostering problems, such approaches may, for example, enable unravelling structural under- or overstaffing within a ward. So far, as we know, such research has never had full attention and will extend the range of possible applications.

Section 2 introduces a generic model to nurse rostering problems covering the most relevant constraints identified by practitioners. The mathematical model serves as an unambiguous description of nurse rostering constraints and their contribution to the objective function. This integer model incorporates all the elements of the nurse rostering datasets available at present.

Section 3.1 of the present chapter brings a comprehensive overview of nurse rostering research. The main contributions hold the centre state, focussing on those breakthroughs leading to new theory, while at the same time underscoring the main causes of the evolution at hand. The section then moves over into questions on generality trying to identify characteristics, modelling techniques and algorithms, capable to be applied in a broader context.

Section 3.2 elaborates on public sets of nurse rostering instances serving as benchmarks for algorithm developers. Some of the benchmark datasets under discussion have been generated automatically while others have been derived from actual hospital situations.

A case study discussing the modelling of six different wards in two Belgian hospitals is presented in Sect. 3.3. This section tries to identify challenging characteristics of real-world nurse rostering problems and show how these can be included in modelling approaches.

A brief discussion and directions for future research closes this chapter in Sect. 4.

2 Mathematical Formulation

A mathematical formulation has been developed for a generic personnel rostering model and is introduced in this section as an integer model. It represents abstract adaptable nurse rostering components to meet various specific problem descriptions. In contrast to the majority of models discussed in the academic literature, the present model is devised in such a way that it is applicable in a large number of real hospital scenarios. The model covers problems matching $ASBCI|RVNO|PL$ in the $\alpha|\beta|\gamma$ notation of [De Causmaecker and Vanden Berghe, 2011].

In an effort to close the gap between theory and practice, the elements defined in the mathematical formulation presented in this chapter, are also described in an XML schema definition file [Smet et al., to appear].

2.1 Nurse Rostering Model

Academic research on personnel rostering has mostly been focused on new solution techniques. These approaches are then used to find solutions for instances based on often simplified models. Many real-world requirements concerning the problem definitions are therefore ignored during the conception of the novel approaches. The result is that only a small fraction of the academic solution methods are used in practice [Kellogg and Walczak, 2007].

The present model aims to overcome this issue by enabling to represent real-world requirements accurately and completely. Therefore, several extensions have been made to standard academic models. The purpose of the presented model is for it to be utilised in real-world personnel rostering applications. Researchers having to address rostering problems with real-world complexity can represent their problems using this generic model with few or no adjustments.

The KAHO dataset (Section 3.2) illustrates an application of this generic nurse rostering model. Its instances incorporate several complex regulations from practice with which the model can cope.

Nurse rostering problems present highly constrained combinatorial optimisation problems with both hard and soft constraints. A solution needs to satisfy all hard constraints in order to be considered feasible, while soft constraints should be attempted to be satisfied as much as possible. This is accomplished by minimising an objective function (Eq. (1)), which is a linear combination of the weight (w_c) and the number of violations (n_c) of each soft constraint ($c \in C$). The quality of a roster is inversely proportional to its objective function value, i.e. the lower the objective function value, the higher the quality of the roster.

$$\sum_{c \in C} w_c n_c \tag{1}$$

Soft constraints in the present model include rest time between two consecutive shifts, penalties with regard to skill types, coverage requirements, collaborations and absence requests. Furthermore, there are three general types of soft quantitative constraints [Bilgin et al., 2012].

Counters. These constraints restrict the number of instances of a specific subject (e.g. days worked, weekends idle, hours worked, etc.) over a specified period.

Series. These constraints restrict the number of consecutive occurrences of instances of a specific subject (e.g. consecutive shifts of the same type, consecutive idle days, etc.)

Successive series. These constraints restrict the succession of two series (e.g. days worked - days idle). Any occurance of the first series implies the second series to follow.

Threshold values are foreseen for counters, series, successive series and coverage constraints. A threshold value can be either a minimum, a maximum, or an interval defined by a minimum and a maximum value.

The constraint sets and the weights of the constraints show differences among sectors, countries and organisations, and even among different employees in the same unit. Therefore, the model does not employ one predefined constraint set with fixed weights that apply to all employees in a problem instance. Instead, it provides a manner to construct constraints and constraint sets that can be specific to each employee and problem instance. Consequently, the model does not foresee a predefined objective function, but an objective function that is constructed according to the constraint sets that are defined by the planner using the model, as expressed in Eq. (1) [Bilgin, 2012].

2.2 Definitions and Variables

The following parameters are defined:

E The set of employees

D The set of days in the current schedule period and in the relevant parts of the previous and upcoming schedule period

S The set of shift types

K The set of skill types

W The set of weekends in the current schedule period and in the relevant parts of the previous and upcoming schedule period

C The set of constraints

The following decision variables are defined $\forall e \in E, d \in D, s \in S, k \in K$:

$$x_{e,d,s,k} = \begin{cases} 1 & \text{if employee } e \text{ is assigned on day } d\text{, in shift type } s \text{ and skill type } k \\ 0 & \text{otherwise} \end{cases}$$

The auxiliary variable $p_{e,d}$ in Eq. (2) denotes the presence of employee $e \in E$ on day $d \in D$. $p_{e,d}$ takes the value 1 if employee $e \in E$ is assigned on day $d \in D$, otherwise it is 0.

$$\begin{aligned} -|S|\,|K|\,p_{e,d} + \sum_{s \in S} \sum_{k \in K} x_{e,d,s,k} \leq 0 \\ -p_{e,d} + \sum_{s \in S} \sum_{k \in K} x_{e,d,s,k} \geq 0 \end{aligned} \tag{2}$$

$p_{e,d,S'}$ in Eq. (3), another auxiliary variable, denotes the presence of employee $e \in E$ on day $d \in D$, in any of the shift types $s \in S' \subset S$. $p_{e,d,S'}$ takes the value 1 if employee $e \in E$ is assigned on day $d \in D$, in any of the shift type $s \in S' \subset S$. It takes the value 0 otherwise.

$$\begin{aligned} -|S'|\,|K|\,p_{e,d,S'} + \sum_{s \in S'} \sum_{k \in K} x_{e,d,s,k} \leq 0 \\ -p_{e,d,S'} + \sum_{s \in S'} \sum_{k \in K} x_{e,d,s,k} \geq 0 \end{aligned} \tag{3}$$

$p_{e,D',S',K'}$ in Eq. (4) denotes the presence of employee $e \in E$ on any of the days $d \in D' \subset D$, in any of the shift types $s \in S' \subset S$ and with any of the skill types $k \in K' \subset K$. $p_{e,D',S',K'}$ takes the value 1 if employee $e \in E$ has at least one assignment on any of the days $d \in D' \subset D$, in any of the shift types $s \in S' \subset S$, with any of the skill types $k \in K' \subset K$, otherwise it takes the value 0.

$$\begin{aligned} -|D'|\,|S'|\,|K'|\,p_{e,D',S',K'} + \sum_{d \in D'} \sum_{s \in S'} \sum_{k \in K'} x_{e,d,s,k} \leq 0 \\ -p_{e,D',S',K'} + \sum_{d \in D'} \sum_{s \in S'} \sum_{k \in K'} x_{e,d,s,k} \geq 0 \end{aligned} \tag{4}$$

Similarly, $q_{e,w}$ denotes the presence of employee $e \in E$ in weekend $w \in W$ (Eq. (5)). This variable becomes 1 if employee e works at least one shift in weekend w, and 0 otherwise. l refers to the length of the weekend. $d_{w,i}$ denotes day i of weekend w.

$$\begin{aligned} &\forall e \in E, w \in W : \\ &-l q_{e,w} + \sum_{i=1}^{l} p_{e,d_{wi}} \leq 0 \\ &-q_{e,w} + \sum_{i=1}^{l} p_{e,d_{wi}} \geq 0 \end{aligned} \tag{5}$$

2.3 Hard Constraints

2.3.1 One Assignment Start Per Day Per Employee

A maximum of one assignment can start for each employee on each day.

$$\forall e \in E, d \in D \ : \sum_{s \in S} \sum_{k \in K} x_{e,d,s,k} \leq 1 \tag{6}$$

2.3.2 Schedule locks

In some cases, manual planners can fix the assignments of an employee $e \in E$ on a day $d \in D$, for a shift type $s \in S$ and a skill type $k \in K$ in advance. These preset assignments are represented by a tuple $\langle e, d, s, k \rangle$ in V_{on} and are modelled using Eq. (7). Days off can also be fixed in advance for employee $e \in E$ on a day $d \in D$ using Eq. (8).

$$\forall (e, d, s, k) \in V_{on} \ : x_{e,d,s,k} = 1 \tag{7}$$

$$\forall (e, d) \in V_{off} \ : \sum_{s \in S} \sum_{k \in K} x_{e,d,s,k} = 0 \tag{8}$$

2.3.3 Honour Skill Types

Let the skill types of an employee $e \in E$ be $K_e \subseteq K$. In that case, Eq. (9) forbids any assignment to e with a skill type k that e does not have.

$$\forall e \in E \ : \sum_{d \in D} \sum_{s \in S} \sum_{k \in K \setminus K_e} x_{e,d,s,k} = 0 \tag{9}$$

2.3.4 Defined Assignments Only

If an assignment on day $d \in D$, with shift type $s \in S$ and skill type $k \in K$ is not defined in any of the coverage constraints $cc_{d,S',K'} \in CC \subseteq C$, then such an assignment cannot be made to an employee. Assignments that are provided in the input schedule cannot be modified, deleted or reassigned to another employee if no coverage constraint is defined for that day, skill and shift type.

The motivation behind this constraint can be found in the fact that the coverage constraints are considered soft constraints. For example, an assignment can be made even if the maximum threshold of the corresponding coverage constraint is equal to zero. This is necessary in practice when an employee does not meet the required amount of working hours specified in his or her contract. However, there are situations when the assignment of a particular shift type to an employee with a particular skill type cannot be considered at all. For example, in some wards, the head nurse is never assigned to a night shift and therefore such a coverage constraint is never defined.

$$\begin{aligned} &\forall d \in D, s \in S, k \in K \ : \\ &\neg\exists \left(cc_{d,S',K'} \in CC | s \in S', k \in K'\right) \Rightarrow \sum_{e \in E} x_{e,d,s,k} = 0 \end{aligned} \tag{10}$$

2.3.5 Overlapping Shift Types

Let T be the set of shift type pairs (s_i, s_j) such that s_i and s_j overlap if they are assigned on consecutive days d and $d+1$, respectively. Equation (11) ensures that overlapping shift type assignments are never made.

$$\forall e \in E, d \in D, (s_i, s_j) \in T : \sum_{k \in K} x_{e,d,s,k} + \sum_{k \in K} x_{e,(d+1),s,k} \leq 1 \tag{11}$$

2.4 Soft Constraints

Let $v(x)$, $w(x)$, $m(x)$, $n(x)$, $p(x)$ refer to the value, weight, maximum threshold, minimum threshold and the penalty of a constraint x from the set of soft constraints $X \subset C$. The objective value of a candidate solution is the sum of the penalties of all soft constraints X in the problem instance (Eq. (12)).

$$\sum_{x \in X} p(x) \tag{12}$$

This section describes various soft constraints in the model and how their penalty $p(x)$ can be calculated.

2.4.1 Rest Times between Shift Types

Let R be the set of shift type pairs (s_i, s_j) such that at least one of s_i and s_j violates the rest time of the other if they are assigned on consecutive days d and $d+1$, respectively. Let $w(rest)$ be the weight for the constraint restricting the rest time between shift types. $p(rest)$ in Eq. (13) refers to the total rest penalty, which is the number of violations of this constraint multiplied by the corresponding weight.

$$p(rest) = \sum_{e \in E} \sum_{d \in D} \sum_{(s_i, s_j) \in R} \sum_{k \in K} w(rest) \cdot x_{e,d,s_i,k} \cdot x_{e,(d+1),s_j,k} \tag{13}$$

2.4.2 Employee Skill Type Penalties

Let the skill types of an employee $e \in E$ be $K_e \subseteq K$. Each skill type $k \in K_e$ has a penalty $w(skill_{e,k})$ that is added to the objective value, in case an assignment is made with that skill type. $p(skill)$ in Eq. (14) refers to the total penalty due to skill type assignments.

$$p(skill) = \sum_{e \in E} \sum_{d \in D} \sum_{s \in S} \sum_{k \in K} w(skill_{e,k}) \cdot x_{e,d,s,k} \tag{14}$$

2.4.3 Coverage Constraints

Coverage constraints are defined as soft constraints. The coverage constraint $cc_{d,S',K'}$ restricts the number of employees assigned on day $d \in D$, for shift type set $S' \subset S$ and for skill type set $K' \subset K$. As it can be understood from its formulation, any assignment with any shift type $s \in S' \subset S$ and any skill type $k \in K' \subset K$ on day $d \in D$ is counted by the coverage constraint. $v(cc_{d,S',K'})$ refers to the total number of assignments on day d, with any shift type $s \in S' \subset S$ and any skill type $k \in k' \subset K$. The total penalty $p(cc_{d,S',K'})$ of coverage constraint $cc_{d,S',K'}$ is calculated using Eq. (16).

$$v(cc_{d,S',K'}) = \sum_{e \in E} \sum_{s \in S'} \sum_{k \in K'} x_{e,d,s,k} \tag{15}$$

$$\begin{aligned} p(cc_{d,S',K'}) = \\ & w(cc_{d,S',K'}) \cdot max\{0, v(cc_{d,S',K'}) - m(cc_{d,S',K'})\} + \\ & w(cc_{d,S',K'}) \cdot max\{0, n(cc_{d,S',K'}) - v(cc_{d,S',K'})\} \end{aligned} \tag{16}$$

2.4.4 Collaboration

Let $E' \subset E$ be the employee set, $D' \subset D$ be the day set, $S' \subset S$ be the shift set, $K' \subset K$ be the skill set and the triple $\langle D', S', K' \rangle$ be the domain of the collaboration constraint $l_{E',D',S',K'}$. $v(l_{E',D',S',K'})$ refers to the total number of assignments made to the members of the employee set E' on day $d \in D'$, for shift type set S' and skill type set K'. The total penalty $p(l_{E',D',S',K'})$ is calculated using Eq. (18).

$$\forall d \in D' : v(l_{E',D',S',K'}) = \sum_{e \in E'} \sum_{s \in S'} \sum_{k \in K'} x_{e,d,s,k} \tag{17}$$

$$\begin{aligned} p(l_{E',D',S',K'}) = & \\ & \sum_{d \in D'} w(l_{E',D',S',K'}) \cdot max\left\{0, v(l_{E',d,S',K'}) - m(l_{E',D',S',K'})\right\} + \\ & \sum_{d \in D'} w(l_{E',D',S',K'}) \cdot max\left\{0, n(l_{E',D',S',K'}) - v(l_{E',d,S',K'})\right\} \end{aligned} \tag{18}$$

2.4.5 Absence Requests

Let $D' \subset D$ be the day set, $S' \subset S$ the shift type set and $K' \subset K$ be the skill type set of the domain $\langle D', S', K' \rangle$ of the absence request $ar_{e,D',S',K'}$ of employee $e \in E$. The handling of the day set in the domain of the absence request determines how the absence request is evaluated.

The *handling* of absence request can be either *individual* or *complete.* In case the handling is individual, the days in the day set D' are handled individually. A complete handling requires all days in the day set to be considered as one block. Whenever a single request cannot be fulfilled, a penalty is incurred for all other requests on days in D' as well.

If the handling is individual, then the total penalty $p(ar_{e,D',S',K'})$ of the absence request is calculated using Eq. (19).

$$p(ar_{e,D',S',K'}) = w(ar_{e,D',S',K'}) \sum_{d \in D'} \sum_{s \in S'} \sum_{k \in K'} x_{e,d,s,k} \tag{19}$$

If the handling is complete, then the total penalty $p(ar_{e,D',S',K'})$ of the absence request is calculated using Eq. (20).

$$p(ar_{e,D',S',K'}) = w(ar_{e,D',S',K'}) \cdot p_{e,D',S',K'} \tag{20}$$

2.4.6 Counters

Let $csv(c)$ be the counter start value and $crv(c)$ the counter remainder value of counter $c \in C$. These values are used to define a starting value of a specific counter (e.g. to account for occurence in the previous period), and to define how many occurences are present in future scheduling periods. By including these values, the model better corresponds to real-world practice where assignments from previous and future periods influence the current scheduling period.

The penalty $p(c)$ of counter c is calculated using Eq. (21).

$$p(c) = \\ w(c) \cdot max\{0, v(c) - m(c)\} + \\ w(c) \cdot max\{0, n(c) - v(c) - crv(c)\} \tag{21}$$

Days worked counter
Let $D' \subseteq D$ be the day set of the days worked counter $dwc_{e,D'}$ of employee $e \in E$. The counter value $v(dwc_{e,D'})$ of days worked counter $dwc_{e,D'}$ is calculated using Eq. (22).

$$v(dwc_{e,D'}) = csv(dwc_{e,D'}) + \sum_{d \in D'} p_{e,d} \tag{22}$$

Days idle counter
Let $D' \subseteq D$ be the day set of the days idle counter $dic_{e,D'}$ of employee $e \in E$. The counter value $v(dic_{e,D'})$ of days idle counter $dic_{e,D'}$ is calculated using Eq. (23).

$$v(dic_{e,D'}) = csv(dic_{e,D'}) + \sum_{d \in D'} (1 - p_{e,d}) \tag{23}$$

Shift types worked counter
Let $D' \subseteq D$ be the day set and $S' \subseteq S$ be the shift types set of the shift types worked counter $swc_{e,D',S'}$ of employee $e \in E$. The counter value $v(swc_{e,D',S'})$ of shift types worked counter $swc_{e,D',S'}$ is calculated using Eq. (24).

$$v(swc_{e,D',S'}) = csv(swc_{e,D',S'}) + \sum_{d \in D'} \sum_{s \in S'} \sum_{k \in K} x_{e,d,s,k} \tag{24}$$

Weekends worked counter
Let $W' \subseteq W$ be the weekends in the counter period of the weekends worked counter $wwc_{e,W'}$ of employee $e \in E$. The counter value $v(wwc_{e,W'})$ of weekends worked counter $wwc_{e,W'}$ is calculated using Eq. (25).

$$v(wwc_{e,W'}) = csv(wwc_{e,W'}) + \sum_{w \in W'} q_{e,w} \tag{25}$$

Weekends idle counter
Let $W' \subseteq W$ be the weekends in the counter period of the weekends idle counter $wic_{e,W'}$ of employee $e \in E$. The counter value $v(wic_{e,W'})$ of the weekends idle counter $wic_{e,W'}$ is calculated using Eq. (26).

$$v(wic_{e,W'}) = csv(wic_{e,W'}) + \sum_{w \in W'} (1 - q_{e,w}) \tag{26}$$

Hours worked counter
Let $D' \subseteq D$ be the day set of the hours worked counter $hwc_{e,D'}$ of employee $e \in E$. Let $D_{ar} \subseteq D$, $S_{ar} \subseteq S$, and $K_{ar} \subseteq K$ be the day set, shift type set

and skill type set of absence request ar, respectively. Let $D'_{ar} = D' \cap D_{ar}$. Let AR_e be the set of all absence requests of an employee $e \in E$. Let $JobTime(s)$ be the net job time of a shift type $s \in S$ and $JobTime(ar)$ the net job time of an absence request ar.

$$\begin{aligned} AR_e^c &= \{ar \in AR_e | D'_{ar} \neq \emptyset \wedge \text{ handling of } D_{ar} \text{ is complete}\} \\ AR_e^i &= \{ar \in AR_e | D'_{ar} \neq \emptyset \wedge \text{ handling of } D_{ar} \text{ is individual}\} \end{aligned} \tag{27}$$

$$\begin{aligned} v(hwc_{e,D'}) = \\ & csv(hwc_{e,D'}) + \sum_{d \in D'} \sum_{s \in S} \sum_{k \in K} x_{e,d,s,k} \cdot JobTime(s) + \\ & \sum_{ar \in AR_e^c} (1 - p_{e,D_{ar},S_{ar},K_{ar}}) \cdot JobTime(ar) + \\ & \sum_{ar \in AR_e^i} \sum_{d \in D'_{ar}} \sum_{s \in S_{ar}} \sum_{k \in K_{ar}} (1 - x_{e,d,s,k}) \cdot JobTime(ar) \end{aligned} \tag{28}$$

Domain counter

Let $D' \subseteq D$ be the day set, $S' \subseteq S$ be the shift types set, and $K' \subseteq K$ be the skill type set of the domain counter $dc_{e,D',S',K'}$ of employee $e \in E$. The counter value $v(dc_{e,D',S',K'})$ of domain counter $dc_{e,D',S',K'}$ is calculated using Eq. (29).

$$v(dc_{e,D',S',K'}) = csv(dc_{e,D',S',K'}) + \sum_{d \in D'} \sum_{s \in S'} \sum_{k \in K'} x_{e,d,s,k} \tag{29}$$

2.4.7 Series

In what follows, the calculation of the penalty $p(s)$ of a series constraint $s \in C$ is presented for different types for series constraints.

Days worked series

The calculation of the total penalty $p(dws_e)$ of days worked series dws_e with a maximum threshold $m(dws_e)$ of an employee $e \in E$ is given in Eq. (30).

$$p(dws_e) = w(dws_e) \sum_{d \in D} max \left\{ \left(\sum_{i=0}^{m(dws_e)} p_{e,d+i} \right) - m(dws_e), 0 \right\} \tag{30}$$

The calculation of the total penalty $p(dws_e)$ of days worked series dws_e with a minimum threshold $n(dws_e)$ of an employee $e \in E$ is given in Eq. (31).

$$p(dws_e) = \\ w(dws_e) \sum_{d \in D} (1 - p_{e,d}) \cdot \\ p_{e,d+1} \cdot max_{i=1}^{n(dws_e)} \{(n(dws_e) - i)(1 - p_{e,d+i+1})\} \tag{31}$$

Days idle series
The calculation of the total penalty $p(dis_e)$ of days idle series dis_e with a maximum threshold $m(dis_e)$ of an employee $e \in E$ is given in Eq. (32).

$$p(dis_e) = w(dis_e) \sum_{d \in D} max \left\{ \left(\sum_{i=0}^{m(dis_e)} (1 - p_{e,d+i}) \right) - m(dis_e), 0 \right\} \tag{32}$$

The calculation of the total penalty $p(dis_e)$ of days idle series dis_e with a minimum threshold $n(dis_e)$ of an employee $e \in E$ given in Eq. (33).

$$p(dis_e) = w(dis_e) \sum_{d \in D} p_{e,d} \cdot (1 - p_{e,d+1}) \cdot max_{i=1}^{n(dis_e)} \{(n(dis_e) - i)(p_{e,d+i+1})\} \tag{33}$$

Weekends worked series
The calculation of the total penalty $p(wws_e)$ of weekends worked series wws_e with a maximum threshold $m(wws_e)$ of an employee $e \in E$ is given in Eq. (34).

$$p(wws_e) = w(wws_e) \sum_{w \in W} max \left\{ \left(\sum_{i=0}^{m(wws_e)} q_{e,w+i} \right) - m(wws_e), 0 \right\} \tag{34}$$

The calculation of the total penalty $p(wws_e)$ of weekends worked series wws_e with a minimum threshold $n(wws_e)$ of an employee $e \in E$ is given in Eq. (35).

$$p(wws_e) = \\ w(wws_e) \sum_{w \in W} (1 - q_{e,w}) \cdot q_{e,w+1} \cdot \\ max_{i=1}^{n(wws_e)} \{(n(wws_e) - i)(1 - q_{e,w+i+1})\} \tag{35}$$

Weekends idle series
The calculation of the total penalty $p(wis_e)$ of days idle series wis_e with a maximum threshold $m(wis_e)$ of an employee $e \in E$ is given in Eq. (36).

$$p(wis_e) = w(wis_e) \sum_{w \in W} max \left\{ \left(\sum_{i=0}^{m(wis_e)} (1 - q_{e,w+i}) \right) - m(wis_e), 0 \right\} \tag{36}$$

The calculation of the total penalty $p(wis_e)$ of days idle series wis_e with a minimum threshold $n(wis_e)$ of an employee $e \in E$ given in Eq. (37).

$$\begin{aligned} p(wis_e) = & \\ & w(wis_e) \sum_{w \in W} q_{e,w} \cdot (1 - q_{e,w+1}) \cdot \\ & max_{i=1}^{n(wis_e)} \{(n(wis_e) - i)(q_{e,w+i+1})\} \end{aligned} \tag{37}$$

Shift types worked series

The calculation of the total penalty $p(sws_{e,S'})$ of shift types worked series $sws_{e,S'}$ defined on $S' \subset S$ with a maximum threshold $m(sws_{e,S'})$ of an employee $e \in E$ is given in Eq. (38).

$$p(sws_{e,S'}) = w(sws_{e,S'}) \sum_{d \in D} max \left\{ \left(\sum_{i=0}^{m(sws_{e,S'})} p_{e,d+i,S'} \right) - m(sws_e), 0 \right\} \tag{38}$$

The calculation of the total penalty $p(sws_{e,S'})$ of shift types worked series $sws_{e,S'}$ defined on $S' \subset S$ with a minimum threshold $n(sws_{e,S'})$ of an employee $e \in E$ is given in Eq. (39).

$$\begin{aligned} p(sws_{e,S'}) = & \\ & w(sws_{e,S'}) \sum_{d \in D} (1 - p_{e,d,S'}) \cdot p_{e,d+1,S'} \cdot \\ & max_{i=1}^{n(sws_{e,S'})} \{(n(sws_{e,S'}) - i)(1 - p_{e,d+i+1,S'})\} \end{aligned} \tag{39}$$

2.4.8 Successive Series

As stated in Sect. 2.1, a penalty for a successive series constraint $ss \in C$ is incurred when an occurrence of the first series $s1$ is not followed by the second series $s2$. Let $p(ss)$ be the total penalty for a successive series constraint and $w(ss)$ the weight. Eq. (40) shows how $p(ss)$ is calculated. α_{s1} is a binary variable, whereas β_{s2} is a non-negative integer variable. For every $s1 - s2$ succession in the schedule, if $s1$ satisfies the condition of the first series α_{s1} is 1, and otherwise α_{s1} is 0. Again, for every $s1 - s2$ succession in the schedule, β_{s2} corresponds to the amount of the deviation between the actual schedule and the second series $s2$.

$$p(ss) = sum_{s1-s2}(w(ss).\alpha_{s1}.\beta_{s2}) \tag{40}$$

3 Solution Approaches and Datasets for Nurse Rostering

3.1 Algorithmic Progress

Domains of applied research in optimisation tend to go through a fixed set of stages. Initial modelling efforts produce simplified versions of the problem statement accessible for standard approaches, leading to efficient solution methods or to conclusions about the intrinsic complexity of the problem. When confronted with practice, the simplified models cannot catch the complications of the real application and must be extended to accommodate new characteristics. The aforementioned approaches will only partly grasp these extensions. The resulting optimisation problems may either be more or less complex. In the latter case, a deeper study of the problem characteristics may reveal essential complicating factors which were not observed initially.

In nurse rostering, the initial models covered assignment of nurses to shifts with the main goal of satisfying the nursing demand in the hospital. Work regulations would typically reduce to best practice conventions used by the operators in the hospital. An example is cyclic scheduling, where the working schedules of the nurses follow a cyclic pattern [Rocha et al., to appear]. As the studies evolved, and depending on the country, work regulations for nurses turned out to be as extremely diverse as extensive. The number of shift structures is of the same order of magnitude as the number of hospitals. Nurses have regulations allowing very dynamic and detailed tuning to their private lives. Healthy work pattern requirements restrict the number of times a nurse can be set to work in a specific kind of shift [Valouxis and Housos, 2000]. The transition in time from one shift to another is subject to specific constraints. Many of these constraints are considered to be preferably satisfied while some must be met under any condition. The model has shifted from the initial assignment problem with a limited number of constraints to a general optimisation problem taking soft constraints into the objective function, like the one represented by Eq. (1).

Initial approaches to nurse rostering problems used linear and integer programming [Abernathy et al., 1973]. In later times, intrinsic complexities made it a target for heuristic techniques [Burke et al., 2001]. As often with problems that cannot be solved to optimality, benchmarks were set up for approaches to allow themselves to be compared (Sect. 3.2). Benchmarks partially grasp the essential characteristics of problems and, if refined over time, serve to arrive at sharper problem definitions. Being a complex problem with many facets, nurse rostering did not lead to one single problem definition. It presently stands for a collection of problems agreeing on certain common elements and differing on others. Two of the authors recently set up a categorisation for nurse rostering, systematically listing the elements that can make up a nurse rostering problem [De Causmaecker and Vanden Berghe, 2011]. The first International Nurse Rostering Competition (2010 NRC) [Haspeslagh et al.,

2012] (Sect. 3.2.4), took specific properties of this list and effectively produced a new benchmark set.

The results of the competition in fact reflect an interesting evolution in optimisation techniques. The organisers mentioned six 'lessons learned' highlighting the importance and possibilities of dedication, hybridization and decomposition, modelling power, restart, hyperheuristics and time spent. These lessons of course reflect a general feeling in the community of combinatorial optimisation and quite naturally fit into the research focus of the time. Dedication is a natural plus in combinatorial problems: the fact that a solver designed for personnel scheduling problems did well on nurse rostering confirms both the power of the solver and the relevance of the competition's problem setting. A hybrid approach consisting of an iterated heuristic decomposition and an exact solver for the component problems [Valouxis et al., 2012] won all tracks in the competition. This alone confirms good results previously obtained by other hybrid approaches and opens a promising new line of thought to the heuristics community. Modelling power present in constraint based systems and the strength of available solvers allowed one team to concentrate on the model to arrive at a very competitive solution [Nonobe, 2010]. Hyperheuristics are acquiring maturity and could stand up against more standard techniques in this competition. Finally, the effectiveness of random restarts in local search has been around for a long time and showed up again in this setting and the fact that algorithms in general do not succeed to take profit from longer computation times was confirmed here. Details of the above approaches are discussed in Sect. 3.2. Besides insights resulting from the best performing approaches to the 2010 NRC's instances, the increase of exact and hybrid mathematical approaches to the nurse rostering problem is noteworthy [Burke and Curtois, 2011, Della Croce and Salassa, to appear, Glass and Knight, 2010].

The field has evolved from a simplification of a practical problem into a detailed description of this problem, identifying crucial problem properties. Benchmarks and competitions put several approaches to a test and may open up the landscape for new techniques not mainstream in the community at a specific time.

3.2 Benchmark Datasets and Solution Methods

As the previous section demonstrates, several solution approaches have been proposed by various authors for different variants of the nurse rostering problem. In the majority of these publications, researchers focus on a specific ward or hospital which has its own regulations. As a result, the presented solution approaches are often (willingly or not) tailored to one specific instance of the problem. Furthermore, in most cases the data on which the performance of an algorithm is verified is not made publicly available. Due to the nature of the nurse rostering problem, this is an understandable course of action. Hospitals do not want to compromise data concerning their daily operations. However, the research community could benefit from a wider availability of instances for verifying and comparing various solution approaches [Burke et al., 2004]. Consequently, robust approaches that handle more variated instances of the nurse rostering problem would become attractive for practitioners.

In an effort to help close the gap between theory and practice, a collection of benchmark instances have been made publicly available. These benchmark datasets offer a large number of variegated instances which researchers can use to evaluate the performance or robustness of their novel solution approaches. The research community has made an effort to include a large diversity in the available instances. As will be shown in what follows, some datasets present real-world scenarios, while others are artificially generated based on a number of parameter settings. Since the introduction of these different datasets, several new best results have been published. This section aims to give an overview of the existing benchmark datasets for the nurse rostering problem along with some of their defining characteristics. Furthermore, a survey is presented of papers in which the different datasets are targeted.

Four benchmark datasets were selected based on popularity and diversity. In doing so, we aim to present an array of possibilities to anyone wishing to evaluate an optimisation approach to the nurse rostering problem. Table 1 shows an overview of the selected datasets. The second column refers to an elaborate description of the instances. Different approaches verified on the instances of a dataset are shown in the third column. These algorithms are discussed later in this section for each dataset individually.

Table 1 Overview of nurse rostering datasets

Dataset	Description	Relevant publications
KAHO	Bilgin et al. [2012]	Smet et al. [to appear] Smet et al. [2012]
Nottingham	Curtois [2012]	Brucker et al. [2010], Burke et al. [2010a, 2011], Burke and Curtois [2011], Glass and Knight [2010]
NSPLIB	Vanhoucke and Maenhout [2007]	Maenhout and Vanhoucke [2006, 2007, 2008, 2010]
2010 NRC	Haspeslagh et al. [2012]	Awadallah et al. [2011], Bilgin et al. [2010], Burke and Curtois [2011], Geiger [2011], Lü and Hao [2012], Messelis and De Causmaecker [2011], Nonobe [2010], Valouxis et al. [2012]

Looking at the four benchmark datasets, some common properties of the nurse rostering problem in general can be identified. Typically, only few hard constraints are defined while there exists a large set of soft constraints. Common hard constraints include the coverage constraints and the requirement for a single assignment per day and per nurse. However, not all datasets consider the same hard constraints, and in doing so attribute to the variety of instances in the benchmarks.

Due to the public availability of the instances, important contributions have been made to the research community. By presenting new, possibly best, results for the benchmarks, a more thorough comparison can be performed, resulting in a fairer evaluation of an approach. On the downside, researchers often select instances from only one dataset. Combining instances from different datasets is advocated so as to enable more comprehensive studies.

3.2.1 KAHO Dataset

The benchmark instances in the KAHO dataset are constructed based on data collected from six different wards in two Belgian hospitals. The data was provided by companies specialised in decision support software for personnel rostering[1]. The information they provided, combined with feedback from human planners, was used to construct the instances so that they compare very closely to reality.

Table 2 shows the number of employees, shifts, skill types and days for each instance in the benchmark dataset. For each ward, three different scenarios are considered. The *normal* scenario models a standard situation in the ward considering the average coverage requirements as well as the normal availability of the nurses. The *overload* scenario simulates a situation where a large amount of work (i.e. coverage) is to be met by the regular staff. This could be the case in situations where the hospital is required to treat a larger number of patients than normal, e.g. in case of a natural disaster or an epidemic. The last scenario, *absence*, represents a case in which a nurse is absent for some part of the roster horizon. The problem instances, as well as an XSD of the input files, can be downloaded from `http://allserv.kahosl.be/~pieter/nurserostering`.

Table 2 General characteristics of the KAHO problem instances

Instance	Employees	Shift types	Skill types	Days
Emergency	27	27	4	28
Psychiatry	19	14	3	31
Reception	19	19	4	42
Meal preparation	32	9	2	29
Geriatrics	21	9	2	28
Palliative care	27	23	4	91

Bilgin et al. [2012] present a study on various local search neighbourhoods for the nurse rostering problem. Their experiments are conducted on instances from the KAHO dataset. They use a variable neighbourhood search and an adaptive large neighbourhood search (ALNS) to generate the first results for

[1] SAGA Consulting and GPS NV Belgium.

these benchmark instances. To the best of the authors' knowledge, these are the best known results and they represent the state of the art for this dataset.

Smet et al. [to appear] discuss various modelling issues with the existing nurse rostering models from the literature. They present extensions to the KAHO dataset model which are used to model complex constraints such as collaboration between groups of nurses and requirements for training junior nurses. Furthermore, experimental results are presented using a hyper-heuristic algorithm deploying the neighbourhood sets which were identified by Bilgin et al. [2012] as the most effective for these types of nurse rostering problems. The results are compared to the results obtained with the ALNS presented in Bilgin et al. [2012].

The original instances along with a modified version of each normal scenario were used by Smet et al. [2012] to investigate different objective functions for fair nurse rostering. The authors tested three different objectives using a metaheuristic search algorithm and discussed their performance for these real-world instances.

3.2.2 Nottingham Dataset

In an effort to centralise a number of instances presented in the literature, Curtois [2012] converted data described by different authors into a uniform data format. All collected instances, a link to the original paper, lower bounds and best known solutions are provided on a web page. The dataset represents instances collected from all over the world and includes both real-world and artificial instances. The different instances show a great variety in problem properties. Table 3 shows an overview of the characteristics of some selected instances.

Table 3 General characteristics of the Nottingham problem instances

Instance	Employees	Shift types	Skill types	Days
ORTEC	16	4	1	31
QMC	19	3	1	28
Ikegami-3Shift	25	3	7	30
BCV-3.46.2	46	3	1	29

The instances are available at `http://www.cs.nott.ac.uk/~tec/NRP/`. Furthermore, an extensive description of the data format can be found on the same web page together with software to solve problems and verify solutions.

Both exact and (meta)heuristic approaches have been applied to different instances from the Nottingham dataset. Among them are an adaptive constructive heuristic [Brucker et al., 2010], scatter search [Burke et al., 2010a], iterated local search [Burke et al., 2011], branch and price [Burke and Curtois, 2011] and mathematical programming [Glass and Knight, 2010]

3.2.3 NSPLib

NSPLib is a benchmark dataset specifically designed to conduct controlled experiments on nurse rostering problems. The dataset contains a large number of generated instances, based on different parameter settings for nine indicators [Vanhoucke and Maenhout, 2007]. These indicators describe different properties of an instance such as the problem size, the preferences and the coverage constraints.

Problem size is defined by the number of nurses, the number of shifts and the number of days.

The preferences of nurses are characterised by three parameters. First, the *nurse-preference distribution* specifies how the preferences are distributed among all nurses. Second, based on this initial distribution, these preferences are further distributed among the shifts. This is specified by the *shift-preference distribution.* Finally, the preferences are distributed among the different days. By controlling these three parameters, the distribution of preferences is fully manageable.

Constraints on coverage are also determined by three parameters. The *total-coverage constrainedness* defines how many nurses are required in total for the complete rostering period. The *day-coverage distribution* controls how the total coverage requirements are distributed over specific days. Finally, the coverage per shift on each day is determined by the *shift-coverage distribution.* By combining these three sub-indicators, coverage requirements are generated for each shift on each day.

Two sets of instances can be identified in NSPLib: a diverse set and a realistic set. Properties of both subsets are shown in Table 4. For each combination of number of employees, shifts and days, ten instances were generated. This results in a total of 29160 instances in the diverse set and 1960 instances in the realistic set. The benchmark dataset can be downloaded from `http://www.projectmanagement.ugent.be/nsp.html`. The problem generator NSPGEN provides the possibility to generate new instances based on chosen values for the nine indicators.

Several algorithms have been applied to the instances in NSPLib, including scatter search [Maenhout and Vanhoucke, 2006], a hybrid genetic algorithm [Maenhout and Vanhoucke, 2008] and a metaheuristic based on the principles of electromagnetism [Maenhout and Vanhoucke, 2007]. To the best of the authors' knowledge the branch and price framework of Maenhout and Vanhoucke [2010] is the only exact approach used to solve the instances.

Table 4 General characteristics of the NSPLib instances

Subset	Employees	Shift types	Skill types	Days
Diverse set	25, 50, 75, 100	3	1	7
Realistic set	30, 60	3	1	28

3.2.4 The First International Nurse Rostering Competition

In 2010, the first International Nurse Rostering Competition (2010 NRC) was organised. The main goal of the organisers was to encourage the development of new algorithms for the nurse rostering problem. Furthermore, the competition also introduced challenges from practice by incorporating some real-world constraints in the instances. The benchmark dataset contains three types of instances: *sprint*, *middle distance* and *long distance*. These types refer to the size of the instances and the allowed computation time. The problem formulation, i.e. the hard and soft constraints, are the same for all types. Three sets of instances were published for each type: *early*, *late* and *hidden*. Table 5 shows some properties of the benchmark instances. In total the dataset contains 60 instances which can be downloaded from `http://www.kuleuven-kulak.be/nrpcompetition`.

Table 5 General characteristics of the INRC instances

Type	Employees	Shift types	Skill types	Days
Sprint	10	3, 4	1	28
Middle distance	30, 31	4, 5	1, 2	28
Long distance	49, 50	5	2	28

As far as the first aim of the competition is concerned, the goal has certainly been achieved. In the first round, 15 competitors submitted an algorithm. From these 15, five finalists were selected, whose contributions were then compared based on their performance on hidden instances. Among the five finalists, various solution approaches were used.

Nonobe [2010] reformulated the instances as constraint optimisation problems and used a general purpose solver to find solutions. A hyperheuristic combined with a greedy shuffle heuristic which performed particularly well on the long distance instances [Bilgin et al., 2010]. The approach described by Lü and Hao [2012] switches between a local search and an elite solution restart mechanism. Burke and Curtois [2011] applied an ejection chain method and a branch-and-price framework. The winners, Valouxis et al. [2012], decomposed the problem into subsequent subproblems which could be solved to optimality. A two-phase approach was used in which first days-off are scheduled and in the second phase particular shifts are assigned to the nurses.

Other algorithms have been evaluated on the 2010 NRC instances. Awadallah et al. [2011] use a population based metaheuristic called harmony search to solve the *sprint* instances. The algorithm mimics the improvisation by musicians and includes recombinations of solutions, random assignments and local search procedures. However, it does not manage to find new best solutions, or to match the current best solutions. A variable neighbourhood search is presented by Geiger [2011] for the *sprint* instances. First, a constructive approach generates an initial feasible solution based on the coverage requirements. Afterwards, an iterative improvement procedure sequentially explores four neighbourhoods, including one for shaking. The approach finds the optimal solution for 8 out of 20 instances and matches one other best known result. Della Croce and Salassa [to appear] present a matheuristic based on a variable neighbourhood search to solve a real-world nurse rostering problem. The algorithm was capable of finding new best solutions for five *sprint* instances. Messelis and De Causmaecker [2011] also investigated the benchmark instances to evaluate hardness measures for the nurse rostering problem.

3.3 Modelling Real World Scenarios: A Case Study

The KAHO dataset was composed with real data from Belgian hospitals. The actual personnel manager or head nurse assisted in setting the parameter values. In this section, we elaborate on this dataset, on its parameter values and on the challenging real-world elements it incorporates.

The schedule period length of the KAHO benchmarks varies between four and 13 weeks. The same value varies between one week and 29 days in the Nottingham benchmarks. As can be seen in Table 6, it is common practice in Belgian hospitals to organise the work into a high number of shift types. Furthermore, Table 6 shows that in some wards, the sum of the number of employees in each skill category exceeds the total number of employees in the ward. This implies that at least some employees possess multiple skills. This skill structure has a direct impact on the employee, assignment and coverage constraint elements of the problem model.

The flexibility of the employment contracts in Belgian hospitals can be seen in Table 7. Nurses can opt for different working agreements: full time, half time and various ratios in between. Real-world applications require the ability to deal with multiple serial contracts per employee, per schedule period. This happens when an employee switches from one contract type to another within the same scheduling period due to a promotion or for another self-induced reason. Multiple serial contracts are encountered in the meal preparation and geriatrics wards of the KAHO benchmarks. Similarly, the sum of the number of nurses for each contract is therefore greater than the total number of nurses in these wards.

Table 6 The number of shift types and number of nurses with each skill type

Ward	Shift types	Skill 1	Skill 2	Skill 3	Skill 4	Total employees
Emergency	27	1	15	4	26	27
Psychiatry	14	1	17	1	-	19
Reception	19	1	1	3	15	19
Meal Preparation	9	1	31	-	-	32
Geriatrics	9	4	20	-	-	21
Palliative Care	23	1	21	4	1	27

Table 7 The weekly job time and the corresponding number of nurses

Ward	38 hours (100%)	34.2 hours (90%)	30.4 hours (80%)	28.5 hours (75%)	22.8 hours (60%)	19 hours (50%)
Emergency	24	-	-	3	-	-
Psychiatry	13	-	-	2	-	4
Reception	5	-	-	7	-	7
Meal P.	3	2	-	1	-	28
Geriatrics	9	-	-	9	1	3
P. Care	13	-	2	4	1	7

For each ward in the KAHO benchmarks, three different scenarios have been developed in close cooperation with experienced planners: normal, overload and absence. These scenarios stem from situations that arise in the real-world. The normal scenario represents the personnel rostering problem most regularly faced in the wards.

In real-world situations, the workload of a ward is not stable over different schedule periods and varies depending on external events such as epidemics and seasonal diseases. A significant increase in the workload must be matched by a proportional increase in the workforce demand. A higher workforce demand is represented by higher threshold values in the coverage constraints. The overload scenarios represent cases where the workload demand is higher than usual. These instances have been carefully adapted by incrementing the threshold values of the coverage constraints of the normal scenario instances. This modification was performed for all the skill types but one, namely the head nurse. One head nurse is sufficient in any situation and consequently, the corresponding coverage constraints have not been incremented.

In practice, it is possible that a scheduled employee cannot work his or her planned shifts due to an unforeseen event such as sickness. In this case, the complete roster needs to be rescheduled to take into account the unforeseen absence of the employee. Several factors need to be considered when addressing this problem. In contrast to the normal and overload scenarios, the roster was already published to the staff in the ward. That means the employees have already adjusted their private plans according to it. Therefore the changes to the schedule must be kept to a minimum. In order to ensure this, only the affected parts, i.e. the days when the employee is scheduled but cannot be present, should be rescheduled. The remainder of the schedule needs to be fixed to the original one. This practice reduces the size of the problem instance considerably, while at the same time increasing the possibility for generating an alternative roster with a low number of constraint violations.

4 Discussion

The present chapter covered general academic advances in nurse rostering research and provided a new comprehensive mathematical model. This generic nurse rostering model enables incorporating complex real-world constraints into the formulation in order to cover uncommon contractual constraints and varied rostering practices in hospitals. The lack of a common modelling approach is one of the major lacunas in the nurse rostering literature. This has prevented the community from building up knowledge and making clear statements about the academic progress in the field. The authors therefore advocate adoption of the presented model by the research community as well as by software developers, in order to provide a ground for comparison and exchange of data and results. Obviously, extensions to the present model will be necessary whenever additional challenges would emerge in real hospital environments.

Algorithmic developments have often concentrated on dedicated performance for sets of very specific instances. Surprisingly large differences among the nurse rostering problems' descriptions impede thorough analysis and comparison of approaches. The introduction of nurse rostering benchmarks is a major step towards a sound performance assessment of past and future approaches to nurse rostering problems. The present chapter reviewed four benchmark datasets for the nurse rostering problem along with publications based on these instances. In order to stimulate the growth of the collection of benchmark instances, the authors encourage researchers to model new problem instances using the presented generic model and to make these new datasets available to the community.

References

Abernathy, W.J., Baloff, N., Hershey, J.C., Wandel, S.: A three-stage manpower planning and scheduling model - a service-sector example. Operations Research 22, 693–711 (1973)

Awadallah, M.A., Khader, A.T., Al-Betar, M.A., Bolaji, A.L.: Nurse rostering using modified harmony search algorithm. In: Panigrahi, B.K., Suganthan, P.N., Das, S., Satapathy, S.C. (eds.) SEMCCO 2011, Part II. LNCS, vol. 7077, pp. 27–37. Springer, Heidelberg (2011)

Bilgin, B.: Advanced Models and Solution Methods for Automation of Personnel Rostering Optimisation. PhD thesis, KU Leuven (2012)

Bilgin, B., Demeester, P., Misir, M., Vancroonenburg, W., Vanden Berghe, G., Wauters, T.: A hyper-heuristic combined with a greedy shuffle approach to the nurse rostering competition. In: Proceedings of PATAT 2010 (2010)

Bilgin, B., De Causmaecker, P., Rossie, B., Vanden Berghe, G.: Local search neighbourhoods for dealing with a novel nurse rostering model. Annals of Operations Research 194, 33–57 (2012)

Brucker, P., Burke, E.K., Curtois, T., Qu, R., Vanden Berghe, G.: A shift sequence based approach for nurse scheduling and a new benchmark dataset. Journal of Heuristics 16(4), 559–573 (2010)

Burke, E.K., Curtois, T.: New computational results for nurse rostering benchmark instances. Technical report (2011)

Burke, E.K., Cowling, P., De Causmaecker, P., Vanden Berghe, G.: A memetic approach to the nurse rostering problem. Applied Intelligence, Special Issue on Simulated Evolution and Learning 15, 199–214 (2001)

Burke, E.K., Kendall, G., Newall, J., Hart, E., Ross, P., Schulenburg, S.: Hyper-heuristics: An emerging direction in modern search technology. In: Glover, F., Kochenberger, G. (eds.) Handbook of Metaheuristics, vol. 57, pp. 457–474. Springer, New York (2003)

Burke, E.K., De Causmaecker, P., Vanden Berghe, G., Van Landeghem, H.: The state of the art of nurse rostering. Journal of Scheduling 7(6), 441–499 (2004)

Burke, E.K., Curtois, T., Qu, R., Vanden Berghe, G.: A scatter search methodology for the nurse rostering problem. Journal of the Operational Research Society 61(11), 1667–1679 (2010a)

Burke, E.K., Hyde, M., Kendall, G., Ochoa, G., Özcan, E., Woodward, J.R.: A classification of hyper-heuristics approaches. In: Gendreau, M., Potvin, J.-Y. (eds.) Handbook of Metaheuristics. International Series in Operations Research & Management Science, vol. 146, pp. 449–468. Springer (2010b)

Burke, E.K., Curtois, T., Fijn Van Draat, L., Van Ommeren, J.-K., Post, G.: Progress control in iterated local search for nurse rostering. Journal of the Operational Research Society 62(2), 360–367 (2011)

Curtois, T.: Employee scheduling benchmark data sets (2012), `http://www.cs.nott.ac.uk/~tec/NRP/`

De Causmaecker, P., Vanden Berghe, G.: A categorisation of nurse rostering problems. Journal of Scheduling 14, 3–16 (2011)

Della Croce, F., Salassa, F.: A variable neighborhood search based matheuristic for nurse rostering problems. Annals of Operations Research (to appear)

Geiger, M.J.: Personnel rostering by means of variable neighborhood search. In: Hu, B., Morasch, K., Pickl, S., Siegle, M. (eds.) Operations Research Proceedings 2010, pp. 219–224. Springer, Heidelberg (2000) ISBN 978-3-642-20009-0

Glass, C.A., Knight, R.A.: The nurse rostering problem: A critical appraisal of the problem structure. European Journal of Operational Research 202(2), 379–389 (2010)

Haspeslagh, S., De Causmaecker, P., Schaerf, A., Stolevik, M.: The first international nurse rostering competition 2010. Annals of Operations Research, 1–16 (2012)

Kellogg, D.L., Walczak, S.: Nurse scheduling: From academia to implementation or not? Interfaces 37(4), 355–369 (2007)

Lü, Z., Hao, J.-K.: Adaptive neighborhood search for nurse rostering. European Journal of Operational Research 218(3), 865–876 (2012) ISSN 0377-2217

Maenhout, B., Vanhoucke, M.: New computational results for the nurse scheduling problem: A scatter search algorithm. In: Gottlieb, J., Raidl, G.R. (eds.) EvoCOP 2006. LNCS, vol. 3906, pp. 159–170. Springer, Heidelberg (2006)

Maenhout, B., Vanhoucke, M.: An electromagnetic meta-heuristic for the nurse scheduling problem. Journal of Heuristics 13, 359–385 (2007) ISSN 1381-1231

Maenhout, B., Vanhoucke, M.: Comparison and hybridization of crossover operators for the nurse scheduling problem. Annals of Operations Research 159, 333–353 (2008) ISSN 0254-5330

Maenhout, B., Vanhoucke, M.: Branching strategies in a branch-and-price approach for a multiple objective nurse scheduling problem. Journal of Scheduling 13, 77–93 (2010) ISSN 1094-6136

Messelis, T., De Causmaecker, P.: An algorithm selection approach for nurse rostering. In: Proceedings of the 23rd Benelux Conference on Artificial Intelligence, pp. 160–166 (2011)

Misir, M., Verbeeck, K., De Causmaecker, P., Vanden Berghe, G.: Design and analysis of an evolutionary selection hyper-heuristic. Technical report, KAHO Sint-Lieven (2012)

Nonobe, K.: An approach using a general constraint optimization solve. In: Proceedings of PATAT 2010 (2010)

Rocha, M., Oliveira, J.F., Carravilla, M.A.: Cyclic staff scheduling: optimization models for some real-life problems. Journal of Scheduling 16(2), 231–242

Smet, P., Martin, S., Ouelhadj, D., Özcan, E., Vanden Berghe, G.: Investigation of fairness measures for nurse rostering. In: Kjenstad, D., Riise, A., Norlander, T.E., McColumn, B., Burke, E.K. (eds.) Proceedings of the 9th International Conference on the Practice and Theory of Automated Timetabling, PATAT, Son, Norway, pp. 369–372 (August 2012)

Smet, P., Bilgin, B., De Causmaecker, P., Vanden Berghe, G.: Modelling and evaluation issues in nurse rostering. Annals of Operations Research (to appear), doi:10.1007/s10479-012-1116-3

Valouxis, C., Housos, E.: Hybrid optimisation techniques for the workshift and rest assignment of nursing personnel. Artificial Intelligence in Medicine 20, 155–175 (2000)

Valouxis, C., Gogos, C., Goulas, G., Alefragis, P., Housos, E.: A systematic two phase approach for the nurse rostering problem. European Journal of Operational Research (2012) ISSN 0377-2217, doi:10.1016/j.ejor.2011.12.042

Vanhoucke, M., Maenhout, B.: NSPLib - a tool to evaluate (meta-) heuristic procedures. In: Brailsford, S., Harper, P. (eds.) Operational Research for Health Policy: Making Better Decisions, Proceedings of the 31st Meeting of the European Working Group on Operational Research Applied to Health Services, pp. 151–165 (2007)

Warner, M.: Nurse staffing, scheduling, and reallocation in the hospital. In: Hospital & Health Services Administration, 77–90 (1976)

Radiotherapy Scheduling

Dobrila Petrovic, Elkin Castro, Sanja Petrovic, and Truword Kapamara

Abstract. This chapter concerns radiotherapy scheduling problems identified at two cancer centres in the UK. The scheduling of radiotherapy pretreatment and treatment appointments is a complex problem due to various medical and scheduling constraints, such as patient category, machine availability, a doctors' rota, waiting time targets (i.e., the time when a patient should receive the first radiotherapy fraction, etc.), and, also, due to the size of the problem (i.e., number of machines, facilities and patients). Different objectives need to be considered including minimisation of the number of patients who do not meet their waiting time targets, minimisation of usage of overtime slots, minimisation of machines idle time, and so on. Motivated by heuristics developed for production scheduling problems, two novel heuristics-based approaches to scheduling of radiotherapy patients are developed. Both approaches involve priority rules; while one of the approaches applies the a priori selected priority rules, the other one employs a genetic algorithm (GA) to select priority rules which will lead to the best scheduling performance. Different experiments are carried out to analyse the performance of the two radiotherapy scheduling approaches.

Dobrila Petrovic · Truword Kapamara
Faculty of Engineering and Computing,
Coventry University, Priory Street, Coventry, CV15FB, UK
e-mail: d.petrovic@coventry.ac.uk

Elkin Castro
ASAP research group, School of Computer Science,
University of Nottingham,
Jubilee Campus, Wollaton Road,
Nottingham, NG8 1BB, UK
e-mail: edc@cs.nott.ac.uk

Sanja Petrovic
Division of Operations Management and Information Systems,
Nottingham University Business School, Jubilee Campus,
Wollaton Road, Nottingham NG8 1BB, UK
e-mail: sanja.petrovic@nottingham.ac.uk

A.Ş. Etaner-Uyar et al. (eds.), *Automated Scheduling and Planning*,
Studies in Computational Intelligence 505,
DOI: 10.1007/978-3-642-39304-4_7,

1 Introduction

Radiotherapy is often used as an essential means to treat cancer patients. Its application has grown worldwide. According to Delaney et al, (2005), an estimated 52% of cancer patients received radiotherapy at least once in their regimen. Around 40% of cancer cases in the UK require radiotherapy alone or in combination with other treatment modes (National Radiotherapy Advisory Group 2007).

Radiotherapy comprises two phases: pretreatment and treatment on linac (linear particle accelerator) machines which deliver radiation. The purpose of the pretreatment is to define precisely the area to be treated with radiotherapy, and to generate a radiotherapy treatment plan which delivers a uniform dose to tumour cells while minimising the radiation to the surrounding healthy tissues and organs. In radiotherapy, radiation is given on linac machines as a series of small doses, referred to as fractions, over a period of time (days, weeks) or a single fraction.

Cancer patients are typically classified according to the treatment intent and their waiting list status. The treatment intent can be palliative or radical. Palliative treatments are meant to relieve symptoms of cancer, whereas radical treatments are given with the aim to cure. There are three waiting list statuses: emergency, urgent and routine. The waiting list status of a patient is determined according to the site of cancer and a level of progress of the tumour. The site of the cancer and the treatment intent determine the patient pathway in a radiotherapy centre. It defines an ordered sequence of medical procedures and consultations, hereafter referred to as operations, which have to be performed in order to determine the best way of delivering a radiotherapy treatment.

Waiting time targets determine the date by which a patient needs to receive their first radiotherapy fraction. Waiting time is measured in consecutive days including weekends and public holidays, from the time when the decision to treat the patient using radiotherapy is made to the time when the first fraction is administered. Table 1 displays the Joint Council for Clinical Oncology (JCCO) good practice and maximum acceptable waiting time targets in days (separated by "/") for different treatment intents and waiting list statuses (JCCO 1993). The Department of Health (DH) has also set a 31-day target to the start of radiotherapy for all cancers (Department of Health 2000).

Table 1 JCCO recommended waiting time targets given in days (good practice / maximum acceptable)

Intent	Waiting list status		
	Emergency	Urgent	Routine
Palliative	1/2	2/14	2/14
Radical	-	-	14/28

Most radiotherapy centres in the UK use handcrafted appointment booking systems. A radiotherapy patients' scheduling problem is very complex, requiring multi-objective considerations. The challenges in radiotherapy centres are to develop scheduling approaches to improving the waiting times in such a way as to meet the waiting time targets, to minimising percentages of late patients, to maximising the utilisation of the machines involved in the entire process and so on, at minimal overhead costs. Different medical and technical constraints have to be considered. Thus, this area offers new challenges to development of radiotherapy patient scheduling systems.

In this chapter, relevant literature concerning radiotherapy scheduling problems is reviewed. Two novel radiotherapy scheduling systems developed in collaboration with two radiotherapy centres in the UK: Arden Cancer Centre, University Hospitals Coventry and Warwickshire NHS Trust, in Coventry, and Nottingham University Hospitals NHS Trust, City Hospital Campus, in Nottingham are described next. Both radiotherapy departments comprise common radiotherapy operations, while having some characteristics relevant to their departments. Two approaches to generating schedules for radiotherapy patients are presented: (1) heuristics based on priority rules which consider the whole radiotherapy process from the decision to treat the patient to the first treatment session and (2) a GA for scheduling appointments for radiotherapy pretreatment operations. Various tests are carried out using real-life data from the two hospitals to evaluate scheduling performance achieved using the two approaches.

2 Literature Review

Long waiting times for radiotherapy treatments adversely affect the success of the treatment. Diverse approaches to mitigating delays in the start of radiotherapy have been proposed. Effects of increasing the number of both machines and staff [Department of Health 2004; Summers and Williams 2006) and improving the staff skills mix and changing work patterns have been investigated (Dodwell and Crellin 2006). However, investments in the expansion of resources may not be financially nor technically feasible (Agarwal et al. 2008). Therefore, optimal use of existing resources subject to keeping the quality of treatment at satisfactory standards is seen to be a viable option. Challenges and opportunities in development of health care systems for appointment scheduling is given in (Gupta and Denton 2008).

There is a paucity of papers on scheduling radiotherapy patients presented in the literature. A linac management system, implemented using Excel, is presented in (Larsson 1993). It is likely to be one of the seminal papers on radiotherapy scheduling. However, the proposed system does not offer scheduling capabilities; instead it provides improved start and completion date estimations and better characterisation of resource utilisation. A discrete-event simulation model for the radiotherapy planning process is proposed in (Werker et al. 2009). This study is focused on one step of the radiotherapy pretreatment only. Reducing planning time is crucial in order to mitigate delays in radiotherapy. Analysis of several scenarios concerning extending human and physical resources shift hours, adding a new treatment machine and reducing demand for doctors presence using a discrete-event simulation model is presented in (Proctor et al. 2007) and (Kapamara et al. 2007).

The following research studies present approaches to solving radiotherapy scheduling problems using optimisation methods. In (Conforti et al. 2008; Conforti et al. 2010; Conforti et al. 2011), two mathematical programming models are presented. In the basic model, existing appointments are kept unchanged, while the extended model allows changes to the already booked appointments in such a way that these patients are guaranteed to finish their remaining fractions. The scheduling horizon is six days. However, not all patients are guaranteed to get appointments' schedule. In (Conforti et al. 2008; Conforti et al. 2010), the objective is to maximise the number of scheduled patients. It is concluded that the extended model has better performance than the basic model. Experimental results obtained using both models indicate that at most 80% of the patients are scheduled, and for the remaining 20% a schedule is not created. In (Conforti et al. 2011), the objective in the basic model is the maximisation of the sum of the number of scheduled patients and the number of scheduled fractions while the extended model has an additional objective to minimise the delay in starting the treatment of already booked patients.

Heuristic approaches to radiotherapy scheduling problems have been investigated as well. In (Petrovic et al. 2006), two algorithms for scheduling radiotherapy treatment patients that request treatment on a given day, ASAP and JIT, are presented. In both algorithms patients are prioritised according to their categories and due dates. Following the prioritised list, the ASAP algorithm assigns to each patient, , the earliest feasible start date for the first fraction, then subsequent sessions are booked from this date onwards, while the JIT algorithm assigns the latest feasible start date for the first fraction, then subsequent sessions are booked from this date onward. In (Petrovic and Leite-Rocha 2008 a,b) four parameters to be used in a constructive method for generating a schedule for radiotherapy patients are investigated. The constructive method schedules patients one by one from a sorted list. Patients are sorted by their due date, while ties are broken by category, and then by the number of planned fractions. The first parameter considers the patient's release date (the date when the patient completes the pre-radiotherapy treatment) and the patient's due date (the waiting time target of the patient). The constructive method operates in a forward (backward) manner from the release date (due date) of each patient, trying to schedule the required number of sessions prescribed to the patient subject to the given constraints. The second parameter determines the utilisation level of each machine. Patients of a given category are allocated to a machine until the predefined level of the machine utilisation is achieved. The third parameter determines days on which patients are considered for scheduling (everyday, on 3 days or on 2 days per week) to investigate whether the accumulation of patients to be scheduled might lead to better schedules due to the larger search space. The forth parameter determines the number of days from the release date after which the patient is considered for scheduling. Each of these parameters is tested individually and the parameters values that led to the best performance of the schedule are used in the constructive method that incorporates all four parameters. The authors also propose a GRASP metaheuristic which takes, as the starting solution, the solution generated by the constructive method described above. Schedules are evaluated with respect to the weighted tardiness of patients, where tardiness is positive if a patient breaches the waiting time target and it is zero otherwise. All constructive and GRASP-based methods give satisfactory results. However, GRASP does not improve appreciably the quality of the solutions given by the constructive methods.

Multi-objective GAs for scheduling of radiotherapy patients are proposed in (Petrovic et al. 2011). Two objectives are defined including minimisation of the average patients' waiting time and minimisation of average length of breaches of waiting time targets. Three GAs are developed which treat emergency, palliative and radical patients in different ways: (1) a GA which considers all three patient categories equally, (2) a GA with an embedded knowledge on the scheduling of emergency patients and (3) a GA which associate different weights to the patient categories. Statistical analysis is carried out to test performances of the three GAs. The results show that the GA with the embedded knowledge generated the schedules with best performance considering emergency patients and slightly outperforms the other two GAs when all patient categories are considered simultaneously.

In our initial work on radiotherapy pretreatment, a daily scheduling problem is considered in which mathematical programming and priority rules are combined (Castro and Petrovic 2011). Priority rules are used to generate an initial feasible solution that is passed to CPLEX. Then, this solution is improved through sequential phases of optimisation in which different criteria are used in a lexicographic manner. However, the overall performance of schedules is worse than that of the schedules generated by priority rules alone. In our further research work described in this chapter, we investigate the effect of penalizing the idle time on the machines close to the current scheduling date on the schedule performance. The latter is related to allocating idle time on machines available for further patient arrivals.

While the literature on the radiotherapy scheduling presented so far typically considers either a pretreatment or a treatment phase, we have considered the whole process. Four heuristics for four unites that comprise the whole radiotherapy process are developed and integrated into the scheduling system. The heuristics involve various priority rules and novel strategies developed to improve scheduling performance for all radiotherapy patients categories.

3 Heuristics for Radiotherapy Scheduling

3.1 Background

In this chapter, we will describe the radiotherapy process at Arden Cancer Centre; however, the process is similar in other oncology departments. It comprises four units: planning, physics, pretreatment, and treatment. Each unit includes several machines and/or facilities as shown in Figure 1. Each patient visits or revisits at least one of the machines and/or facilities depending on their predetermined pathway. The pathway, which must commence and terminate in the planning and treatment unit respectively, is determined based on the cancer case and/or doctor consultation.

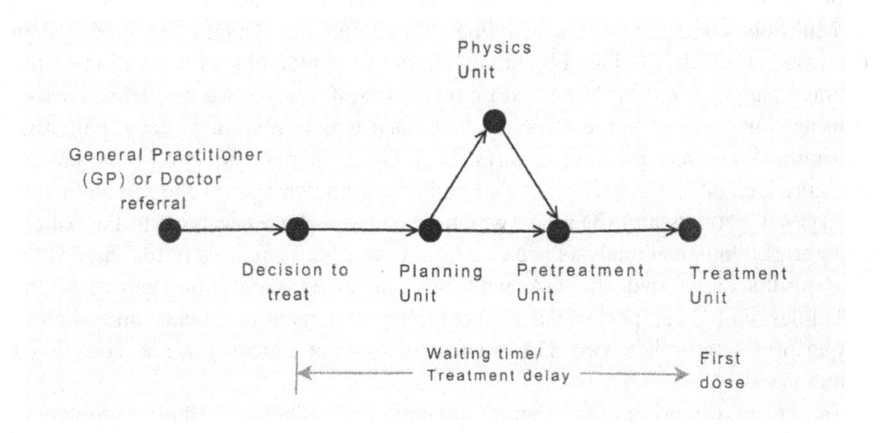

Fig. 1 Units in Arden Cancer Centre

The tumour volume is imaged in the planning unit using a simulator or computed tomography (CT) scanner. The simulator is a machine that takes radiographs, while the CT scanner takes images from different angles and builds a 3D image of the tumour volume. Some patients may require a mask for immobilisation during planning and treatment. The mask is made in the mould room prior to the patient's operations on the simulator or CT scanner. If the mask is not required, the patient proceeds straight to the simulator or CT scanner. Most importantly, the doctor should be available for each operation in the planning unit. A patient visits the simulator or CT scanner and the mould room at most once. The output of the planning unit operations are digital images required for outlining and planning, and dosimetry calculations.

The digital images are sent to either the physics or pretreatment unit depending on the perceived complexity of the dosimetry calculations. Complex dosimetry calculations are handled in the physics unit, while the simple dosimethry calculations are carried out in the pretreatment unit. The physics unit performs two operations: outlining and planning, and dosimetry calculations. Firstly, technicians do the outlining and planning; then the allocated doctor approves and signs their output, before they carry out the dosimetry calculations. After completing the dosimetry calculations, the physics unit technicians hand the treatment plans over to the pretreatment unit.

In the pretreatment unit, a single accuracy calculation check is done if the treatment plans are received from the physics unit; otherwise, if they are received from the planning unit, three dosimetry calculations and checks are performed separately, on three different desks.

The treatment unit comprises linacs machines for dispensing the calculated radiation doses (currently there are 7 linacs at the Arden Cancer Centre). There are complex constraints on scheduling patients on linacs machines. Patients receive the dose fractions daily, except on weekends and bank holidays. Some cancer cases (head and neck, gynaecological, repirator and urinary) should start treatment on Monday. Doctors may prescribe multiple treatment phases for some patients. The number of fractions in each phase is predetermined beforehand. New radiographs of the tumour volume should be obtained on the simulator before the commencement of each phase. Plan checks during the treatment should be done within the last three fractions before the completion of each treatment phase.

The Arden Cancer Centre considers the following three patient categories and the corresponding target waiting times: emergency, palliative and radical patients and 2, 14 and 28 days, respectively.

3.2 *Problem Statement*

Characteristics of the radiotherapy scheduling problem under consideration are as follows:

- Details and pathways of newly arriving patients to be scheduled are released on daily basis.
- The number of patients is uncertain.
- All appointments are scheduled daily at 9.00am.
- A patient's schedule of appointments cannot be altered once scheduled.
- Machines and/or facilities are available continuously, from Monday to Friday from 9.00am to 17.00pm, except for bank holidays.
- It is assumed that there are no machine breakdowns.
- Machines are under periodic maintenance according to a predetermined maintenance plan. During the maintenance period, the machines cannot be used for planning or treatment.
- There is no separation time between processing of operations for two consecutive patients or treatment plans.
- Each doctor is available in specified time periods per week.
- The allocated doctor examines the patient during the first five minutes of the processing time on a planning machine or facility.

The department would like to have schedules generated for each unit. The planning and treatment units involve patients, while the physics and pretreatment units handle treatment plans. Currently, only schedules for the planning and treatment units are generated and the corresponding appointments are booked manually. Furthermore, the patient's treatment appointments are booked once the planning operations are completed. The system proposed is generating the appointments for a patient and the corresponding schedule for treatment plan in advance for the all four units. Therefore, we formulated the radiotherapy scheduling problem as follows; a schedule of appointments is to be generated for the planning and treatment units and for the corresponding operations in the physics and pretreatment units for all patients which arrive within a scheduling horizon in such a way as:

- To minimise the average waiting times of all patients scheduled during the considered time horizon

$$\text{Minimise } z_1 = \frac{1}{N} \cdot \left(\sum_{j=1}^{N} RW_j\right) \tag{1}$$

where RW_j is the waiting time for patient P_j, and N is the total number of patients within a scheduling horizon under consideration.

- To minimise the percentage of patients that do not meet their JCCO due date for their first treatment

$$\text{Minimise } z_2 = \frac{1}{N} \cdot \left(\sum_{j=1}^{N} U_j\right) \cdot 100 \tag{2}$$

where binary variable U_j is equal to 1 if patient P_j exceeds the waiting time target, 0 otherwise.

- To minimise the total overtime penalty calculated as the sum of durations of all overtime slots used

$$\text{Minimise } z_3 = \frac{1}{N} \cdot \left(\sum_{j=1}^{N} O_j\right) \tag{3}$$

where O_j is the penalty for performing an operation for patient P_j after normal working hours.

The planning unit comprises both human (doctors) and machine resources including the CT scanner, simulator and mould room. The scheduling problem of the planning unit can be described as a dynamic (sequences of patients are arriving every day), flexible multi-resource (including doctors and machines simultaneously) two-stage hybrid flowshop problem. The first stage of the problem involves the mould room which patients have to visit before going to the other planning unit machines. The second stage involves the two parallel unrelated alternative planning machines, the CT scanner and simulator. Some of the patients may not visit the first stage of the planning unit, but all the patients visit a machine in the second stage.

The scheduling problem of the physics unit can be described as a flowshop problem that involves two facilities and resources (i.e. the physics desk and doctor). Each treatment plan has to be processed using these two resources in the same order: the physics desk, followed by the doctor and then the physics desk again.

The scheduling problem of the pretreatment unit comprises two subproblems. The first subproblem involves the calculation of treatment plans received from the physics unit and can be described as a multiple parallel machine scheduling problem where machines correspond to three identical desks. The second problem concerns treatment plans received from the planning unit that have to be processed once on each of the three desks using any route possible, as in an open shop problem. Hence, the pretreatment scheduling problem has characteristics of mixed shop scheduling problems.

Since each patient visits and/or revisits a treatment machine specified by the doctor over a given number of consecutive days, the scheduling problem of the treatment unit can be described as a single machine scheduling problem.

3.3 Scheduling Heuristics

In the radiotherapy centre, it is essential to create good schedules of appointments quickly, on daily basis. Therefore, a suitable scheduling method needs to be less computationally intensive compared to some optimisation methods and heuristics that have been designed for the identified scheduling subproblems associated with each unit.

Four constructive heuristics, H1, H2, H3, and H4 are proposed for the planning, physics, pretreatment, and treatment units, respectively. They are connected as presented in Figure 2. The heuristics involve priority rules combined with specially designed mechanisms relevant to the scheduling of patients in the corresponding units. All the priority rules are first tested separately and then in different orders. The most appropriate sequence in which they should be applied is determined empirically.

Input into planning heuristics H1 is a list of patients' booking requests which is created on the daily basis. Heuristics H1 are applied to book operations for the planning unit for all the patients considered on a given day and they calculate for each patient the completion times of their last operations in the planning unit. The list of patients who require complex treatment plans and their completion times are handled first by physics heuristics H2. Then the list of all patients and their completion times are passed to pretreatment heuristics H3. Heuristics H3 schedule operations on the pretreatment desks and pass the list of treatment plans and their completion times to treatment heuristics H4. If a treatment plan needs checking, i.e. needs to go back to the planning unit, it is sent to heuristics H1 which book required slots on the simulator and the allocated doctor. Otherwise, heuristics H4 are applied to book appointments on the treatment machines.

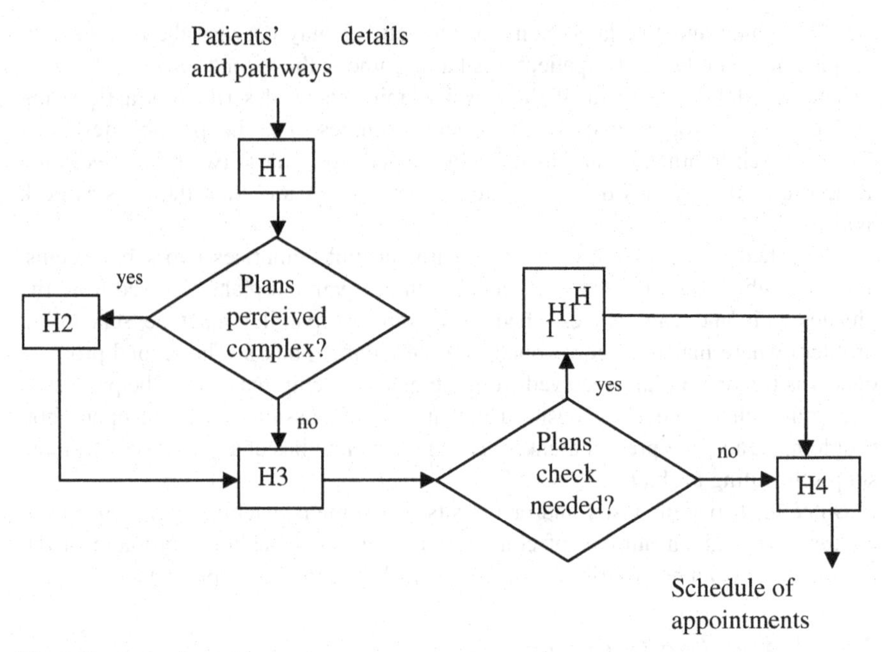

Fig. 2 Heuristics flow chart

The heuristics consider due dates for the completion of required operations in the corresponding unit. The due dates are determined in line with the practice of the radiotherapy centre in such a way as to meet the JCCO waiting time targets.

The planning and treatment units were identified as potential bottleneck for the patient flow. Consequently, two relevant heuristics, H1 and H4, will consider overtime slots. Heuristics H4 use the overtime slots for emergency and palliative patients when they do not meet their respective waiting time targets, while the overtime slots for radical patients are considered only when they breach their waiting time target by an empirically determined threshold. Similarly, H1 uses overtime slots when the lateness of the patient exceeds a predetermined threshold.

3.3.1 Heuristics for the Planning Unit – H1

In the planning unit, it is imperative to consider and synchronise doctor's availability for each operation on the machines or facility, which makes the scheduling problem of this unit very complex. For example, Figure 3 shows a feasible schedule of the appointments for the planning unit resources. Patient 1 is seen by the doctor in the mould room in slot 1, but stays on that machine for 3 more slots afterwards. The same doctor oversees the taking of radiographs for patient 1 on the simulator in slot 5 and so on.

Resource \ Slots	1	2	3	4	5	6	7	8	9	10
Doctor	1	2			1	4				
Mould		1								
Simulator								1		
Scanner			2				4			

Fig. 3 Example of a feasible schedule of the planning unit

The priority rules in H1 reorder the list of arrived patients into a new list by combining the following simple rules in lexicographic manner: the most urgent patient category (MUPC), the most number of steps in the entire radiotherapy process (MNSRP), the most number of operations in the planning unit (MNOP), and the least slack (LS), where the slack for a patient is calculated as the time period between the patient's due date for completion of the operations in the planning unit and the patient's release date in the planning unit. The first rule is applied to order the patients, the second one is used to break the ties, and so on. The aforementioned priority rules generate a sequence of patients which are then scheduled in such a way that the first available appointment slot on the requested machine and doctor in the planning unit is booked.

However, in order to leave some available slots for possible emergency and palliative patient who may arrive in future, a scattering mechanism for radical patients concerning available doctors' slots is introduced. This mechanism skips some available doctor's slots for radical patients, but books the doctor's slot in such a way that the patient's target waiting time for the planning unit could be met. If that is not possible, the first available doctor slot is booked.

3.3.2 Heuristics for the Physics Unit – H2

The first operation in the physics unit is outlining and planning. It has to be performed upon receipt of digital images from the planning unit and has to be approved by the doctor. The time required to produce treatment plans is mostly affected by doctors' availability. The time period between the doctor's availability and the release date of the patient's digital image in the physics unit, referred to as "the doctor delay", is calculated. Thus, the priority rule in H2 employs in lexicographic manner two simple rules: the least doctor delay (LDD) and the most urgent patient category (MUPC). Due to staff shortages, technicians who process operations on the physics desk also perform the mould room operations. Thus, physics unit operations are handled when the mould room is free. After finding a slot in the doctor schedule for the approval of the first operation, a slot for the second operation, i.e. dosimetry calculations, is searched for. For example, in the doctor's schedule in Figure 3, slots 3 and 4 are targeted for approval of the outlining and planning operations which are carried out in the physics unit. The slot for the following second operation performed by technicians is searched based on the date the doctor approves the plan. Figure 4 gives a simplified example of the schedule generated by heuristics H2. Patient 3 has the first operation done, approved by the doctor at a time when the technicians could be working on the first operation of patient 6. Upon the completion of patient 6's outlining and planning operation, the second operation for patient 3 is performed and so on.

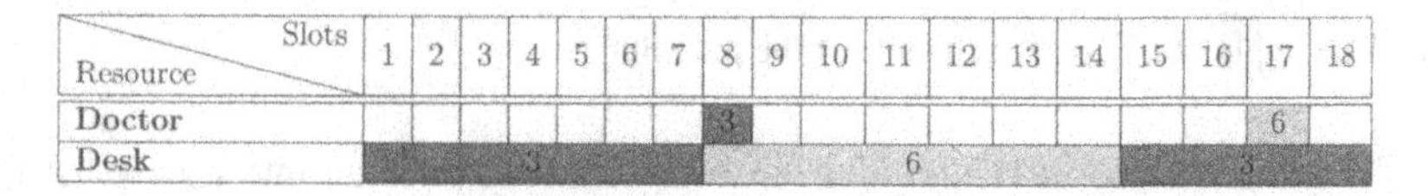

Fig. 4 Example of a feasible schedule of the physics unit

3.3.3 Heuristics for the Pretreatment Unit – H3

Each operation for each treatment plan should be processed on a different desk. An example of the pretreatment unit desks schedule is demonstrated in Figure 5. The treatment plan for patients 2 and 4 undergo dosimetry calculations and checks on the desks in the following order: B → C → A and C → A → B, respectively. However, plans for patients 3, 5, and 6 have complex calculations performed in the physics unit and thus, a single operation – checking, is performed on desks, B, A, and C, respectively.

Desk	Slots			
A	1	4	5	2
B	2	3	4	1
C	4	1	2	6

Fig. 5 Example of a feasible schedule of the pretreatment unit

Heuristics H3 comprise a priority rule which reorders treatment plans received from heuristics H2 using these simple rules: the least number of pretreatment operations (LNPO), the most urgent patient category (MUPC), the least slack (LS), i.e, the difference between the JCCO due date and the due date of completion of pretreatment operations, and the least work on a desk queue (LWINQ) which is employed to find the desk with the earliest free slot for a given date. The above hybrid heuristic greedily allocates the earliest possible slots on the desks for the most urgent treatment plans. To free slots for incoming urgent treatment plans, H3 disperses slots for the treatment plans for the radical patients without affecting their possible treatment start dates by considering their slack time (i.e. the difference between the JCCO due date and the due date of completion of pretreatment operations). Plans with slack times longer than a predetermined threshold are 'scattered' across three consecutive days so as to reduce the amount of work on each of the desks per day and make some slots free.

3.3.4 Heuristics for the Treatment Unit – H4

Constructive heuristics H4 use the following simple rules to reorder the sequence generated by heuristics H3: the most urgent patient category (MUPC), the least number of prescribed treatment phases (LNPTP), the least number of prescribed fractions (LNPF), and the earliest treatment due date (ETDD). Some cancer cases such as head and neck, lung, and respiratory cancers may require an initial plan verification check before the start of treatment. Thus, heuristics H1 is used to book their appointments on the simulator, before commencing booking their treatments. These plan verification checks and other plan checks which should be done within the last three fractions to the completion of a phase, hinder the search for treatment slots. The plan checks are conducted solely on the simulator and a free slot on the simulator within the last three consecutive days of the treatment phase should be found before proceeding with the search for the next phase's treatment slots.

One critical requirement in radiotherapy is that all the fractions must be delivered on the same machine. Heuristics H4 search and book for each patient the earliest available slot on a treatment machine of the requested machine type, such as a low or high energy linac. The difference between the date of the earliest feasible starting slot and the JCCO target start date is the number of days the target is breached. In order to free some slots for incoming patients (especially urgent patients), parameters which denote the maximum allowed number of days the JCCO target could be breached for different patient categories are incorporated into heuristics H4. Values of these parameters are determined empirically. Heuristics H4 finds the first available treatment start date using normal working hours. If the first available treatment start date for an emergency or palliative patient is breached by more than a prespecified threshold, H4 uses overtime slots. In the case of a radical patient, if the first available treatment start date is late by more than a predetermined number of days, the slot for treatment is searched again considering overtime slots. However, if the treatment for the radical patient is late by less than or equal to the predetermined number of days, the patient is 'retained', i.e. put at the end of list of patients to be scheduled. Such patients are scheduled last considering overtime slots also. The main aim of this strategy is to minimise the lateness of the patients requiring radical treatments while creating free slots that can be used for emergency or palliative patients.

3.4 Analysis of Test Results

A discrete-event simulation model of the Arden Cancer Centre was developed based on historical data with details of over 2000 patients collected between September 2005 and January 2007 (Kapamara et al. 2007). The number of patients arriving on a given day was modelled using a Poisson distribution with the expected rates 8.88, 7.76, 7.47, 6.59 and 11.6 for 5 days (Monday to Friday) in a week respectively. This probability distribution was proven to be a good model for new cancer referral rates in several studies (for example, Thomas et al. 2001). The simulation model was used to generate probability distributions for all data describing newly arriving patents such as category of the patient, the date of the arrival to the Cancer Centre, allocated doctor, type of cancer, due date of the patient's first fraction operation planned for the patient treatment, the treatment machine prescribed by the doctor, number of prescribed treatment phases per cancer type etc. For example, according to the historical data, 67%, 31% and 2% of the newly arriving patients were radical, palliative and emergency patients, respectively.

The simulation model is used to carry out various tests to analyse performance of the proposed scheduling heuristics. In this chapter, results of the analyses of impact of: (1) maximum allowed JCCO target breaches, (2) numbers of reserved slots on treatment machines and (3) number of overtime slots are presented. In each test, all slots of machines and facilities are initially free and doctors are available according to the rota. A 'warm-up' period of 3 months is used during which the machines, facilities and doctors appointment slots are booked resulting in a partially booked timetable. Then, patient scheduling is carried out for a consecutive year period during which the scheduling performance is recorded. Each test is repeated 10 times using different data generated by the simulation model. Average waiting times per patient category, the average percentages of late patients with respect to each category and the corresponding standard deviations are presented.

3.4.1 Maximum Allowed JCCO Target Breach

There is no medically established threshold below which treatment delays are safe (Mackillop 2007) and thus, the JCCO recommended waiting time guidelines deemed to be short (Joint Council of Clinical Oncology 1993), but reasonably achievable . However, in order to prevent creating schedules in which a patient could have had an unacceptably very long delayed waiting time, a constraint which set the maximum JCCO target breach is included in the developed heuristics. Based on some studies, (for example, Huang et al. 2003), different combinations of the maximum allowed JCCO target breaches for each patient category are suggested, as shown in Table 2. For example, in test 2, patients requiring emergency treatments have to adhere to the target JCCO waiting time, while 3 and 7 days are set to be the maximum number of days for breaching the waiting time targets for palliative and radical patients, respectively.

Introducing the maximum allowed target breaches produces slightly better average waiting times for emergency and palliative patients, as shown in Table 3. By allowing JCCO target breaches, some additional slots, that otherwise would have been booked for radical patients, become available for incoming urgent patients. However, the average waiting times for emergency and palliative patients are reduced at the expense of higher number of late radical patients (Table 4).

JCCO waiting time targets are reached, but the percentages of late emergency and palliative patients are high; for example, in test 2, 22.9% of all emergency patients are late and 13.5% of all palliative patients are late.

Table 2 Maximum number of days JCCO target breaches are allowed

	JCCO target breach (in days)		
Test	Emergency	Palliative	Radical
1	0	0	0
2	0	3	7
3	0	3	14
4	0	7	14

Table 3 Average waiting times (standard deviations) obtained using different maximum allowed JCCO target breaches

	Average waiting time (in days)			
Test	Emergency	Palliative	Radical	All
1	1.2 (0.21)	10.0 (0.20)	20.5 (0.05)	16.8
2	1.0 (0.19)	9.7 (0.17)	20.5 (0.05)	16.6
3	1.0 (0.19)	9.7 (0.19)	20.5 (0.05)	16.6
4	1.0 (0.19)	9.7 (0.16)	20.5 (0.05)	16.6

Table 4 Average percentages (standard deviations) of late patients obtained using different maximum allowed JCCO target breaches

	Average percentage of late patients (%)			
Test	Emergency	Palliative	Radical	All
1	25.0 (7.85)	17.0 (2.14)	1.0 (0.49)	7.0
2	22.9 (8.78)	13.5 (1.57)	1.1 (0.23)	5.5
3	22.9 (8.78)	13.7 (1.81)	1.0 (0.18)	5.5
4	22.9 (8.78)	13.8 (0.13)	1.1 (0.13)	5.6

As test 2 produced the same average waiting times for all patient categories as tests 3 and 4 (Table 3), but slightly better percentage of late palliative patients who have priority over radical patients (Table 4), the maximum JCCO target breaches set in test 2 are used in further tests, which aim at reducing the percentages of late patients.

3.4.2 Reserved Slots on Treatment Machines

Most cancer centres use the block/slot approach to create schedules of appointments in the planning and treatment units. In this study, the size of a slot for a machine or facility is estimated as the average time taken to treat a patient on the machine or facility. For example, it takes approximately 15 minutes to treat a patient on a high energy linac. Thus, on a normal working there are 29 slots available for bookings of the high linac, assuming that the work (i.e. clinical treatments) starts at 9.20am and ends at around 4.30pm. Similar approach is used to determine the number of slots on other treatment machines.

In this study, reserved appointment slots are only used in heuristics H4 to restrict the number of slots available for certain categories of patients on the treatment machines, as shown in Table 5. Reserved slots are allocated in such a way that emergency patients that need treatment have access to the entire capacity of the machine for the day. Palliative patients have access to the full machine's capacity excluding the number of slots reserved for emergency patients on that machine. Finally, the difference between the machine's full capacity for the day and the sum of number of slots reserved for emergency and palliative patients is the number of slots that are made available to radical patients.

Table 5 Appointment slots reserved on the treatment machines

	Number of reserved slots on a treatment machine per day						
	Emergency		Palliative		Radical		
	DXR	Linac	DXR	Linac	DXR	Linac	
Test						High	Low
5	0	0	0	0	0	0	0
6	1	1	3	3	9	25	32
7	1	1	3	6	9	22	29
8	1	1	6	6	6	22	29
9	1	1	6	12	6	16	23

The introduction of reserved slots on the treatment machines does not considerably improve the average waiting time of emergency patients, as shown in Table 6. However, the reserved slots in test 8 produce a slight improvement on the average waiting time for palliative patients. Although, the reserved slots in test 9 produce good average waiting times for emergency and palliative patients, the average waiting time for radical patients is worse compared to test 8. This is a consequence of reserving more appointment slots for palliative patients at the expense of radical patients. In addition, the percentage of late radical patients is improved in test 8 compared to test 9, as shown in Table 7. Therefore, in the tests to follow the numbers of reserved slots suggested in test 8 are used.

Table 6 Average waiting times (standard deviations) obtained using different combinations of reserved slots

Average waiting time (in days)				
Test	Emergency	Palliative	Radical	All
5	1.0 (0.19)	9.7 (0.17)	20.5 (0.05)	16.6
6	1.0 (0.18)	9.8 (0.19)	20.8 (0.04)	16.9
7	1.0 (0.18)	9.8 (0.18)	20.9 (0.04)	17.0
8	0.9 (0.17)	9.4 (0.15)	20.8 (0.08)	16.8
9	1.0 (0.20)	9.4 (0.11)	21.5 (0.25)	17.2

Table 7 Average percentages (standard deviations) of late patients obtained using different combinations of reserved slots

Average percentage of late patients (%)				
Test	Emergency	Palliative	Radical	All
5	22.9 (8.78)	13.5 (1.57)	1.1 (0.23)	5.5
6	22.9 (8.78)	15.5 (2.34)	1.1 (0.20)	6.2
7	22.9 (8.78)	15.4 (1.87)	1.2 (0.15)	6.2
8	22.5 (8.76)	13.0 (1.64)	0.6 (0.16)	5.0
9	22.9 (8.75)	13.0 (1.62)	1.1 (0.43)	5.3

3.4.3 Overtime Slots

Extending the working day can allow the machines to be used to full capacity to meet demand for radiotherapy. However, it can also negatively impact the quality of service due to increased staff exhaustion and reduction in other hospital services such as pharmacy, medical and nursing cover, transportation and so on. However, some cancer centres showed that extended working days is cheaper than investing in new linacs (Routsis et al. 2006). Therefore, some centres have extended their working days. In this study, it is assumed that the use of overtime slots offers a short-term practical solution to the problem of meeting the JCCO targets and reducing the percentage of late patients. It is therefore decided to evaluate the effect of using overtime slots on each treatment machine for certain categories of patients and to explore the minimum amount of time that added to normal working hours improves the scheduling performance. Heuristics H4 allocate overtime slots to emergency and palliative patients. Table 8 shows different overtime slots considered. In test 11, thirty minutes of overtime is accrued on each treatment machine (that is, the DXR, low and high energy linacs). This means that the treatment unit would be working until about 5.00pm. In tests 12 and 13, the machines would be working until about 5.30pm and 6.30pm, respectively.

Table 8 Overtime appointment slots

	Number of overtime appointment slots		
	Emergency and palliative		
	DXR	Linacs	
Test		High	Low
10	0	0	0
11	2	2	2
12	4	4	5
13	8	8	10

In tests 10 to 13, different numbers of overtime slots are used while the maximum allowed JCCO target breaches and reserved slots are set as in tests 2 and 8, respectively. Extending working hours by half an hour (test 11) is not improving waiting times (Table 9), but it improves proportions of late emergency and palliative patients at the expense of late radical patients, i.e., 6.7% and 78% of all late patients are emergency and palliative patients, while in test 10, when overtime slots are not used, these proportions are 7.9% and 84.0%, respectively (Table 11). In this case, when half an hour overtime is added, more patients that need palliative treatment are booked on dates immediately close to their target dates. Thus, on most of these dates, only one slot is left available for emergency treatments. Any other incoming patients (either palliative or radical) are booked on later dates, because there are no available slots on the treatment machines. Hence, the average waiting times and percentage of late patients are worse. However, the results obtained in tests 12 and 13 (Table 10) are slightly improved compared to test 11, while the average waiting time obtained are similar (Table 9). This means that if the centre's treatment unit worked until 5.30pm, the average waiting times obtained would be the same as when the working hours are extended to about 6.30pm. Average percentages of late patients (Table 10) show that extending the working hours by 2 hours is slightly better for both palliative and radical patients. However, Table 11 shows that the proportion of total number of late radical patients is slightly higher when 2 hours of overtime work is introduced compared to 1 hour. These results demonstrate that the centre can obtain the same average waiting times and similar percentages of late patients for each patient category and proportions of total late patients when working hours on the treatment machines are extended by one or two hours. Given the costs associated with longer overtime working hours, the results show that the centre should only add up to an hour of overtime on the DXR, low and high energy linacs.

Table 9 Average waiting times (standard deviations) obtained using different overtime

	Average waiting time (in days)			
Test	Emergency	Palliative	Radical	All
10	0.9 (0.17)	9.4 (0.15)	20.8 (0.08)	16.8
11	1.0 (0.17)	9.8 (0.19)	21.1 (0.08)	17.2
12	1.0 (0.21)	9.3 (0.15)	20.8 (0.05)	16.8
13	1.0 (0.21)	9.3 (0.15)	20.8 (0.08)	16.8

Table 10 Average percentages (standard deviations) of late patients obtained using different overtime

Percentage of late patients (%)				
Test	Emergency	Palliative	Radical	All
10	22.5 (8.76)	13.0 (1.64)	0.6 (0.16)	5.0
11	23.8 (7.94)	15.5 (1.88)	1.5 (0.25)	6.4
12	22.9 (8.75)	13.0 (1.70)	0.7 (0.24)	5.0
13	22.9 (8.75)	12.9 (1.76)	0.6 (0.16)	4.9

Table 11 Average proportions (standard deviations) of late patients obtained using different overtime

Average proportions of total late patients (%)			
Test	Emergency	Palliative	Radical
10	7.9 (2.48)	84.0 (3.28)	8.1 (3.22)
11	6.7 (2.00)	78.0 (3.80)	15.3 (2.86)
12	8.1 (2.45)	83.6 (3.45)	8.3 (3.06)
13	8.1 (2.46)	82.5 (5.55)	9.4 (5.35)

3.5 *Summary*

The results obtained from these tests show that the Arden Cancer Centre requires at least 1 and 6 reserved slots on each type of the treatment machines (DXR and linacs) for emergency and palliative patients, respectively. In addition, waiting time targets for emergency patients should not be breached, while those for palliative and radical patients can be breached by a maximum of three and seven days, respectively. The tests also showed that further improvements to the performance measures can be obtained if the centre's working day is extended by an hour.

It might be interesting to note that the average waiting times of palliative and radical patients is improved by 34% and 41%, respectively, compared to the waiting times achieved in the practice, in 2008. However, it is worth noting that in the tests conducted, machine breakdowns are not considered.

4 Heuristics for Radiotherapy Pretreatment Scheduling

4.1 *Background*

The scheduling of radiotherapy patients can be divided into two consecutive phases: scheduling of patients for radiotherapy pretreatment and scheduling of patients on linacs machines; the latter can commence once the patients complete their pretreatment. In radiotherapy pretreatmentscheduling, the patients are allocated due dates which correspond to their waiting time targets, i.e. dates when they have to start their treatment on linac machines. We consider these two scheduling phases separately in order to search solution spaces of smaller sizes which will ultimately lead to the higher quality of schedules of both phases. Once the schedules of both phases are generated, they can be coordinated.

The radiotherapy pretreatment scheduling problem is defined in collaboration with the Nottingham University Hospitals NHS Trust, City Hospital Campus, in Nottingham, UK. The problem is very similar to the radiotherapy pre-treatment in Arden Cancer Centre described in Section 3. Some specifics relevant to City Hospital Campus are outlined below.

Two types of resources required in radiotherapy pretreatment scheduling: doctors and machines are considered including five types of machines and facilities which are in use: the mould room, the CT scanner, the physics unit, the simulator and the verification system.

Machines are continuously available throughout the clinic opening hours, in the City Hospital, from Monday to Friday, from 8:30 to 18:00, and weekends from 9:00 to 13:00. On the other hand, each doctor has three availability shifts: for planning, for simulation, and emergency-urgent availability. Planning and simulation availability shifts are different for each doctor. The emergency-urgent availability is from Monday to Friday from 9:00 to 16:00, and on weekends from 9:00 to 13:00 on an on-call basis. Doctors see emergency and urgent patients within the emergency-urgent availability.

A pretreatment pathway, i.e., an ordered sequence of medical procedures and consultations are determined for each patient. The pathway depends on the site of the cancer and the waiting list status of the patient (radical or palliative). For illustration purposes a radical head and neck pathway is displayed in Fig. . This figure displays the required operations, processing and lead times on the left side, and the sequence of resources on the right side. This pathway is chosen because it is the most complex one as it has the largest number of operations and includes most possible operations and features that may be present in any other radical pathway. In the head and neck radical pretreatment, the patient first goes to the mould room where the beam direction shell (BDS) is made and fit. This procedure takes one hour; however, the shell will only be ready one day after this operation is completed. Once this lead time has elapsed, the patient attends the CT (computed tomography) scanner. When the images from the CT scanner are ready, the doctor and physics unit define an appropriate arrangement of beams, produce a dose distribution (an isodose plan) and calculate the appropriate linac settings to deliver the treatment. The isodose plan usually takes two working days to complete, because it includes discussion between the physics unit staff and the doctor. During this time, both the physics unit and doctor are not exclusively devoted to this operation, hence they both can perform other tasks. The patient is then referred to the simulator for the treatment verification. Depending on the patient, the doctor may like to be present for this operation. After the treatment verification, the doctor approves the verification and prescribes the radiation dose. Then, the physics unit checks the generated treatment plan. Patient identification, dose data and linac parameters are entered into the verification system. Finally, these data are checked by a second person. Radical patients are seen from Monday to Friday (not on weekend days) and each patient is seen by exactly one doctor who is a specialist for the cancer site of the patient. Doctors are scheduled depending on the operation they are going to perform. Planning is scheduled within the planning availability of the corresponding doctor, whereas the treatment verification, and approval of the verification and prescription of the dose are scheduled within their simulation availability.

4.2 Problem Statement

The performance indicator of the City Hospital is the percentage of patients exceeding the DH 31-day waiting time target which is reported every month. The hospital aims not to exceed this waiting time targets by more than 5%. However, patients are scheduled with respect to their treatment intent and waiting list status according to the JCCO waiting time targets given in Table 1. The problem is to find a schedule of radiotherapy pretreatment processes in such a way that patients meet their JCCO waiting time targets.

4.3 A Genetic Algorithm for a Radiotherapy Pretreatment Scheduling Problem

The proposed genetic algorithm (GA) is built on two ideas. First, we concluded that scheduling patients using priority rules can generate schedules with good performance (Castro and Petrovic 2011). Second, computational studies indicate that a further improvement of an initial solution gives better results, when compared with schedules generated with the initial solution without the improvement.

The first idea is realised by encoding schedules with a priority rule-based representation (Dorndorf and Pesch 1995; Hart and Ross 1998). In (Dorndorf and Pesch 1995), this representation is used for static job-shop benchmark problems, in which all jobs are simultaneously available for processing. On the other hand, a modification of the representation is proposed for a dynamic job-shop problem, in which jobs have different arrival times (Morton and Pentico 1993). The second

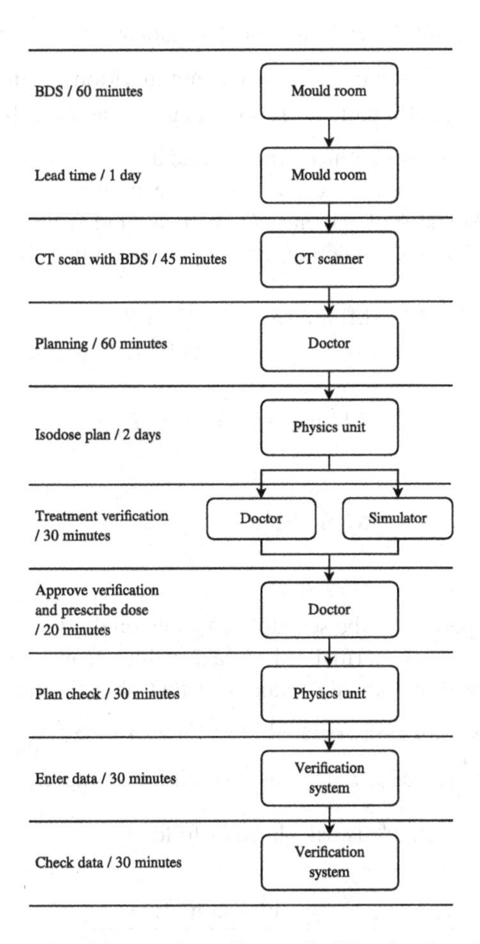

Fig. 6 The radiotherapy pretreatment pathway for a radical head and neck patient

idea is implemented by penalising the early resource idle time in the fitness function (Branke and Mattfeld 2000). The influence of the early idle time in the GA is regulated by two parameters: $\alpha \in [0,1]$ which denotes the importance of the early idle time term in the fitness function, and ω which denotes the number of days during which the idle time is to be penalised, starting from the earliest release date.

Fitness Function. A given schedule is evaluated with respect to a fitness function that incorporates two terms. The first term penalises breaches of the maximum acceptable JCCO waiting time targets, while the second one penalises early resource idle time within a given time window (Branke and Mattfeld 2000).

The first term is the sum of three objectives defined in (1)-(3). Objective (1) is the minimisation of the weighted numbers of patients exceeding the JCCO waiting time targets. Priority weights of patients w_j depend on the waiting list status of patients. Binary variable U_j is equal to 1 if patient P_j exceeds the waiting time target, 0 otherwise. Objective (2) is the minimisation of the maximum JCCO lateness. The lateness of a patient, L_j, is calculated as the difference between the completion day of the pretreatment and the due date as given by the corresponding JCCO waiting time target. Objective (3) is the minimisation of the sum of the weighted JCCO lateness of patients. Objectives (1)-(3) are calculated using the maximum acceptable JCCO waiting time targets given in Table 1.

$$\text{Minimise } z_1 = \sum_{P_j \in \mathrm{P}} w_j U_j \tag{4}$$

$$\text{Minimise } z_2 = \max_{P_j \in \mathrm{P}} \{L_j\} \tag{5}$$

$$\text{Minimise } z_3 = \sum_{P_j \in \mathrm{P}} w_j L_j \tag{6}$$

where P is set of patients to be scheduled on a given day.

Objectives (4)-(6) are normalised to take values from the [0,1] interval with respect to the largest and smallest values of that objective in a given population. Let v^k be the value associated to chromosome k to be normalised $(v^k \in \{z_1^k, z_2^k, z_3^k\})$, and l is the l-th chromosome in the population. The normalised value $\bar{v}^k$ of v^k is calculated as follows:

$$\bar{v}^k = \frac{v^k - \min_l \{v^l\}}{\max_l \{v^l\} - \min_l \{v^l\}} \tag{7}$$

The normalised values of (4)-(6) are summed up and correspond to the first term of the fitness function

$$z^k = W_1 \bar{z}_1^k + W_2 \bar{z}_2^k + W_3 \bar{z}_3^k \tag{8}$$

where weights $W_i\ (i = 1, 2, 3)$ represent the relative importance of objectives (4)-(6). This value is normalised to take values from the [0,1] interval before it is incorporated in the fitness function: $\bar{z}^k = z^k / (W_1 + W_2 + W_3)$.

The second term of the fitness function corresponds to the early resource idle time. The idle time is calculated within a predefined interval of days, denoted by ω. The idle time is normalised to take values from the [0,1] interval in the same way as given in (7). Let $\overline{\zeta}^k(\omega)$ be the normalised resource idle time of chromosome k within an interval of ω days, and $\alpha \in [0,1]$ its importance in the fitness function, then the fitness function of chromosome k is:

$$f^k = \overline{z}^k + \alpha \overline{\zeta}^k(\omega) \tag{9}$$

Parameters α and ω control the balance between the performance of the schedule in terms of the JCCO waiting time targets ($\overline{z}^k$) and the early resource idle time ($\overline{\zeta}^k(\omega)$).

Encoding and Decoding. Schedules are encoded by implementing a variation of the priority rule-based representation (Dorndorf and Pesch 1995; Hart and Ross 1998). Chromosomes are decoded using the modified Giffler and Thompson's algorithm given in (Storer et al. 1992). In this algorithm, parameter $\delta \in [0,1]$ determines the size of the search space by defining the number of the schedulable operations. In the proposed GA, δ is evolved (Hart and Ross 1998), as opposed to having a fixed value (Bierwirth and Mattfeld 1999; Branke and Mattfeld 2000; Mattfeld and Bierwirth 2004). The motivation for evolving δ is based on complex interactions between this parameter, other GA parameters and problem instances which are difficult to characterise. The length of the chromosome in our GA is equal to the total number of schedulable operations plus one. This additional gene stores parameter $\delta \in [0,1]$ which regulates the size of the search space.

Chromosome k is represented by the string $(\pi_1^k, \pi_2^k, \ldots, \pi_M^k, \delta^k)$ in which M is the total number of operations and π_d^k is a rule from a set of priority rules. We consider a comprehensive set of 44 priority rules which can be classified as in (Blackstone Jr. et al. 1982). Rules that involve waiting time targets such as earliest waiting time target, slack-based rules, etc; rules involving the processing time: shortest processing time, least work remaining, fewest remaining operations, etc; rules that depend on characteristics other than processing time and waiting time targets such as: random selection and arrived at queue first, and rules that depend on two or more characteristics such as: slack per number of operations remaining, slack per work remaining, etc. If $\delta = 1$ the algorithm produces active schedules, while if $\delta = 0$ the algorithm generates non-delay schedules. For values of δ from the (0,1) interval, schedules from a subset of active schedules, including all non-delay ones, are generated (Storer et al. 1992).

The pseudo-code of the modified Giffler and Thompson's algorithm is given in Algorithm 1. Iteration t of the algorithm is explained as follows. Let Γ_t be the cut set (set of operations ready to be scheduled in iteration t). Operation $O_t \in \Gamma_t$ with earliest completion time is selected; ties are broken arbitrarily. A conflict set

H_t is defined with all operations from Γ_t which require the same resource as O_t and whose processing time overlap with that of O_t. A smaller (respectively larger) value of δ gives a smaller (respectively larger) size of the conflict set H_t. Priority rule π_t is used to select an operation from H_t to be scheduled next.

Algorithm 1. Modified Giffler and Thompson's algorithm, adapted from (Storer et al. 1992)

1. $t \leftarrow 1$
2. Γ_t is the set of operations ready to be scheduled
3. **while** $\Gamma_t \neq \emptyset$ **do**
4. $O_t \leftarrow$ operation from Γ_t with earliest completion time ϕ_t
5. $m_t \leftarrow$ a resource required by O_t.
   If O_t requires multiple resources, m_t is the least available resource
6. $\sigma_t \leftarrow$ earliest start time on resource m_t
7. $H_t \leftarrow \{O \in I(m_t) : \sigma(O) \leq \sigma_t + \delta(\phi_t - \sigma_t)\}$, where $I(m_t)$ is the set of operations ready to be scheduled on m_t and $\sigma(O)$ is the earliest start time of operation O
8. Choose operation $O^* \in H_t$ by using priority rule π_t
9. Schedule O^*
10. $\Gamma_t \leftarrow \{\Gamma_t \setminus O^*\} \cup O$, where O is the immediate successor of O^*
11. $t \leftarrow t+1$
12. **end while**

Variation Operators

Figure 7 displays the implemented one point crossover (Sastry et al. 2005). In the GA proposed here, genes δ are combined by means of two convex combinations, one for each offspring. The combined values of δ are: $\delta_1^{'} = \delta^1\lambda + \delta^2(1-\lambda)$ and $\delta_2^{'} = \delta^1(1-\lambda) + \delta^2\lambda$. Value λ is chosen randomly from the [0,1] interval. A convex combination is chosen because its definition guarantees that $\delta_1^{'}, \delta_2^{'} \in [0,1]$.

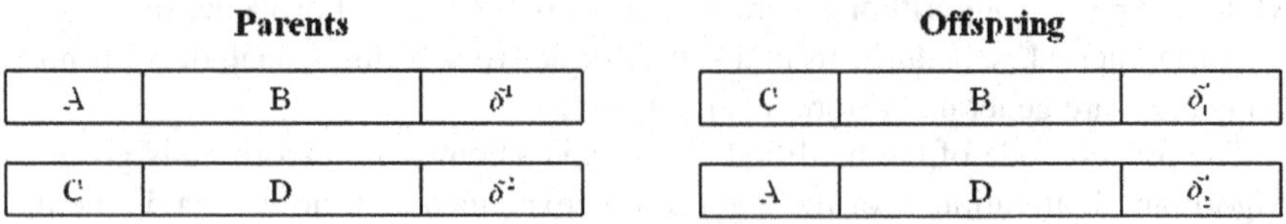

Parents

A	B	δ^1
C	D	δ^2

Offspring

C	B	$\delta_1^{'}$
A	D	$\delta_2^{'}$

Fig. 7 Implemented one point crossover

In the mutation operator introduced here, a chromosome is mutated by randomly selecting one of its genes. If this gene is a priority rule, it is replaced with a randomly selected rule from the set of predefined rules. If the selected gene is δ, a new value is randomly chosen from the $[0,1]$ interval.

4.4 *Experimental Design and Results*

4.4.1 Experimental Design

In this Section, an experimental design will be described together with the results of investigating the effects of changing GA parameters on the schedule performance and run time of the algorithm.

The experiments are designed to compare the performance of different combinations of α and ω in solving the radiotherapy pretreatment scheduling problem. First, all combinations are compared with respect to the monthly percentage of patients exceeding the DH waiting time target as this is the main measure of performance as the City Hospital. The secondary performance indicator is the monthly percentage of patients exceeding the JCCO waiting time targets. In case when the comparison based on these performance measures is not statistically significant, schedules performances are compared with respect to objectives (4)-(6).

The following factors affect the performance of the proposed GA: the intake of patients as a problem factor, and two GA factors, α and ω. The intake of patients is tested at two levels: current intake of patients (Scenario 1) and 10% increase in the current intake (Scenario 2) that the City Hospital expressed interest to investigate. GA levels for factors α and ω are set as follows: $\alpha \in \{0.0, 0.2, 0.4, 0.6, 0.8, 1.0\}$ and $\omega \in \{0,\ 1,\ 4,\ 7\}$. These values give 16 combinations of levels. When ω is 0 days, then the fitness function is JCCO driven (equivalent to α =0). We decide the maximum penalisation window to be 7 days. We also tested value 1 (patients are scheduled on daily basis) and value 4, which is the midpoint in the range.

Priority weights for emergency, urgent and routine patients are set to 5, 2, and 1, respectively. Weights W_1, W_2, W_3 which represent the relative importance of objectives (4)-(6) are set to 6, 3, and 1, respectively. The defined weight values are set in consultation with the City Hospital, but are of subjective nature.

The City Hospital provided two sets of data on patients. The first dataset gives information on the number of patients who arrived every day in a period of five years, 2001-2005. For each patient their treatment intent, waiting list status and admission date are known. This dataset provides us with the daily arrival profile of patients by their type (treatment intent and waiting list status). The second dataset provides information on 188 patients. For each patient the waiting list status, treatment intent, site, doctor, and all pretreatment operations dates are known. These datasets are the input to a problem instance generator described in (Leite-Rocha 2011). The instance generator preserves the observed seasonality of real-world arrivals, namely week of the year and day of the week.

All the parameter combinations proposed are run on the same set of instances which is referred to as the blocking on instances technique (Rardin and Uzsoy 2001), and are run with the same random seed on the same instance (Chiarandini et al. 2007). Blocking on instances and random seeds associated to each instance are a documented variance reduction techniques (Chiarandini et al. 2007; Johnson 2002). Therefore, for each level of the patient intake, a GA with specific values of α and ω is run once on 30 different instances. An instance corresponds to one year of patient arrivals, with a one-year warm-up period within which the GA is used to fill in the booking system. In summary, each combination of values for α and ω is run on the same set of 60 instances: 30 instances for Scenario 1 (current patient intake) and 30 instances for Scenario 2 (10% more than the current patient intake). Performance data are collected in the year following the warm-up period. All together, (16 GA level combinations) $\times$ (60 instances) gives 960 years of experiments.

The values of the GA parameters are set as follows. The crossover probability is 0.6, and the mutation probability is 0.1. These values give satisfactory results in (Bierwirth and Mattfeld 1999; Branke and Mattfeld 2000; Mattfeld and Bierwirth 2004). The GA is allowed to run after a given number of generations has been reached with no improvement in the value of the fitness function. This number is equal to the number of patients on a given day (Bierwirth and Mattfeld 1999). Accordingly, larger instances (in terms of the number of patients) are allowed to run for more generations than smaller ones. A fitness proportionate selection with an elitist strategy of one individual is used. Chromosomes of the initial population are randomly chosen. Specifically, each gene representing a priority rule is initialised by randomly selecting a rule from the set of rules, while δ is randomly selected from the [0,1] interval. In pilot studies, it is found that having a population of 50 individuals does not affect solution quality and halves the run time, compared to 100 individuals.

4.4.2 Experimental Results

Table 12 gives a summary of the results obtained. There are two columns under each Scenario. Entries in the first and third columns are the average over 30 realisations (one for each instance). Each realisation is the average over 12 observations where each observation is the monthly percentage of patients exceeding the JCCO and the DH waiting time targets, respectively. Elements in the second and fourth columns are the average percentages (one for each instance) of the total number of patients exceeding the JCCO and the DH waiting time targets over 30 realisations, respectively. The bottom rows give the overall average, best, worst and sample standard deviation values observed across all instances in each scenario.

By examining Table 12 we can see that there is no apparent difference between GA parameter combinations with respect to the average percentage of patients exceeding the JCCO and DH waiting time targets. However, all combinations are sensitive to a 10% increase in the current patient intake. The JCCO and DH indicators are increased almost three and over fourfold from Scenario 1 to 2, respectively. Also, 10% more patients than the current intake still leads to an average DH indicator and its worse value within the acceptable 5% limit.

Table 12 Average monthly percentage of patients exceeding the JCCO and the DH waiting time targets

α	ω	% of patients exceeding JCC / % patients exceeding DH targets	
		Scenario 1	Scenario 2
0.0	0	0.74 / 0.15	2.14 / 0.66
0.2	1	0.76 / 0.16	2.13 / 0.67
	4	0.76 / 0.16	2.18 / 0.69
	7	0.77 / 0.15	2.16 / 0.68
0.4	1	0.77 / 0.16	2.15 / 0.68
	4	0.78 / 0.16	2.16 / 0.68
	7	0.77 / 0.17	2.12 / 0.65
0.6	1	0.78 / 0.15	2.09 / 0.67
	4	0.77 / 0.16	2.13 / 0.68
	7	0.78 / 0.18	2.15 / 0.68
0.8	1	0.75 / 0.15	2.12 / 0.67
	4	0.81 / 0.15	2.19 / 0.69
	7	0.77 / 0.17	2.16 / 0.69
1.0	1	0.78 / 0.14	2.12 / 0.69
	4	0.76 / 0.15	2.17 / 0.66
	7	0.78 / 0.16	2.16 / 0.68
Overall average		0.77 / 0.16	2.14 / 0.68
Overall best		0.08 0.00	0.41 / 0.00
Overall worst		2.98 / 1.45	5.18 / 3.26
Sample st. dev.		0.58 / 0.28	1.15 / 0.99

Descriptive statistics given in Table 12 show that there is no evident distinction between GA parameter combinations of α and ω with respect to both the average percentage of patients exceeding the JCCO and the DH waiting time targets. We use analysis of variance (ANOVA) to confirm or refute these hypotheses. The response variables (monthly percentage of patients exceeding the JCCO and the DH waiting time targets) can be expressed in terms of the problem parameters levels (Scenario 1 and 2), the GA parameter levels, $\alpha \in \{0.0, 0.2, 0.4, 0.6, 0.8, 1.0\}$ and $\omega \in \{0, 1, 4\}$, and the interactions between these factors. That is, the effect of GA parameter level α (or ω) may vary with the level of the problem factor - intake of patients. Table 13 shows the ANOVA results for the monthly percentage of patients exceeding the DH waiting time target. This table reveals that the only main factor that has statistical significance is the intake of patients, and that no interaction between factors is significant. The same conclusion is reached for the monthly percentage of patients exceeding the JCCO waiting time targets, the monthly weighted number of patients exceeding the JCCO waiting time targets, the monthly maximum JCCO lateness, and monthly sum of weighted JCCO lateness.

Table 13 ANOVA results for the monthly percentage of patients exceeding the DH waiting time target

Source	Df	Sum sq	Mean Sq	F value	Pr(>F)	
α	5	0.02	0.004	0.0069	1.000	
ω	2	0.01	0.003	0.0060	0.994	
Scenario	1	64.11	64.107	118.4295	<2e-16	***
$\alpha * \omega$	8	0.03	0.004	0.0077	1.000	
α * Scenario	5	0.002	0.004	0.0076	1.000	
ω * Scenario	2	0.01	0.004	0.0080	0.992	
Residuals	936	506.67	0.541			

Signif. Codes: 0 '***' 0.001 '**' 0.01 '*' 0.05 '.' 0.1 ' ' 1

Combinations of GA parameter levels are not statistically significant with respect to the performance measures of interest. However, our experiments reveal that run times are sensitive to a 10% increase in the number of patients and to certain GA level combinations (the CPU times are shown in Table 14). This Table indicates that for a given value of α the run time becomes shorter as ω becomes larger. On the other hand, there are mixed results on the effect of α on the run time. However, the run time is shortest at $\alpha = 0$ and $\omega = 0$.

4.5 *Summary*

Sixteen parameter combinations of the weight of the early resource idle time in the fitness function (α) and the length during which idle time is penalised (ω) are tested. It is observed that combinations of these parameters do not generate schedules with significantly different performance measures. However, combinations are differentiable with respect to the run time.

The performance of the proposed GA may be enhanced. A different representation can be implemented. The priority rule-based representation may lead to false competition among chromosomes (Hart et al. 2005). That is, different chromosomes can represent the same schedule. Furthermore, in radiotherapy pretreatment, different schedules may give the same fitness function value as the level of resolution of the lateness is a day. Also, it is observed that for minimax objectives (objective (5)) different solutions give the same objective function value (Garfinkel and Gilbert 1978). We suggest to investigate a preference list-based representation (Davis 1985) in order to mitigate the false competition, and possibly minimise the issues that arise from the mathematical definition of the objectives. In the preference list-based representation each resource keeps its own list of patients, thus patterns of sequences of patients may emerge on each resource and can be investigated.

Table 14 Run times of the experiments

α	ω	Total run time / Run time per day instance	
		Scenario 1	Scenario 2
0.0	0	6:55:11 / 0:01:41	9:50:56 / 0:02:24
0.2	1	7:47:01 / 0:01:54	11:40:55/ 0:02:51
	4	7:47:36 / 0:01:54	11:24:41/ 0:02:47
	7	7:34:10 / 0:01:51	10:33:12/ 0:02:34
0.4	1	8:07:41 / 0:01:59	11:06:33/ 0:02:43
	4	7:53:43 / 0:01:55	10:57:30/ 0:02:40
	7	7:23:05 / 0:01:48	10:33:33/ 0:02:35
0.6	1	7:59:02 / 0:01:57	11:37:17/ 0:02:50
	4	7:41:32 / 0:01:52	11:07:27/ 0:02:43
	7	7:22:43 / 0:01:48	10:18:25/ 0:02:31
0.8	1	7:48:07 / 0:01:54	11:19:55/ 0:02:46
	4	7:51:39 / 0:01:55	10:58:37/ 0:02:41
	7	7:40:57 / 0:01:52	10:23:43/ 0:02:32
1.0	1	7:44:35 / 0:01:53	11:11:54/ 0:02:44
	4	7:42:58 / 0:01:53	10:48:24/ 0:02:38
	7	7:42:10 / 0:01:53	10:37:11/ 0:02:35
Overall average		7:41:23 / 0:01:52	10:54:23/ 0:02:40
Overall best		5:13:29 / 0:01:17	7:06:24 / 0:01:43
Overall worst		11:44:47 / 0:02:53	16:18:53/ 0:03:59
Sample st. dev.		1:09:22 / 0:00:17	1:59:25 / 0:00:29

5 Conclusions

Various heuristics have been developed and applied to a wide range of scheduling problems, and, in particular, production scheduling. In this chapter, it is shown that radiotherapy scheduling problems can be typified as production scheduling problems with some unique characteristics relevant to the medical domain.

Novel heuristics for scheduling radiotherapy patients are proposed and analysed. In the first scheduling system presented, the scheduling of the whole radiotherapy process, from the decision to treat a patient using radiotherapy to the administration of the first fraction, is considered as four scheduling subproblems, corresponding to the four units of the centre. Priority rules are developed for each unit and combined with special strategies proposed to improve scheduling performance of urgent patients, such as emergency and palliative patients. The strategies include: (1) allowing breaches of the target waiting times for different patient categories, (2) reserving slots on treatment machines for certain patient categories and (3) introducing overtime slots on the treatment machines. The heuristics created schedules of good performance with respect to waiting times and percentages of late patients of different categories.

The heuristics presented are developed for specific real-life radiotherapy centres. However, they are generic in the sense that they consider typical objectives and constraints of any radiotherapy centre.

The future work will be carried out in the following directions. The heuristics will be extended to consider various aspects of real-life radiotherapy scheduling problems, such as considering forecasts of uncertain numbers of arriving patients and the corresponding patient categories, anticipating failures of certain patients to attend their appointments, rescheduling of patients that missed their appointments, including preferences of certain patients categories for being treated in a certain period of the day, etc. A comparison of the two different approaches to radiotherapy scheduling based on priority rules and metaheuristics will be carried out.

Our further research work into development of genetic algorithms for preradiothearpy scheduling will be focused on investigation of a different chromosome representation. Namely, a priority rule-based representation may lead to false competition among chromosomes That is, different chromosomes may represent the same schedule. Furthermore, different schedules may give the same fitness function value because lateness is measured in days. We will investigate a preference list-based representation in which each resource maintains its own list of patients. Such a representation may be both more appealing to the City Hospital in terms of an easier explanation and also it would be interesting to investigate if patterns of patient sequences could be identified on each resource.

We will also investigate the coordination of radiotherapy pretreatmentand radiotherapy scheduling. For instance, a patient who has to wait for their treatment on a linac machine, due to its unavailability for a required number of consecutive sessions, can give priority to patients who are late in their pretreatment phase, but can start radiotherapy treatment earlier.

Acknowledgments. The authors would like to thank the Engineering and Physics Science Research Council (EPSRC), UK, for supporting this research (Ref. No. EP/NB2076949511/1 and EP/C54952X/1). The authors would also like to acknowledge the support of the Nottingham University Hospitals NHS Trust, City Hospital Campus, UK and to thank the staff from the Arden Cancer Centre, University Hospitals Coventry and Warwickshire NHS Trust, UK for their help, support and involvement in this project.

References

Agarwal, J., Ghosh-Laskar, S., Budrukkar, A., Murthy, V., Mallick, I.: Finding solutions for the endless wait–reducing waiting times for radiotherapy. Radiotherapy and Oncology 87(1), 153–154 (2008)

Bierwirth, C., Mattfeld, D.: Production scheduling and rescheduling with Genetic Algorithms. Evolutionary Computation 7(1), 1–17 (1999)

Blackstone Jr., J., Phillips, D., Hogg, G.: A state-of-the-art survey of priority rules for manufacturing job shop operations. International Journal of Production Research 20(1), 27–45 (1982)

Branke, J., Mattfeld, D.: Anticipation in dynamic optimization: The scheduling case. In: Deb, K., Rudolph, G., Lutton, E., Merelo, J.J., Schoenauer, M., Schwefel, H.-P., Yao, X. (eds.) PPSN 2000. LNCS, vol. 1917, pp. 253–262. Springer, Heidelberg (2000)

Castro, E., Petrovic, S.: Combined mathematical programming and heuristics for a radiotherapy pretreatment scheduling problem. Journal of Scheduling (2011), doi:10.1007/s10951-011-0239-8

Chiarandini, M., Paquete, L., Preuss, M., Ridge, E.: Experiments on metaheuristics: Methodological overview and open issues. Technical report University of Southern Denmark (2007)

Conforti, D., Guerriero, F., Guido, R.: Optimization models for radiotherapy patient scheduling. 4OR: A Quarterly Journal of Operations Research 6(3), 263–278 (2008)

Conforti, D., Guerriero, F., Guido, R.: Non-block scheduling with priority for radiotherapy treatments. European Journal of Operational Research 201(1), 289–296 (2010)

Conforti, D., Guerriero, F., Guido, R., Veltri, M.: An optimal decision making approach for the management of radiotherapy patients. OR Spectrum 33(1), 123–148 (2011)

Davis, L.: Job Shop Scheduling with Genetic Algorithm. In: Grefenstette, J. (ed.) Proceedings of the International Conference of Genetic Algorithms (ICGA), pp. 136–240 (1985)

Delaney, G., Jacob, S., Featherstone, C., Barton, M.: The role of radiotherapy in cancer treatment: Estimating optimal utilization from a review of evidence-based clinical guidelines. Wiley InterScience 104(6), 1129–1137 (2005)

Department of Health, The NHS cancer plan: a plan for investment, a plan for reform (2000)

Department of Health, The NHS cancer plan and the new NHS: Providing a patient-centred service (2004)

Dodwell, D., Crellin, A.: Waiting for radiotherapy. British Medical Journal 332(7533), 107–109 (2006)

Dorndorf, U., Pesch, E.: Evolution based learning in a job shop scheduling environment. Computers & Operations Research 22(1), 25–40 (1995)

Garfinkel, R., Gilbert, K.: The bottleneck traveling salesman problem: Algorithms and probabilistic analysis. Journal of the ACM 25(3), 435–448 (1978)

Gupta, D., Denton, B.: Appointment scheduling in health care: Challenges and opportunities. IIE Transactions 40, 800–819 (2008)

Hart, E., Ross, P.: A heuristic combination method for solving job-shop scheduling problems. In: Eiben, A.E., Bäck, T., Schoenauer, M., Schwefel, H.-P. (eds.) PPSN 1998. LNCS, vol. 1498, pp. 845–854. Springer, Heidelberg (1998)

Hart, E., Ross, P., Corne, D.: Evolutionary Scheduling: A Review. Genetic Programming and Evolvable Machines 6(2), 191–220 (2005)

Huang, J., Barbera, L., Brouwers, M., Browman, G., Mackillop, W.J.: Does delay in starting treatment affect the outcomes of radiotherapy? Journal of Clinical Oncology 21, 555–563 (2003)

Johnson, D.: A theoretician's guide to the experimental analysis of algorithms. In: Goldwasser, M., Johnson, D., McGeoch, C. (eds.) Data Structures, near Neighbour Searches, and Methodology: Fifth and Sixth DIMACS Implementation Challenges, vol. 59, pp. 215–250 (2002)

Joint Council of Clinical Oncology, Reducing delays in cancer treatment: Some targets, 1993 Report. Royal College of Physicians, London (1993)

Kapamara, T., Sheibani, K., Petrovic, D., Haas, O., Reeves, C.R.: A Simulation of a radiotherapy treatment system: A case study of a local cancer centre. In: Proceedings of the ORP3 2007 Conference Guimaraes, Portugal, pp. 29–35 (2007)

Larsson, S.N.: Radiotherapy patient scheduling using a desktop personal computer. Clinical Oncology 5(2), 98–101 (1993)

Leite-Rocha, P.: Novel approaches to radiotherapy treatment scheduling. PhD thesis, University of Nottingham (2011)

Mackillop, W.J.: Killing time: the consequences of delays in radiotherapy. Radiotherapy Oncology 84, 1–4 (2007)

Mattfeld, D., Bierwirth, C.: An efficient genetic algorithm for job shop scheduling with tardiness objectives. European Journal of Operational Research 155(3), 616–630 (2004)

Morton, T., Pentico, D.: Heuristic scheduling systems. Wiley (1993)

National Radiotherapy Advisory Group, Radiotherapy: Developing a world class service for England. 2007 Report to Ministers. National Health Service, England (2007)

Petrovic, D., Morshed, M., Petrovic, S.: Multi-objective genetic algorithm for scheduling of radiotherapy treatments for categorised cancer patients. Expert Systems with Applications 38, 6994–7002 (2011)

Petrovic, S., Leite-Rocha, P.: Constructive approaches to radiotherapy scheduling. In: Ao, S., Douglas, C., Grundfest, W., Schruben, L., Burgstone, J. (eds.) World Congress on Engineering and Computer Science (WCECS), pp. 722–727 (2008a)

Petrovic, S., Leite-Rocha, P.: Constructive and GRASP approaches to radiotherapy scheduling. In: Ao, S. (ed.) Advances in Electrical and Electronics Engineering (IAENG) Special Edition of the World Congress on Engineering and Computer Science 2008 (WCECS), pp. 192–200. IEEE Computer Society (2008b)

Petrovic, S., Leung, W., Song, X., Sundar, S.: Algorithms for radiotherapy treatment booking. In: Qu, R. (ed.) Proceedings of the Workshop of the UK Planning and Scheduling Special Interest Group (PlanSIG), pp. 105–112 (2006)

Proctor, S., Lehaney, B., Reeves, C.R., Khan, Z.: Modelling patient flow in a radiotherapy department. OR Insight 20, 6–14 (2007)

Rardin, R., Uzsoy, R.: Experimental Evaluation of Heuristic Optimization Algorithms: A Tutorial. Journal of Heuristics 7(3), 261–304 (2001)

Routsis, D., Thomas, S., Head, J.: Are extended working days sustainable in radiotherapy? Journal of Radiother. Practice 5, 77–85 (2006)

Sastry, K., Goldberg, D., Kendall, G.: Genetic algorithms. In: Burke, E., Kendall, G. (eds.) Search Methodologies. Introductory Tutorials in Optimization and Decision Support Techniques, pp. 97–126. Springer (2005)

Storer, R., Wu, S., Vaccari, R.: New search spaces for sequencing problems with application to job shop scheduling. Management Science 38(10), 1495–1509 (1992)

Summers, E., Williams, M.: Re-audit of radiotherapy waiting times. Royal College of Radiologist London UK (2005)

Thomas, S.J., Williams, M.V., Burnet, N.G., Baker, C.R.: How much surplus capacity is required to maintain low waiting times? Clinical Oncology 13, 24–28 (2001)

Werker, G., Sauré, A., French, J., Shechter, S.: The use of discrete-event simulation modelling to improve radiation therapy planning processes. Radiotherapy and Oncology 92(1), 76–82 (2009)

Recent Advances in Evolutionary Algorithms for Job Shop Scheduling

Bahriye Akay and Xin Yao

Abstract. Scheduling decides the order of tasks to efficiently use resources considering criteria such as minimization of the number of late tasks, minimization of the completion time, minimization of the idle times of the machines, etc. Approaches for solving scheduling problems can be divided into three broad groups: (a) exact methods that produce exact optimal solutions, (b) approximation methods that find high quality near optimal, and (c) hybrid methods based on the first two. Approximate methods can be easily combined with other types of heuristics and can be applied to a wide range of problems.

In the category of approximation algorithms, evolutionary algorithms (EAs) are very promising tools for the problems with dynamic characteristics, contradicting multi-objectives and highly nonlinear constraints. For EAs to be effective and efficient for a combinatorial optimisation problem like scheduling, the structure of an EA needs to be designed carefully to exploit the problem structures. An appropriate representation for the problem and the type of search operators suitable for the representation should be studied because they directly affect the search efficiency of the EA.

In this chapter, our focus will be on EAs for job shop scheduling problems (JSP). First, JSP will be formulated as an optimization problem and approaches for JSP will be given briefly. Second, EAs will be introduced and the key issues in the application of EAs for JSP will be emphasized. Third, various representations used in EAs for handling JSP will be described and advantages and drawbacks of

Bahriye Akay
Dept. of Computer Engineering, Erciyes University,
38039, Melikgazi, Kayseri, Turkey
e-mail: bahriye@erciyes.edu.tr

Xin Yao
Center of Excellence for Research in Computational Intelligence and Applications,
School of Computer Science, University of Birmingham,
Birmingham B15 2TT, U.K
e-mail: x.yao@cs.bham.ac.uk

A.Ş. Etaner-Uyar et al. (eds.), *Automated Scheduling and Planning*,
Studies in Computational Intelligence 505,
DOI: 10.1007/978-3-642-39304-4_8,

different representations will be described based on the results from the literature. Forth, crossover and mutation operators designed for particular representations will be illustrated and their strength and limitations will be discussed. Almost all successful applications of evolutionary combinatorial optimisation include some kind of hybrid algorithms, where both EAs and local search were used. The seventh topic of this chapter is devoted to local search strategies which are frequently integrated into EAs.

1 Job Shop Scheduling

Combinatorial optimization deals with problem of finding the optimal subset from a finite set of subsets. Some common problems in combinatorial optimization are maximum matching, minimum spanning, travelling salesman, knapsack and scheduling. Scheduling decides the order of tasks to use resources efficiently. I/O scheduling, CPU scheduling,time table scheduling, project scheduling, job shop scheduling, etc., are some examples of scheduling problems. The objective of scheduling may be minimization of the number of late jobs, minimization of the makespan, minimization of the idle times of processors, minimization of the completion time of a project etc.,.

In a job shop scheduling problem, given n jobs are assigned to m machines to minimize a criterion such as completion time. In a job shop scheduling problem, each job is composed of a set of operations and has its predefined job sequence. Each operation is processed by the required machine and a machine can process each job at different speeds, so, processing time of each job is different at different machines.

Job Shop Scheduling (JSP) problems have characteristics given below (Yamada, 2003):

- All the jobs J_j must follow the pre-defined technological sequence.
- Each machine M_r can process only one job at a time.
- An operation O_{jr} must be processed on its machine M_r for its processing time p_{jr} without being preempted by another operation.
- A job does not visit the same machine twice.
- Let the starting time be s_{jr} and the completion time be c_{jr} of an operation O_{jr}. A schedule is a set of completion times for each operation $\{c_{jr}\}_{1<j<n,1<r<m}$ that satisfies above constraints, and where n is the number of jobs and m is the number of machines.
- Makespan (C_{max}) is the time required to complete all the jobs. $C_{max} = \max_{1\leq j\leq n,1\leq r\leq m} c_{jr}$

An example of a 3x3 JSP is given in Table 1. The data for the first job means that its first operation (O_{11}) is processed on machine 1 for its processing time, p_{11}, on machine 1 (3), and then second operation (O_{12}) is processed on machine 2 for its processing time, p_{12}, on machine 2 (3) and finally its third operation (O_{13}) processed on machine 3 for its processing time, p_{13}, on machine 3(3). Similarly, first operation of the second job (O_{21}) is processed on machine 1 for its processing time, p_{21}, on machine 1 (2), and then second operation of the second job (O_{22})is processed on machine 3 for its processing time, p_{23}, on machine 3 (3), and finally third operation (O_{23})processed on machine 2 for its processing time, p_{22}, on machine 2 (4). The same process is repeated for the third job.

Table 1 An Example 3x3 JSP (Yamada, 2003)

J_j	$M_r(p_{jr})$		
1	1(3)	2(3)	3(3)
2	1(2)	3(3)	2(4)
3	2(3)	1(2)	3(1)

The aim is to find a schedule that minimizes the makespan or another objective without violating the constraints that any two jobs can not be processed on a machine at the same time and an operation can not start before the its previous operation is completed.

A solution to a JSP can be visually represented by a Gantt-chart (Gantt, 1910) as shown in Figure 1 or a disjunctive graph (Roy and Sussmann, 1964) as in Figure 2. x-axis of the Gantt chart shows time units, and the machine numbers are shown along the y-axis. Each box corresponds to an operation O_{jr}. The left edge of the box is aligned at starting time s_{jr} of O_{jr} and the length of the box is the processing time p_{jr} of O_{jr}. Therefore, the makespan of the schedule, $C_{max} = \max_{1 \leq j \leq n, 1 \leq r \leq m} c_{rj}$ is equal to 12 for the example schedule given in Figure 1.

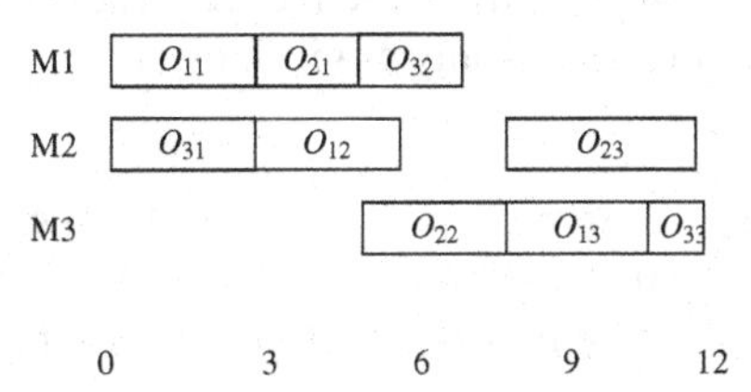

Fig. 1 Gantt-Chart of a 3x3 JSP

A schedule S which does not violate any of the constraints related with the machine sequences and the predefined job sequences is called feasible. However, a feasible schedule may not be an optimal schedule because there may be some other alternative schedules with less makespan. A feasible schedule may be semi-active or active. In a semi-active schedule, there is no operation that can be started earlier without altering the order of jobs in schedule. In order to reduce the makespan of a semi-active schedule, some operations may be shifted to the left without delaying other jobs, which is called a permissible left shift. A schedule whose makespan can not be reduced by permissible left shifts is called an active schedule. Giffler-Thompson algorithm is a heuristic that creates active schedules with a bias towards non-delay schedules depending on various rules; for example, priority dispatching rules, shortest operation time, most work remaining and first come first serve rule.

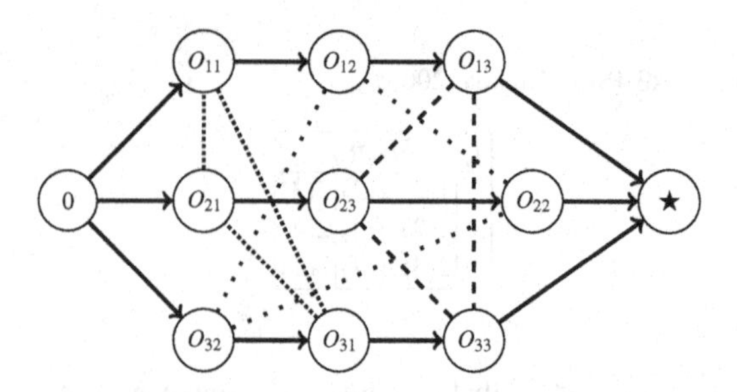

Fig. 2 Disjunctive Graph of a 3x3 JSP

Although Gantt-chart is easy to define a schedule, it does not give any information if the schedule is feasible or not. A more informative disjunctive graph representation (Roy and Sussmann, 1964) can be used to identify a schedule. A disjunctive graph $G = (N, A \cup E)$ is a set of nodes N representing operations of the jobs together with source and sink with zero cost to indicate the beginning and the end of the schedule where A is a set of ordinary conjunctive arcs representing technological sequences of machines for each job and $E = \bigcup_{r=1}^{m} E_r$ is a set of disjunctive arcs representing pairs of operations that must be performed on the same machine (Adams et al., 1988; Yamada, 2003). Solid lines represent constructive arcs (technological sequence) and dotted lines represent disjunctive arcs. The scheduling problem based on graph can be defined mathematically as given by (1):

$$\begin{aligned} \text{minimize}: \; & s_* \\ \text{subjectto}: \; & s_w - s_v \geq p_v, && (v,w) \in A, \\ & s_v \geq 0, && v \in N, \\ & s_w - s_v \geq p_v \vee s_v - s_w \geq p_w && (v,w) \in E_r, 1 \leq r \leq m. \end{aligned} \tag{1}$$

Scheduling is fixing all undirected (disjunctive) arcs into directed ones. A selection is defined as a set of directed arcs selected from the set of disjunctive arcs E. A schedule S from $G = (N, A \cup E)$ (Phan, 2000)

- S is a partial selection iff $i \rightarrow j \in S$ implies $j \rightarrow i \notin S$ for all $i \leftrightarrow j \in E$.
- S is a complete selection iff either $i \rightarrow j \in S$ or $j \rightarrow i \in S$ for all $i \leftrightarrow j \in E$.
- A complete selection S is acyclic iff the directed graph G_s is acyclic.

The objective to be minimized is a function of the completion times of the jobs. Let p_{ij} be processing time of job i on machine j, r_i is the release time of job i at which job i can start its processing, d_i is due date of job i at which job i is expected to finish, $\hat{d}_i$ is deadline of job i by which job i must be completed and w_i is the weight of job i, C_i denotes the completion time of job i. Lateness of a job is given by i $L_i = C_i - d_i$ and the tardiness of job i is defined as $T_i = max(L_i, 0)$. Alternative objective functions to be minimized are as follows:

- Makespan ($Cmax$)=$max(C_1 \dots C_n)$
- Maximum lateness ($Lmax$)=$max(L_1 \dots L_n)$
- Total weighted completion time: ($\sum w_i C_i$)
- The total (unweighted) completion time: $\sum C_i$
- Total weighted tardiness ($\sum w_i T_i$)
- The total (unweighted) tardiness : $\sum T_i$
- Weighted number of tardy jobs ($\sum w_i U_i$)
- The total (unweighted) number of tardy jobs is denoted by $\sum U_i$ where the unit penalty of job i, $U_i = 1$ if $C_i > d_i$; otherwise, $U_i = 0$.

2 Approaches for JSP

Approaches for solving JSPs can be divided into three broad groups: exact methods, approximation methods and methods based on the hybridization of the first two.

Exact algorithms are guaranteed to find an optimal solution, but they need exponential computation time in the worst-case for even small problem instances. Finding an exact solution may be very difficult due to searching an exponential number of possible solutions, and can be inappropriate because the problems may have large size, the dynamic character, constraints difficult to formulate in mathematical terms, contradictory objectives (Widmer and Costa, 2008). In these conditions, the approximate methods are possible alternatives which provide good quality solutions in a reasonable amount of time. Response time of the approximation methods is often much faster than the exact methods, and the approximation methods can be easily adapted or combined with other types of methods, so that they can be applied to a wide range of problems.

- Complete (exact) methods (integer programming, Branch and Bound)
- Approximation methods
  - Constructive methods
    - List scheduler algorithms
    - The shifting bottleneck procedure
    - Insertion techniques and beam search
    - NEH Heuristic
  - Local search methods
    - Simulated annealing
    - Threshold accepting methods
    - Tabu search
  - Evolutionary algorithms
- Hybrid Methods

Constructive methods reduce the size of the problem and the search space into a smaller subset of whole search space, at each step. Although they are fast and simple, the quality of the solutions obtained are not satisfactory. List schedulers, in the category of constructive methods, finds an operation to be processed next depending on some rules called priority rules and dispatching rules. Giffler and Thompson, Non-Delay algorithms are typical examples of list scheduler. Shifting bottleneck heuristic identifies the machine with the longest makespan, schedules this machine and then the other machines are reoptimized. NEH algorithm proposed by Nawaz et al. searches a minimal length sequence for JSP problem. Local search methods exploit the neighbourhood of a solution and move another solution in the search space iteratively. $\mathrm{N(s)} = \{\mathrm{s}' \in \mathrm{X} | \exists \delta \in \Delta : \mathrm{s}' = \mathrm{s} \oplus \delta\}$. Their performance depends on the size and structure of the considered neighborhood $N(s)$ and they can stuck to the local optima far from the optimum. Due to the disadvantages of exact methods and constructive methods on scheduling and local search methods, researchers attempt to find high quality sub-optimal solutions in a reasonable time by using evolutionary algorithms.

3 Introduction to Evolutionary Algorithms

Evolutionary Algorithms (EA) based on biological evolution operate on a population of solutions applying evolution operators: selection, mutation, recombination. They obtain well approximating solutions because they do not make any assumption about the search space and this makes them useful for solving continuous and combinatorial hard problems.

```
begin
    Initialization;
    Evaluation;
    while Termination Criteria is not Met do
        Selection of parents;
        Recombination;
        Mutation;
        Evaluate new solutions;
        Selection to form new population;
    end
end
```

Algorithm 1: Evolutionary Algorithm

A population is constructed by generating a set of randomly generated solutions and each solution is assigned a fitness values. Once all solutions are evaluated, the evolution cycle is repeated until a termination criterion, such as reaching the maximum number of generations, is satisfied. In an evolution cycle, some solutions are selected as parents to produce offsprings by recombination. After recombination, the offsprings are mutated by with a certain probability and a new population is generated. In traditional EAs, solutions are represented by binary strings and operators operating in the binary space to produce a candidate solution.

Although it is easy to adapt a problem to be solved by an EA, the main difficulty is related to deciding the structure of EA. Because the JSP is an ordering problem which tries to find the sequence of the jobs on machines in order to minimize makespan time, classical EA representation and search operators may not be directly applied or they may produce illegal or infeasible solutions. Therefore, the structure of an EA, the encoding and representation and the suitable search operators should be decided carefully to solve JSP in order to obtain efficient results.

- Representation of schedule(phenotype) by suitable genotype
- Decoding the individual to a schedule
- Type of Crossover
- Type of Mutation

Many studies have been presented to the literature based on various representations, recombination operators, fitness functions and local search methods. Cheng et al. (1999) presents a good survey for the encoding schemes, crossover and mutation operators, and hybrid genetic algorithms. In addition to those given in Cheng et al. (1999), this chapter also contains more recent schemes and results.

4 Representation

Representation should be decided carefully because an inappropriate representation may increase the dimension of search space and may worsen the approximation ability of the algorithm. In order to decide the representation of the problem variables (phenotype) and the format of the variables (genotype) which the EA will work through the search process. Deciding the phenotype and genotype and the mapping from genotype to phenotype determines the type of evolutionary operators and hence directs the evolution.

In order to solve JSPs, there are some different representations: binary, real and ordering representations. These main categories vary what they encode such as jobs, operations, machines, completion times, etc. Choosing the appropriate representation is especially important for JSPs due to the precedence constraints they have. In the initialization or after applying a genetic operator, the output solution may not be feasible if precedence constraints are violated. Therefore, the infeasible solution needs to be repaired and this may affect EA's performance. Moreover, a representation that needs many mappings and decodings to be evaluated leads to high running time.

There are two types of encoding in representation, indirect and direct encoding. In the indirect approach, solutions are encoded in a data structure and these genotypes are passed to the operators. To evaluate a solution, the solution at the genotype level is mapped to the phenotype level. In the direct approach, search operators are applied to the solutions which are already in the format of the problem variables (Rothlauf, 2008). The direct representation includes operation-based representation, job-based representation, job pair relation-based representation, completion time-based representation, and random keys representation. In the indirect approach, a schedule is constructed with dispatching rules as in the priority rule-based representation. Preference list-based representation, priority rule-based representation, disjunctive graph-based representation, and machine-based representation are examples of indirect representations (Gen et al., 2008).

4.1 *Binary Representation*

Nakano and Yamada (1991) proposed a GA using binary representation based on the disjunctive graphs. The directed arcs on the graph are labeled as 0 or 1 according to the direction and all the bits corresponding to these labels form a one dimensional bit-string with the length of $mn(n-1)/2$. An arc connecting O_{ij} and O_{kj} is labeled as 1 if the arc is directed from O_{ij} to O_{kj} and $(i < k)$, otherwise it is labeled as 0. The Hamming distance defines the similarity between two bit strings.

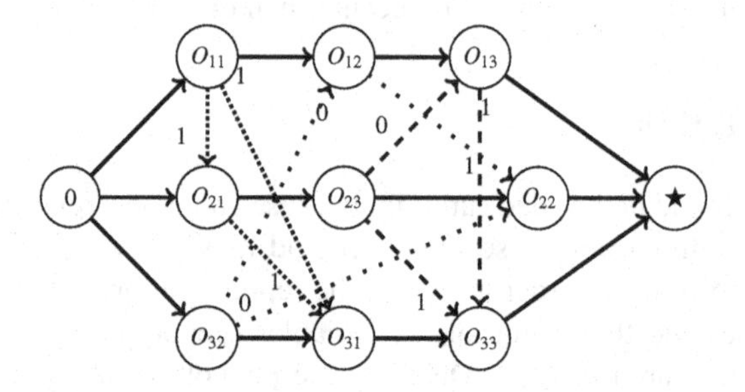

Fig. 3 Labelled Disjunctive Graph

The binary string corresponding to the directed disjunctive graph in Figure 3 is given in Figure 4.

1	1	1	1	0	0	0	1	1
O_{11}	O_{11}	O_{21}	O_{12}	O_{12}	O_{22}	O_{13}	O_{13}	O_{23}
O_{21}	O_{31}	O_{31}	O_{22}	O_{32}	O_{32}	O_{23}	O_{32}	O_{33}

Fig. 4 Binary string representation for the directed disjunctive graph

The approach can be used with the conventional genetic operators, such as one-point, two-point and uniform crossovers without any modification. However, it has a drawback that an offspring generated by the genetic operators may not be legal or feasible. Binary representation used in conventional EAs needs a repairing mechanism to keep the individuals feasible. For this reason, they employed a harmonization approach to deal with infeasible or illegal solutions. The approach was applied to solve ft06 (6x6), ft10 (10x10) and ft20(20x5) problems and found the optimum solution for ft06 and near optimum solutions for ft10 and ft20.

4.2 Priority Rule-Based Representation

Dorndorf and Pesch (1995) proposed a priority rule-based representation in which a chromosome is a sequence of rules for job assignment and a schedule is created based on the priority rules. There is an array of predefined rules such as select the operation with the shortest processing time, select the operation with the longest processing time, select the operation of the job with the shortest remaining time etc., and the chromosome $[\pi_1, \pi_2, \ldots, \pi_i, \ldots, \pi_{nm}]$ holds nxm indices of the rules defined in the array (Gen et al., 2008). Table 2 shows an example of the rules that can be used in the encoding :

Table 2 An example of the rules table

Rule Index	Rule
1	Shortest processing time
2	Longest processing time
3	Shortest remaining time
4	Longest remaining time
5	Latest start time
6	Latest finish time

Fig. 5 A Chromosome encoded based on Priority-rule based Representation

Mattfeld and Bierwirth (2004) defined the GA solutions by priorities between any two operations to solve job shop scheduling problems with release and due-dates. The tardiness was considered as objective function. They aimed to reduce the complexity by narrowing the scope at the machine level through the schedule builder and to decompose a problem through a multi-stage approach to focus on the long-term planning at the shop-floor level. Precedence preservative crossover operator (PPX) and the delete-insert mutation operators were used. Morton and Pentico's 48 scheduling problems provided with the Parsifal software package was used to validate the approach. Proposed GA model was compared to probabilistic scheduling. Comparing both GA variants, the active GA performed superior for three criteria (weighted mean tardiness, the maximum tardiness, the weighted number of tardy jobs). For minimizing the weighted mean flowtime of jobs the non-delay GA is clearly advantageous. It was stated that GA failed to explore the larger space of active schedules. The standard deviation of the GA was small compared to with probabilistic scheduling. The active GA was said to be more time consuming on average than its non-delay counterpart.

4.3 Preference-List Based Representation

A chromosome encoded with preference-list based representation proposed by Davis (1985) consists of *m* sub-chromosomes with length *n* for an *nxm* JSP. A sub-chromosome is a preference list rather than an operation sequence on the machine as shown in Figure 6. An example 3x3 JSP is given in Table 3 (Ponnambalam et al., 2001):

Table 3 An Example 3x3 JSP (Ponnambalam et al., 2001)

J_j	$M_r(p_{jr})$		
1	1(3)	2(3)	3(2)
2	1(1)	3(5)	2(3)
3	2(3)	1(2)	3(3)

Fig. 6 A Chromosome encoded based on Preference-List based Representation

The first triple is the preference-list for the first machine, and the second triple belongs to the second machine and, so forth. In order to create a schedule, the first preferred operations on each machine are considered. That is the J_2 for machine 1, J_1 for machine 2 and J_2 for machine 3. Among these operations only J_2 on machine 1 can be dispatched due to the operator precedence constraints given in Table 3. Once J_2 on machine 1 is scheduled on machine 1, the next operation to be dispatched is J_2 on machine 3. Now, J_3 on machine 1 and J_1 on machine 2 are in the preference list queue. However, they can not start due to the constraints. Hence, the next operations on the machines are taken into preference list. J_1 on machine 1 and J_3 on machine 2 are scheduled and then, J_3 on machine 1 and J_1 on machine 2 can be dispatched now since their job precedences have been completed. And J_1 on machine 3 and J_2 on machine 2 are dispatched and finally J_3 is scheduled on machine 3 (Ponnambalam et al., 2001). The schedules dispatched according to the priority list always lead to a feasible schedule since no illegal move is allowed.

4.4 Completion Time Based Representation

Completion time-based representation proposed by Yamada and Nakano (1992) consists of the completion times of the operations as shown in Figure 7 where each

O_{ijk} represents the completion time of the associated operation k of job i on machine j. It may require to use a specially designed genetic operators and extra computational computation to obtain valid solutions.

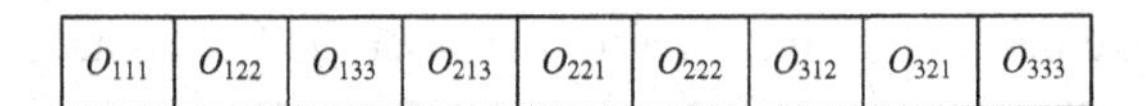

O_{111} O_{122} O_{133} O_{213} O_{221} O_{222} O_{312} O_{321} O_{333}

Fig. 7 A Chromosome encoded based on Completion Time based Representation

4.5 Random Keys Representation

Random keys representation proposed by Bean (1994) can lead genetic operators to produce feasible and valid solutions without a repairing mechanism. Bean and Norman (1993) developed a GA based random keys representation for JSPs. In a chromosome encoded by random keys representation, each gene consists of an integer number in $1, m$ which corresponds to the machine number and a fractional part in $(0, 1)$. The indices of fractional parts after a sort in ascending order gives sequences on the machines. Assuming that jobs are processed at their earliest possible time, Norman and Bean (1995) incorporated non-zero ready times and tool availability procedures into their GA, and in order to handle with constraints they presented an algorithm which schedules a job by avoiding any possible local left-shift on its machine.

Consider the chromosome given in Figure 8. Sorting the keys for machine 1 in ascending produces the job sequence [3 2 1], the job sequence [2 1 3] for machine 2, and the job sequence [1 2 3] for machine 3. As seen from the Figure 8, the job sequences may violate the precedence constraints.

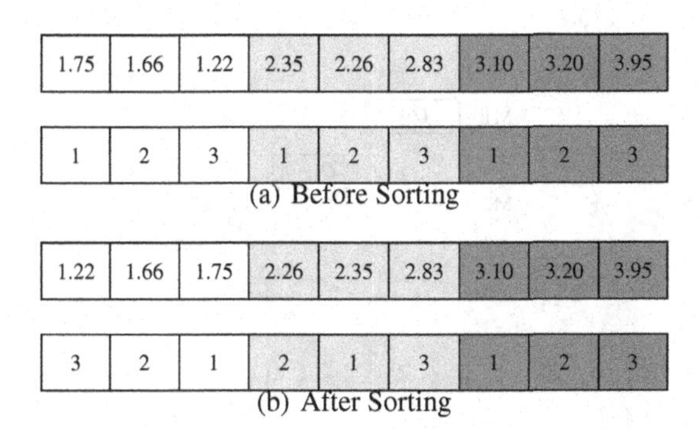

Fig. 8 A Chromosome encoded based on Random Keys Representation

4.6 *Job Permutation Representation*

In the job permutation representation (Giffler and Thompson, 1960), a chromosome of n jobs is scheduled by first assigning the operations of first job in the chromosome and then the the operations of second job, so forth. An operation is assigned to a machine to make best processing time for the associated machine. An example of job-based representation for a 3x3 JSP is shown in Figure 9.

Table 4 An Example 3x3 JSP

J_j	$M_r(p_{jr})$		
1	1(3)	2(3)	3(3)
2	1(2)	3(3)	2(4)
3	2(3)	1(2)	3(1)

J_1	J_3	J_2

Fig. 9 A chromosome encoded based on job permutation representation

According to the chromosome given in Figure 9 for a 3x3 JSP given in Table 4, first job is dispatched and then the third job and finally the second job. The operation precedence of the first job is [M1 M2 M3] according to the Table 4. All operations of the first job are dispatched depending on their processing times given in Table 4 is shown in Figure 10. Then the third job is scheduled on machines [M2 M1 M3] based on the precedences given in Table 4. The Gantt chart is shown in Figure 11 after the third job was scheduled. Finally, the second job is dispatched based on the machine sequence [M1 M3 M2] and corresponding processing times [2 3 4] as shown in Figure 12. As seen from the Gantt chart in Figure 12, the makespan of the schedule is 16.

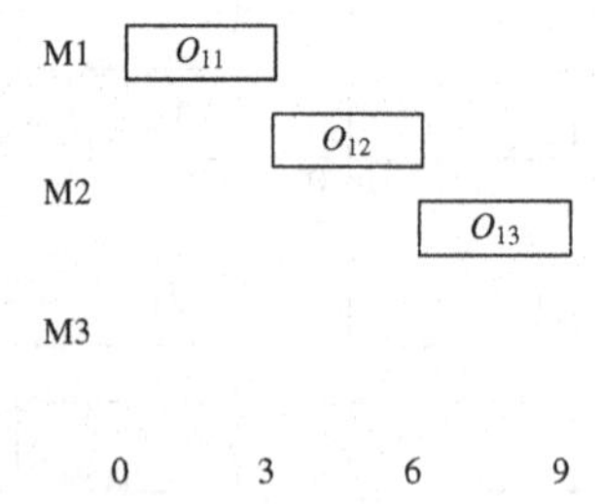

Fig. 10 Gantt Chart after the first job was dispatched

Bierwirth (1995) proposed a generalised-permutation genetic algorithm and Bierwirth et al. (1996) analysed three crossover operators preserving the relative, position and absolute permutation order of operations.

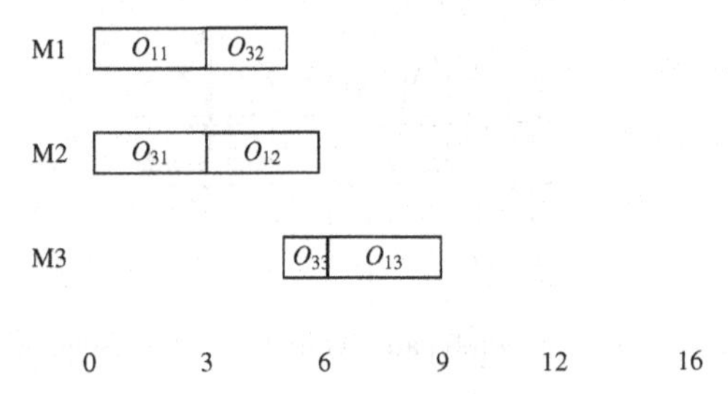

Fig. 11 Gantt Chart after the third job was dispatched

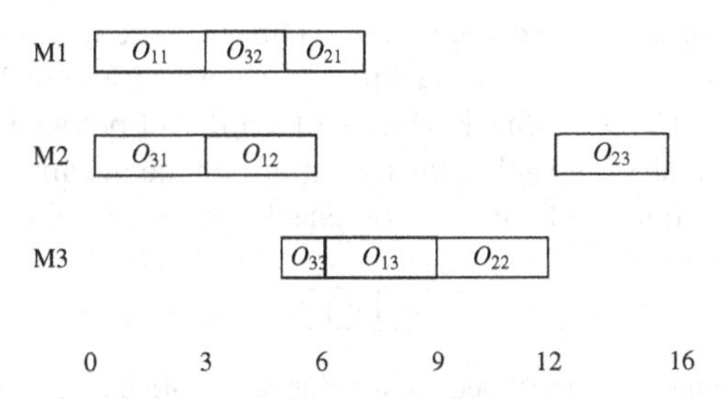

Fig. 12 Gantt Chart after the second job was dispatched

4.7 Job Pair Relation Based Representation

In the job pair relation-based representation, to encode a schedule a solution is represented by a binary string where each entry is the order of a job pair (i, j) for a machine (Cheng et al., 1999; Ponnambalam et al., 2001). Each value in the matrix is determined as follows (Eq. 2):

$$x_{ijm} = \begin{cases} 1, & \text{if job } \mathrm{J_i} \text{ is processed before job } \mathrm{J_j} \text{ on machine m} \\ 0, & \text{otherwise} \end{cases} \qquad (2)$$

If x_{ijm} is 1, it means that the job J_i must be processed before J_j on machine m. A job with the maximum number of ones has the highest priority for the machine. The length of a chromosome is $M\text{x}(N-1)\text{x}N/2$ where N is the number of jobs and M is the number of machines. When traditional crossover and mutation operators are used, this representation may be helpful (Hassan et al., 2009). However, it is complex and redundant, and a repair function or a penalty function is needed due to the illegality of the chromosomes produced (Kleeman and Lamont, 2007). An example of a chromosome encoded by job pair relation based representation is given in Figure 13:

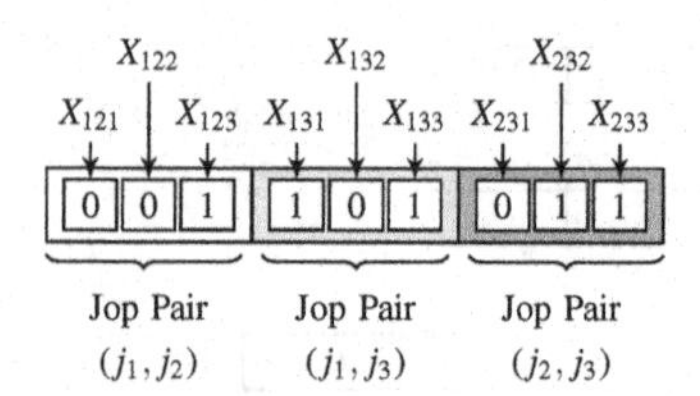

Fig. 13 A chromosome encoded by job pair relation based representation

4.8 Machine Permutation Representation

In the machine permutation representation, a chromosome with a length of the number of machines is a sequence of machines as shown in Figure 14 for a three machine JSP. The shifting bottleneck heuristic identifies a bottleneck machine in the sequence and optimally dispatches the operations on the machine based on the time to process all operations and the next bottleneck machine is determined.

M_1	M_3	M_2

Fig. 14 A chromosome encoded based on machine permutation representation

4.9 Operation-Based Representation

In the operation-based representation, a chromosome is a sequence of operations which leads to a schedule. All operations of the same job are labeled with the same symbol and its converted to an operation sequence according to the number of occurrences of a symbol (Gen et al., 1994). An example of operation based representation is shown in Figure 15 for a 3x3 JSP. The job sequence is interpreted as an operation sequence considering the order of occurrences. For example, the first symbol belongs to job 2 and its the first occurrence of an operation of job 2, it is considered as O_{21}. If we use the problem given in Table 4, depending on the technological sequence matrix, O_{21} is assigned to machine 1. The second symbol belongs to job 3 and it is the first occurrence of job 2, the second symbol is considered as $O31$ and assigned to machine 2. The third symbol is an operation of job 3 and it is the second occurrence of job 3, it is interpreted as O_{32} and assigned to machine 1. All the operations are interpreted in the same manner and scheduled on the machines. It is obvious that a chromosome encoded based on operation based representation always yields a feasible schedule and it is easy to implement.

4.10 Parallel Jobs Representation

Mesghouni et al. (1997) applied evolutionary programs to minimize the makespan of a job shop scheduling. Since binary representation does not consider some constraints such as precedence and resources constrains, they offered a new representation for the chromosomes, which respects the different constraints of the flexible job

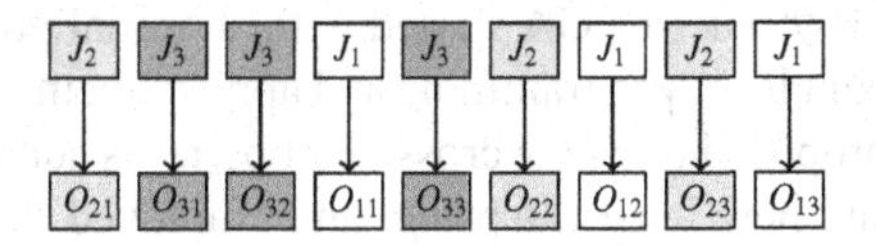

Fig. 15 A chromosome encoded based on operation based representation

shop problem. This representation, called Parallel Jobs Representation(PJsR), was defined by a matrix where each row is an ordered series of the operating sequence of this job. Each element of the row contain two terms: the machine index and the starting time of the operation if the assignment of this machine on this operation is definitive.

Table 5 Parallel Jobs Representation

J_1	(M_{k1}, T_{k1})	(M_{k2}, T_{k2})	...
...	(M_{k3}, T_{k3})	(M_{k1}, T_{k1})	...
J_n	(M_{k4}, T_{k4})	...	...

Mesghouni et al. (1997) proposed two crossover operators called row crossover and column crossover. The mutation operator used in the approach selects an operation at random and reassigns it to another machine. The minimization of makespan was the objective function. The initial population is generated from the solutions that do not violate the precedence constraints. They solved a problem with 10 machines and 10 jobs each with 3 operations. The proposed parallel genetic operators are claimed to be suitable to JSP and effective on this kind of problems.

4.11 Parallel Machines Representation

Mesghouni et al. (1999) proposed parallel machine encoding which provides feasible schedule. A chromosome represented by parallel machine representation (Table 6) consists of a set of machines each consisting of the operations represented by three numbers: the job number, the operation order and starting time of the associated operation which is calculated considering the precedence and resources constraints.

Table 6 Parallel Machines Representation

M_1	$(i^1, j^1, t^1_{i,j,M_1})$	...
M_i	$(i^i, j^i, t^i_{i,j,M_i})$	...
M_m	$(i^m, j^m, t^m_{i,j,M_m})$	...

Mesghouni et al. (1999) used a GA started with an initial population of the solutions given by constraint logic programming and applied them genetic operators and priority rules. They proposed two new crossover operators adapted to the encoding, which always generate new legal offspring, and employed the assigned mutation and the swap mutation. They solved a problem with 10 machines and 10 jobs each with 3 operations and concluded that parallel machines and parallel jobs representations combined with proposed genetic operators are suitable and effective for the job-shop scheduling problem.

4.12 Substring Representation

Wu and Li (1996) used an encoding in which a solution was a large string (Figure 16) made up by several sub strings each of which stands by a machine. In Figure 16, the substring for kth machine is shown where PRT_{ki} is a step of a part is processed by the kth machine ,and $IDTM_{ki}$ is the waiting time of the kth machine to processed a part. For a problem with m machines, the whole string is composed by m substrings.

PRT_{k1}	PRT_{k2}	$\cdots$	PRT_{ki}	$\cdots$	PRT_{kn}	$IDTM_{k1}$	$IDTM_{k2}$	$IDTM_{ki}$	$\cdots$	$IDTM_{ki}$	$\cdots$	$IDTM_{kn}$

Fig. 16 A substring for a machine in substring representation

Wu and Li (1996) used some genes of idle time of machines to narrow the solution space in order to get a completed determined solution. Because the solutions may not be feasible, a penalty term was introduced into the cost function to handle with the constraints. Chen et al. (1999) used two string representation where one string contains a list of all operations of all jobs and machines selected for corresponding operations while the second string contains a list of operations on each machine. The approach requires a feasibility maintenance mechanism because after crossover and mutation decoded solutions may not be feasible due to their representation. In the encoding, each individual consists of two substrings, substring A which defines the routing policy of the problem and substring B which defines the sequence of the operations on each machine, as shown in Figure 17.

4.13 Operations Machines Coding

Kacem et al. (2002) employed an assignment table which provides feasible solutions after crossover and mutation, integrates the notion of the assignment schemata and enables the exchange of information contained in current good solutions. The coding gives the execution of the operations from the rows and the tasks of each machine with the starting and completion times from the columns. They utilized the domain knowledge in the mutation operation. One drawback of the assignment table is its space complexity.

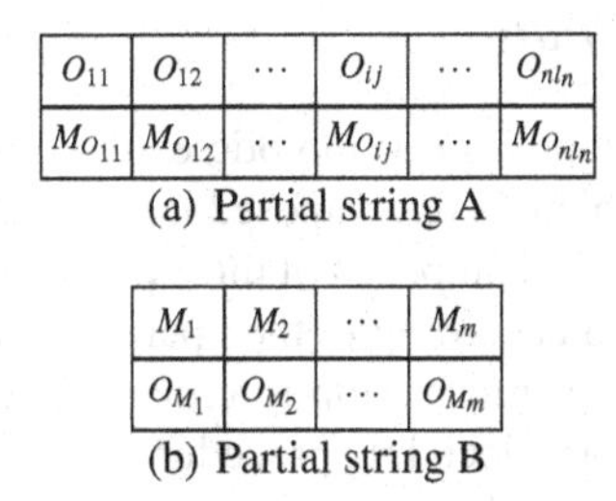

Fig. 17 Chen et al. (1999)'s substring representation

4.14 Complex Number Representation

Gu et al. (2009) and Gu et al. (2010) presented a parallel and co-evolutionary quantum genetic algorithms for stochastic job shop scheduling problem. Q-bit representation based on complex numbers was used to represent a linear superposition of solutions. Because of the complex numbers, the representation cannot be used directly and is converted into binary, decimal and job shop sequence, respectively. The state of a qubit can be represented as given in (3):

$$|\psi\rangle = \alpha\,|0\rangle + \beta\,|1\rangle \tag{3}$$

where α and β are complex numbers that specify the probability amplitudes of the corresponding states. As a string of l Q-bits, a Q-bit individual is defined by (4)

$$\begin{bmatrix} \alpha_1 & \alpha_2 & \dots & \alpha_l \\ \beta_1 & \beta_2 & \dots & \beta_l \end{bmatrix} \tag{4}$$

where $|\alpha_i^2| + |\beta_i^2| = 1$ and $i = 1, 2, \dots, l$.

Using a conversion mechanism, q-bit individual is converted to a job permutation string. Let η be a random number generated from the uniform distribution [0,1]. If α_i from Q-bit individual $P_Q^i(t)$ satisfies $|\alpha_i^2| > \eta$, then a bit of the binary string $X_i(t)$ is set to 1, otherwise set to 0. Every bits of binary string formed is transferred into a decimal number, and then a decimal string D(t) of length n is obtained. Permutation of D(t) is ordered in ascending and the job shop code is obtained.

Gu et al. (2009) employed cyclic crossover while Gu et al. (2010) employed two-point crossover and a not gate mutation operator. They also used a catastrophe operator to avoid premature convergence which occurs when the evolution of the best solution gets stuck in some consecutive generations. They aimed to minimize the makespan of mt06 (6x6), mt10(10x10) and mt20(20x5) problems and compared the proposed algorithm with GA and quantum GA, and the proposed algorithm with its variants using two and three universes. The proposed algorithm was said to be able to generate optimal or near-optimal solutions with fast convergence speed and to be suitable for stochastic JSPs with large number of machines, parts and operations.

4.15 Hybrid Representation

Yan and Hongze (2009) proposed a symbiotic evolutionary algorithm in which a flexible job scheduling problem is decomposed into two sub different problems. In the approach, a chromosome is composed of three parts: operation permutation, machine permutation generated according to the operation permutation and successor sequence which contains the information to choose successors. In decoding of a solution to a schedule, a hybrid decoding procedure was adopted. A new neighbourhood multi-parent crossover(NMX) operator was proposed in this paper because the traditional crossover cannot be applicable to inherit the characteristics of the chromosomes in the neighborhood. Different mutation operators were applied for operation and machine parts of a chromosome. They solved 24 problems from the literature and and the results of the algorithm on the problems were compared to those of other studies. The proposed approach was said to be suitable for solving FJSPs.

4.16 Three Dimensional Encoding

Wang et al. (2009) introduced a three-dimensional chromosome syntax, to solve the job shop problem, Each chromosome consists of a number of m square (*nxn*) matrices called a jobjob matrix, where m is the number of machines and n is the number of jobs as shown in Figure 18. Remaining elements are filled with -1 and then If job n1 is processed before job n2, then the cell at row n1, column n2 is filled with a 1, and the cell at row n2 column n1 is filled with a 0. The column sums for each machine are summarized in a three-dimensional matrix, with the first (horizontal) dimension representing the jobs, the second (vertical) representing the machines, and the third representing the chromosomes. This chromosome is used for creating the schedule. A union crossover operator takes two parent chromosomes and creates two child chromosomes in three steps. The three-dimensional encoding genetic algorithm (3DGA) was compared to standard branch and bound (BB), shifting bottleneck (SB), and tabu search (TS) in literature and obtained the minimum makespan on five instances of JSP (10x10, 20x15, 50x10).

5 Crossover Operators

In EAs, a pair of parent solutions are selected from the population and new offspring solutions are generated using genetic operators. The crossover operator is required to inherit features of the parents as much as possible but it should also explore new patterns of permutation. In this section crossover operators proposed for combinatorial problems and used with JSPs will be described.

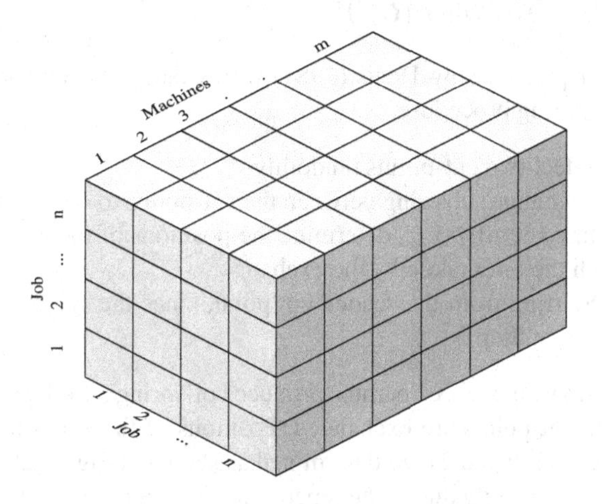

Fig. 18 Three Dimensional Encoding

5.1 Partial-Mapped Crossover (PMX)

PMX (Goldberg and Lingle, 1985) was initially proposed to solve Travelling Salesman Problem. It works like a two-cut-point crossover, which has a repairing procedure based on mapping in order to resolve the illegality of the offsprings.

- Step 1: Select two cut-points randomly
- Step 2: Exchange substring between the cut-points to create offsprings and determine the mapping relations
- Step 3: Apply the mapping relations to other parts of the offsprings to legalize the offspring

Figure 19 shows an example of the crossover operator which creates offsprings using PMX operator. In the example, (6 $\rightleftarrows$ 3), (4 $\rightleftarrows$ 9) and (5 $\rightleftarrows$ 2) are mapped, and the offsprings are legalized by applying this mapping after exchanging the substrings.

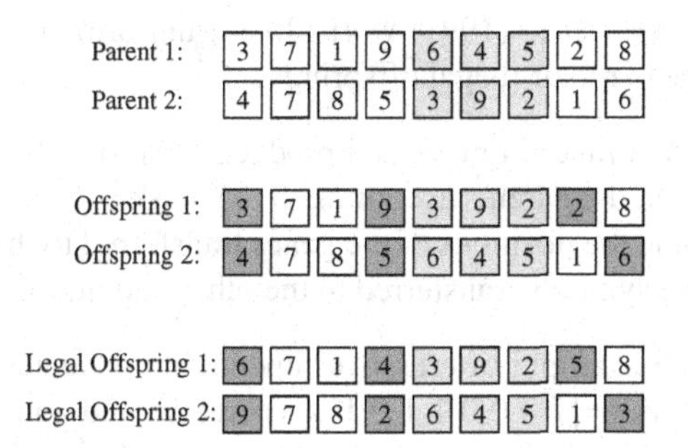

Fig. 19 An example of how PMX operator works

5.2 *Order Crossover (OX)*

OX operator proposed by Davis (1985) works like PMX operator but it applies a different repairing procedure.

- Step 1: Select two cut-points randomly
- Step 2: Exchange substring between the cut-points to create offsprings
- Step 3: In each offspring, determine the positions of the symbols which are the same with the ones taken by the exchange.
- Step 4: Starting from the second cut point, place the symbols into the positions determined in Step 3.

Figure 20 shows how OX operator produces offsprings. In the example, substrings between the cut-points are exchanged. Positions of the genes transferred from the second parent (3, 9 and 2) are determined as shown in Figure 20 labelled with X in order to avoid illegality due to the repetition of the same symbols. To these positions labelled with X, the genes transferred to the other individual (6, 4 and 5 for the first offspring) are placed as in the order of the substring starting from the second cut point. This repairing procedure is repeated for the second procedure, as well.

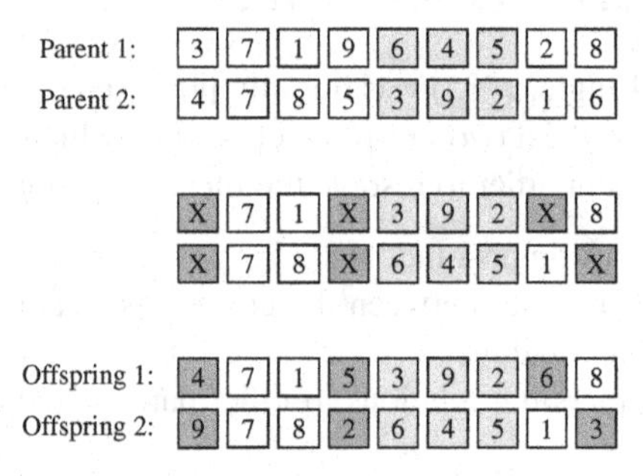

Fig. 20 An example of how OX operator works

5.3 *Position-Based Crossover (PBX)*

PBX proposed by Syswerda (1991) works like a uniform crossover operator and applies a procedure to repair illegal offsprings.

- Step 1: Generate a random mask and produce offsprings by applying multi-cut exchange between the parents according to the mask
- Step 2: Determine the positions of the genes transferred to the other individual
- Step 3: Place the symbols transferred to the other parent, from left to right

Figure 21 shows an example of crossover operation conducted by PBX operator. In the example, a mask=[1 0 0 0 1 1 0 1 0] is created and a multi-cut crossover is applied to the parents in the positions where the mask element is 1. The symbols in the positions where the mask is 0 are labelled with X. In the repairing procedure, the symbols transferred from the other parent (4, 3, 9 and 1 for the first offspring) are deleted. [371964528]− > [76528]. The remaining symbols (7, 6, 5, 2 and 8) are inserted instead of Xs in the same order to preserve the ordering in the parent.

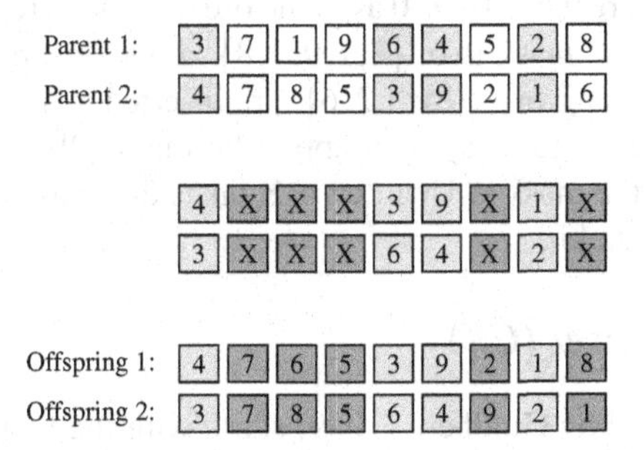

Fig. 21 An example of how PBX operator works

5.4 Order-Based Crossover (OBX)

Order-based crossover proposed by Syswerda (1989) is a slight variation of PBX in that the order of symbols in the selected position in one parent is imposed on the corresponding ones in the other parent.

- Step 1: Generate a random mask
- Step 2: Take symbols from parent 2 where the mask is 1 to produce the first offspring and take symbols from parent 2 where the mask is 0 to generate the second child.
- Step 3: Place the missing symbols in child 1 and 2 in the order they were found in the parent 1 and 2, respectively.

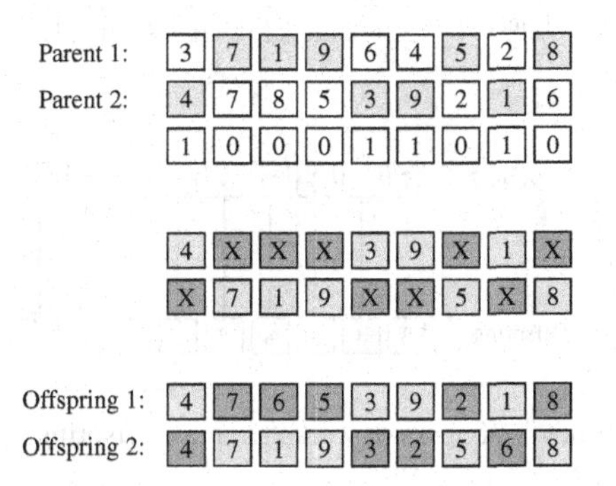

Fig. 22 An example of how OBX operator works

Figure 22 shows an example of crossover operation using by OBX operator. In the example, a mask=[1 0 0 0 1 1 0 1 0] is created and the symbols where the mask is 1, $\{4,3,9,1\}$, are taken from parent 2 to create the first child. The first child's missing symbols at the positions where the mask is 0 are $\{7,6,5,2,8\}$ are inserted into the positions where the mask is 0 as in the order in the parent 1 from left to right. The symbols where the mask is 0, $\{7,1,9,5,8\}$, are taken from parent 1 to create the second child. Missing ones, $\{4,3,2,6\}$, are inserted as in the order of parent 2. OBX and PBX differ in the way they apply the mask. PBX swap the symbols at the positions where the mask is 1 while OBX takes the symbols of only one parent depending on the mask bit.

5.5 Cycle Crossover (CX)

CX operator proposed by Goldberg (1989) starts with the first position of parent1. The entry in the position is copied to the first offspring and the entry in the position of parent2 points out the next symbol in parent1. This cycle continues until parent2 points out an already placed symbol. Once a cycle is completed, a new cycle starts with the first unused symbol of the other parent (Kleeman and Lamont, 2007).

Steps given below are repeated for both parent 1 and parent 2 until the offsprings are fully constructed.

- Step 1: Assign 1 to position p1 of parent1.
- Step 2: Let s1 be the symbol in position p1. If s1 is not used, copy s1 into position p1 of offspring1. Otherwise assign the position of the first unused symbol of the other parent to p1 and the symbol to s1.
- Step 3: Go to the position p1 of parent2. Let s2 be the symbol in position p1 of parent2.
- Step 4: Find the position p2 where s2 is in the parent1.
- Step 5: Assign p2 to p1 and goto Step 2.

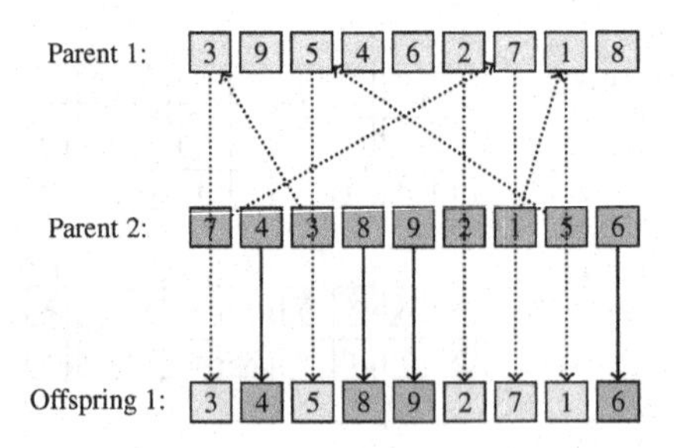

Fig. 23 An example of how CX operator works to create offspring 1

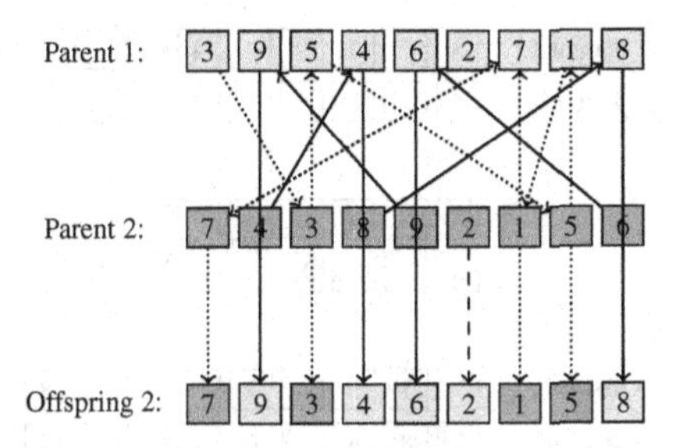

Fig. 24 An example of how CX operator works to create offspring 2

Figure 23 and 24 show how offsprings are created based on CX operator. In Figure 23, the operation starts with the symbol (s1=3) in the first position (p1=1). Since s1 is not used before, it is copied to the position p1 of offspring1. The position of the symbol s2 (s2=7, p1=1) which is in the position p1 of parent2 is assigned to p2. Because s2=7 is in seventh position, p2 is set to 7. In the seventh position of parent2, symbol 1 stands. Symbol 1 points out eighth position of parent1 and symbol 1 is copied into eighth position of the offspring1. Because the symbol 5 in eighth position of parent2 was used before, the cycles again starts with the first unused symbol of the other parent. Parents are switched when a used symbol is met. In order to create the second offspring, CX operator starts with the first symbol of parent2 as shown in Figure 24.

5.6 Linear Order Crossover (LOX)

Linear order crossover (LOX) (Falkenauer and Bouffouix, 1991) works similar to the order crossover operator but LOX does not treat the chromosome in a cyclic fashion unlike OX and handles a chromosome as a linear entity.

- Step 1: Select two cut-points randomly
- Step 2: Exchange substring between the cut-points to create offsprings
- Step 3: Place the other symbols of parent 1 into the other positions of offspring 1 from left to right to preserve the relative ordering, start and end points. Repeat Step 3 for offspring 2 by taking the symbols from parent 2.

Figure 25 shows an example of how LOX operator works. In the example, the symbols between the positions 5 and 7 are exchanged. The symbols [7, 1, 6, 4, 5, 8] coming from parent 1 are inserted into offspring 1 from left to right and the symbols [7, 8, 3, 9, 2, 1] of parent 2 are inserted into offspring 2 as in the same manner.

5.7 Subsequence Exchange Crossover (SXX)

SXX operator proposed by Kobayashi et al. (1995) is an extension of exchange operator proposed for TSP, and it is used with job sequence matrix encoding in

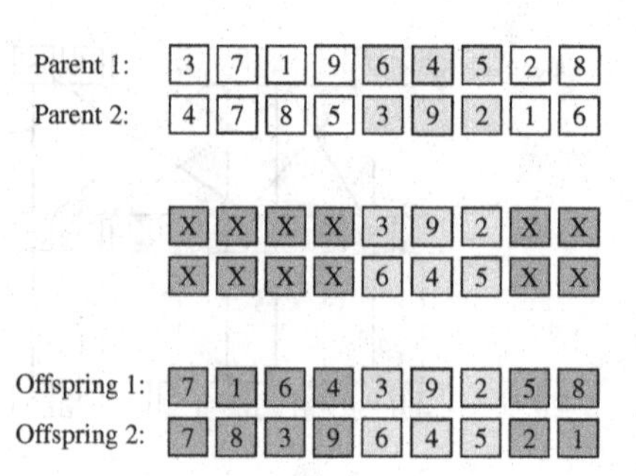

Fig. 25 An example of how LOX operator works to create offsprings

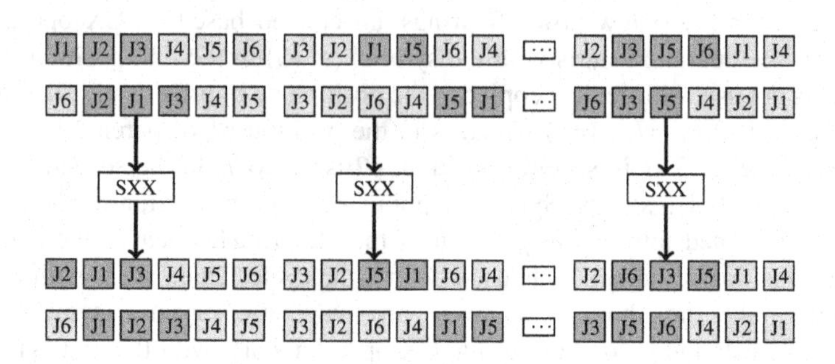

Fig. 26 An example of how SXX operator works to create offsprings

which each row is an operation sequence for each machine. In order to produce offsprings, if the symbols of a substring in parent1 occurs consequently in parent2, these substrings are exchanged between two parents.

- Step 1: Identify subsequences one for one machine for the parents.
- Step 2: Exchange these subsequences among parents to create offspring.

Figure 26 shows how SXX operator works. In the figure, each parent holds the operation sequences on each machine. SXX operator is applied to each operation sequence separately. In the first operation sequences of parent1 and parent2, the subsequence $J_1J_2J_3$ of parent1 consequently occurs in the parent2 as the sequence $J_2J_1J_3$. Therefore, these subsequences are exchanged to form the first substrings of the offsprings. The SXX operator is repeated on the operation sequences of other machines, and then the offsprings are produced. It is obvious that, the offsprings created do not need repairing procedure after SXX crossover.

5.8 Partial Schedule Exchange Crossover (PSXX)

PSXX operator proposed by Gen et al. (1994) is used with operation-based encoding. Partial sequences randomly selected from the parents are exchanges and a repairing mechanism is applied to make the offsprings legal. Steps of the how the operator is applied are given below:

- Step 1: Randomly select partial operation subsequences from parent1 and parent2
- Step 2: Swap these partial subsequences
- Step 3: Delete the exceeding genes out of the cut points
- Step 4: Insert missing genes after the second cut point

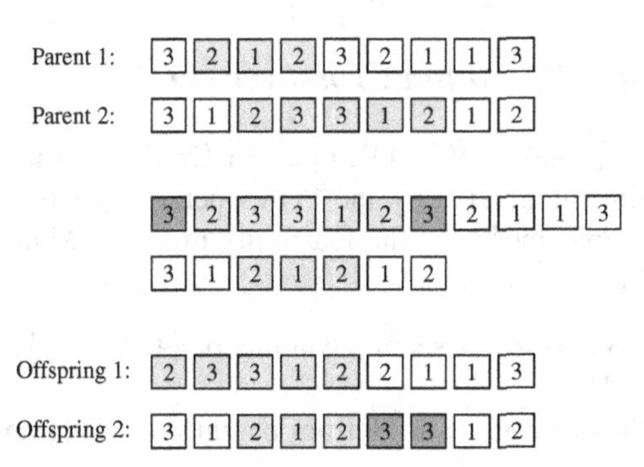

Fig. 27 An example of how PSXX operator works to create offsprings

How offsprings are created by PSXX operator is illustrated in Figure 27. In the figure each parent is a sequence of operations for a 3-jobs and 3-machines problem. In the figure, the partial subsequence $[2,1,2]$ from parent1 and the partial sequence $[2,3,3,1,2]$ are randomly selected. These partial subsequences are swapped between parent1 and parent2 and the offsprings [3,2,3,3,1,2,3,2,1,1,3] and [3,1,2,1,2,1,2] are created. But these offsprings are not valid because the first offspring has two exceeding genes $\{3,3\}$ and the second offspring has two missing genes $\{3,3\}$. In the first offspring, symbols $\{3,3\}$ are deleted out of the cut points and missing symbols in the second offspring $\{3,3\}$ are inserted after second cut point to make the offsprings valid.

5.9 *Precedence Preservative Crossover (PPX)*

PPX is suggested by Bierwirth et al. (1996) to be used for JSPs. A mask is generated randomly with the elements of set 1,2. This mask decides the parent whose unused leftmost gen will be transferred. The main steps of the PPX operator are given below:

- Step 1: Generate a random mask with the elements of set 1,2
- Step 2: Take unused leftmost symbols from parent 1 where the mask is 1 to from parent 2 where the mask is 2.

An example of crossover using PPX operator is shown in Figure 28.

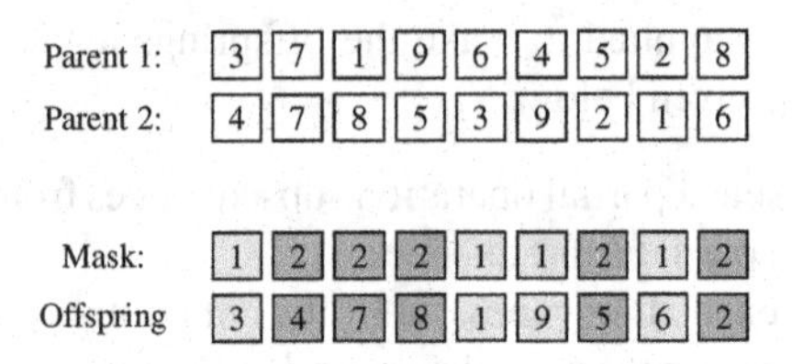

Fig. 28 An example of how PPX operator works

5.10 Precedence Operation Crossover (POX)

Zhang et al. (2005) proposed the POX operator for the operation-based representation. The offsprings generated by the POX operator satisfy the characteristics-preservingness, completeness and the feasibility properly. Main steps of the POX operator are as follows:

- Step 1: Randomly choose the set of job numbers, $\{1,2,\ldots,n\}$, into one nonempty exclusive subset J1.
- Step 2: Copy those numbers in J1 from parent1 to child1 and from parent2 to child2, preserving their locus.
- Step 3: Copy those numbers in J1, which are not copied at step 2, from parent2 to child1 and from parent1 to child2, preserving their order.

Figure 29 gives an example of how the POX operator works.

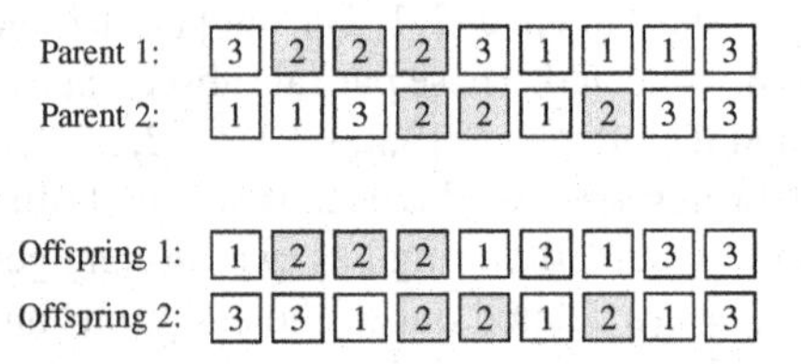

Fig. 29 An example of how POX operator works

6 Mutation Operators

Mutation operator which is a random walk through the search space provides diversity in genetic information of the population in order to prevent the premature convergence, and assures all the points in search space to be likely searched.

6.1 Swap Mutation

The swap mutation operator randomly selects two random positions (p1,p2) and swaps the symbols in these positions. Figure 30 shows how swap operator works. The symbols $\{1,5\}$ in the positions randomly selected are swapped.

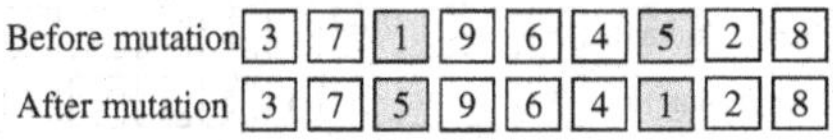

Before mutation 3 7 1 9 6 4 5 2 8
After mutation 3 7 5 9 6 4 1 2 8

Fig. 30 An example of how swap mutation works

6.2 *Shift Mutation*

The shift mutation operator randomly selects a symbol and then shifts left or right it s times where s is a random integer.

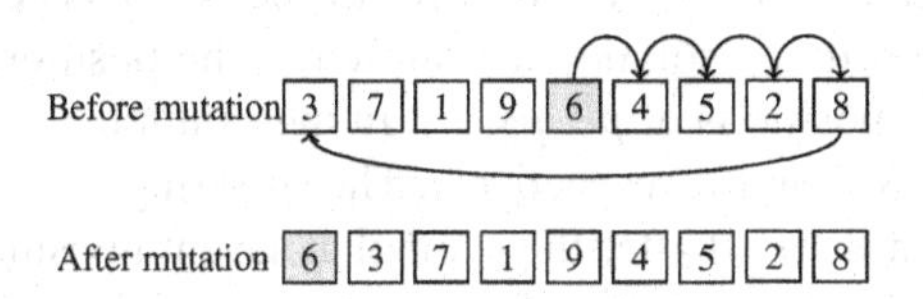

Fig. 31 An example of how shift mutation works

6.3 *Inversion Mutation*

The inversion mutation operator selects two points in a solution and changes the order of the symbols in the mutant solution. Symbols before and after the selected points are copied from the original solution to the mutant solution. Figure 32 shows an example of inversion mutation working on 9-jobs chromosome encoded based on job based representation. In the example third and sixth positions are randomly determined and the symbols between these positions [1 9 6 4] are inverted in the mutant solution as [4 6 9 1]. The other symbols out of the selected points remain the same.

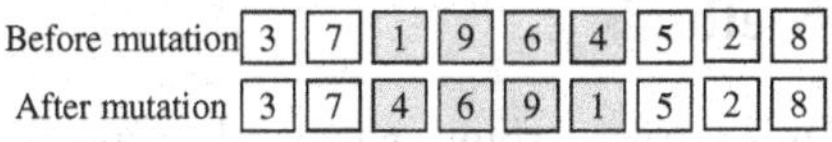

Before mutation 3 7 1 9 6 4 5 2 8
After mutation 3 7 4 6 9 1 5 2 8

Fig. 32 An example of how inversion mutation works

6.4 *Insertion Mutation*

The insertion mutation operator randomly selects two positions (p1,p2). The symbols between the positions p1+1 and p2 are shifted to the previous position and the symbol in the position p1 is assigned to the position p2. The other symbols out of the positions p1 and p2 remain the same. Figure 33 shows an example of insertion mutation working on 9-jobs chromosome encoded based on job based representation. In the example, third (p1) and seventh (p2) positions are determined. The symbols between the positions p1+1 and p2 [9 6 4 5] are shifted to previous positions and the symbol previously in the position p1 [1] is assigned to the position p2.

Before mutation	3	7	1	9	6	4	5	2	8
After mutation	3	7	9	6	4	5	1	2	8

Fig. 33 An example of how insertion mutation works

6.5 Displacement Mutation

The displacement mutation operator randomly selects three points ($p_1 < p_2 < p_3$) in a chromosome and then moves the substring between the positions p_1 and p_2 to the position after the symbol in the position p_3. The symbols between the positions p_2+1 and p_3 in the original solution are shifted to the positions starting from p_1. Figure 34 shows the positions of p_1, p_2, p_3. In the example given in Figure 34, p_1, p_2 and p_3 equals to 3, 5 and 7, respectively. The substring [1 9 6] between the third and fifth positions in inserted after the symbol at seventh position. The symbols [4 5] are shifted to the positions starting from the third position.

Before mutation	3	7	1	9	6	4	5	2	8
After mutation	3	7	4	5	1	9	6	2	8

Fig. 34 An example of how displacement mutation works

6.6 Assigned Mutation

Mesghouni et al. (1999) used an assigned mutation which randomly selects one chromosome and one operation and re-assigns this selected operation to another machine in the same position considering the precedence and resource constraints.

7 Local Search Methods

Local search methods move from a solution to another one in the neighborhood current solution using local changes based on defined rules. Local search methods are also used with EAs for solving JSPs to find improved solutions or active schedules. In this section some local search methods used for JSPs will be explained.

7.1 Giffler-Thompson Algorithm

The Giffler and Thompson (GT) algorithm (Giffler and Thompson, 1960) can generate an active schedule by scheduling operations while avoiding a long idle period. (Yamada, 2003). In EAs, GT algorithm is used as a crossover operator to generate active schedules. The notation used in GT algorithm is given below:

- $PJ(O)$: job predecessor
- $PM(O)$: machine predecessor

- $ES(O) := max\{c(PJ(O)), c(PM(O))\}$: The earliest starting time $ES(O)$ of O
- $EC(O) = ES(O) + p(O)$: The earliest completion time of O
- $O_{*r} = argmin\{EC(O)|O \in D\}$: The earliest completable operation
- $C[M_r, i] = \{O_{kr} \in D | ES(O_{kr}) < EC(O_{*r})\}$: a set of candidate operations for the next processing on M_r (i-1 have already been scheduled, ith operation)
- A subset of G that consists of operations processed on machine M_r is denoted as G_r.

At each step the GT algorithm generates a conflict set and selects the next operation to preserve operation precedence without violating the precedence constraints. The algorithm iterates until all operations are scheduled. Main steps of the GT algorithm are given in Algorithm 2.

begin
  $G = \{O_{1T_{11}}, O_{2T_{21}}, \ldots, O_{nT_{n1}}\}$;
  foreach $O \in G$ **do**
    $ES(O) = 0, EC(O) = p(O)$
  end
  Find the earliest completable operation with machine M_r,
  O_{*r} = arg min {EC(O)|O ∈ D};
  Calculate the conflict set $C[M_r, i] = \{O_{kr} \in D | ES(O_{kr}) < EC(O_{*r})\}$;
  Select one of operations in $C[M_r, i]$ randomly. Let the selected operation be O_{kr}.;
  Schedule $O_k r$ as the i-th operation on M_r; i.e. $S_{ri} = k$, with its starting and completion times equal to $ES(O_{kr})$ and $EC(O_{kr})$ respectively:
  $s(O_{kr}) = ES(O_{kr}); c(O_{kr}) = E(CO_{kr})$;
  foreach $O_{jr} \in G_r \setminus \{O_{kr}\}$ **do**
    $ES(O_{jr}) = max\{ES(O_{jr}); EC(O_{kr})\}$ and $EC(O_{jr}) = ES(O_{kr}) + p(O_{kr})$
  end
  Remove $\{O_{kr}\}$ from G and G_r;
  $(G = G \setminus \{O_{kr}\}, G_r = G_r \setminus \{O_{kr}\})$;
  if $O_{ks} \in G | O_{ks}$ *is next to* O_{kr} *in* T, *i.e.*, $r = T_{ki}$ *and* $i < m$ **then**
    Add O_{ks} to G, $s = T_{ki+1}$, $G = G \setminus \{O_{kr}\} \cup \{O_{ks}\}$
  end
  $ES(O_{ks}) = max\{EC(O_{kr}), EC(PM(O_{ks}))\}$;
  $ECb(O_{ks}) = ES(O_{ks}) + p(O_{ks})$
end

Algorithm 2: Giffler-Thompson Algorithm

7.2 *Variable Neighbourhood Search*

Variable neighborhood search (VNS) is a randomized local search algorithm which systematically changes the neighborhood using multiple trajectories, different neighborhood structures, various perturbations (Hansen and Mladenovic, 2001)

There are different variants of VNS such as forward VNS, backward VNS, VNS that accept worse solutions with some probability.

```
begin
    s0 ← GenerateInitialSolution, choose{N_k}, k = 1,... k_max;
    repeat
        s' ← RandomSolution(N_k(s*)) (shaking);
        s*' ← LocalSearch(s') ;
        if f(s*') < f(s*)(Move or not) then
            s* ← s*';
            k ← 1;
        else
            k ← k+1;
        end
    until termination criteria;
end
```

Algorithm 3: Basic variable neighbourhood search

7.3 Hill Climbing

Hill climbing uses steepest local move which performs a perturbation yielding the best improvement. While generating a neighbor solution, an exhaustive search is conducted to find the best move with the greatest improvement. For this reason, this method is computationally expensive.

7.4 Tabu Search

Tabu search (TS) proposed by Glover (1989) is used as a local search to improve a solution checking its immediate neighbors and using adaptive memory structures and intelligent decisions (Xhafa, 2007). Like the other local search techniques TS uses a neighborhood procedure ($N(s)$ to move current solution (s) to a better solution

```
begin
    s ← Generate Initial Solution, ŝ ← s;
    Determine the tabu criteria and aspiration conditions ;
    repeat
        Generate a subset N*(s) ⊆ N(s) which are not in the tabu list;
        Choose the best s' ∈ N*(s);
        if f(s') < f(ŝ)(Move or not) then
            ŝ ← s';
        end
        Update the recency and frequency lists;
        Update tabu list;
        Perform aspiration if needed;
    until termination criteria;
end
```

Algorithm 4: Tabu Search Algorithm

($\hat{s}$) until a termination criterion is satisfied (Algorithm 4). Unlike the other local search techniques, to avoid getting stuck in local minima and cycling among already visited solutions, TS uses an adaptive memory to memorize these regions and to store them in the tabu list for a certain number of iterations.

8 Conclusion

This chapter considers the principles of different representations, crossover and mutation operators and local search methods especially used for job shop scheduling problems.

Although many representations and operators have been given in the literature for JSPs and EAs using these representations and operators can find optimal solutions for some benchmark problems, there are some problems that can not be solved efficiently by EAs. Choosing the appropriate representation is important because an inappropriate representation will lead to inefficient results, or generated solutions may violate the precedence constraints JSPs have. Therefore, some representations may require repairing mechanisms to make the generated solutions feasible. It should be noted that the repairing mechanism has also affect an EA's performance and running time. Crossover and mutation operators are required to inherit some features of the parents and provide diversity to the population, respectively. The main issues related to them are efficiency and feasibility. EAs with robust representations giving feasible schedules and efficient crossover and mutation operators are still open research topics that should be studied in the future.

Local search techniques together with the search operators of EAs are used to search improved solutions in the neighborhood of solutions. Due to the costs of the local searches using an exhaustive search, local search techniques combined with intelligent decisions are preferred due to the solution quality and the running time.

Based on our detailed review, the scalability of EAs for JSP, which investigates the algorithm performance with respect to problem characteristics, for example, problem size, availability of precedence constraints, is suggested as the most challenging and important gap to fill in the existing literature of EAs for JSPs. From both experimental and theoretical perspective, a rigorous scalability analysis of EA's computation time on different JSP instance classes is needed to quantify the performance characteristics of EAs for JSPs.

Acknowledgement. This work was partially supported by an EPSRC grant (No. EP/I010297/1) and TUBITAK 2219 post-doctoral fellowship.

References

Adams, J., Balas, E., Zawack, D.: The shifting bottleneck procedure for job shop scheduling. Management Science 34(3), 391–401 (1988)

Bean, J.: Genetic algorithms and random keys for sequencing and optimization. ORSA Journal of Computing 6(2), 154–160 (1994)

Bean, J., Norman, B.: Random keys for jos shop scheduling, technical report 93-7. Technical report, Dept. of Industrial and Operations Engineering, University of Michigan (1993)

Bierwirth, C.: A generalized permutation approach to job-shop scheduling with genetic algorithms. OR Spektrum 17(2-3), 87–92 (1995)

Bierwirth, C., Mattfeld, D., Kopfer, H.: On permutation representations for scheduling problems. In: Ebeling, W., Rechenberg, I., Voigt, H.-M., Schwefel, H.-P. (eds.) PPSN 1996. LNCS, vol. 1141, Springer, Heidelberg (1996)

Chen, H., Ihlow, J., Lehmann, C.: A genetic algorithm for flexible job-shop scheduling. In: Proceedings. 1999 IEEE International Conference on Robotics and Automation, vol. 2, pp. 1120–1125 (1999)

Cheng, R., Gen, M., Tsujimura, Y.: A tutorial survey of job-shop scheduling problems using genetic algorithms: part ii. hybrid genetic search strategies. Comput. Ind. Eng. 37, 51–55 (1999)

Davis, L.: Applying adaptive algorithms to epistatic domains. In: Proceedings of the 9th International Joint Conference on Artificial Intelligence, vol. 1, pp. 162–164. Morgan Kaufmann Publishers Inc., San Francisco (1985)

Dorndorf, U., Pesch, E.: Evolution based learning in a job shop scheduling environment. Computers & OR 22(1), 25–40 (1995)

Falkenauer, E., Bouffouix, S.: A genetic algorithm for job shop. In: Proceedings of the 1991 IEEE International Conference on Robotics and Automation, pp. 824–829 (1991)

Gantt, H.L.: Work, Wages and Profits. The Engineering Magazine (1910)

Gen, M., Cheng, R., Lin, L.: Network Models and Optimization: Multiobjective Genetic Algorithm Approach (Decision Engineering). Springer (2008)

Gen, M., Tsujimura, Y., Kubota, E.: Solving job-shop scheduling problem using genetic algorithms. In: Proceedings of the 16th International Conference on Computer and Industrial Engineering, Ashikaga, Japan, pp. 576–579 (1994)

Giffler, J., Thompson, G.: Algorithms for solving production scheduling problems. Operations Research 8, 487–503 (1960)

Glover, F.: Tabu search - part 1. ORSA Journal on Computing 1(2), 190–206 (1989)

Goldberg, D.E.: Genetic Algorithms in Search, Optimization and Machine Learning, 1st edn. Addison-Wesley Longman Publishing Co., Inc., Boston (1989)

Goldberg, D.E., Lingle, J.: Alleles, Loci and the Travelling Salesman Problem. In: Proceedings of the 1st International Conference on Genetic Algorithms and Their Applications. Lawrence Erlbaum Associates, New Jersey (1985)

Gu, J., Gu, M., Cao, C., Gu, X.: A novel competitive co-evolutionary quantum genetic algorithm for stochastic job shop scheduling problem. Comput. Oper. Res. 37, 927–937 (2010)

Gu, J., Gu, X., Gu, M.: A novel parallel quantum genetic algorithm for stochastic job shop scheduling. Journal of Mathematical Analysis and Applications 355(1), 63–81 (2009)

Hansen, P., Mladenovic, N.: Variable neighborhood search: Principles and applications. European Journal of Operations Research 130, 449–467 (2001)

Hasan, S., Sarker, R., Essam, D., Cornforth, D.: A Genetic Algorithm with Priority Rules for Solving Job-Shop Scheduling Problems. In: Chiong, R., Dhakal, S. (eds.) Natural Intelligence for Scheduling, Planning and Packing Problems. SCI, vol. 250, pp. 55–88. Springer, Heidelberg (2009)

Kacem, I., Hammadi, S., Borne, P.: Approach by localization and multiobjective evolutionary optimization for flexible job-shop scheduling problems. IEEE Transactions on Systems, Man, and Cybernetics, Part C: Applications and Reviews 32(1), 1–13 (2002)

Kleeman, M.P., Lamont, G.B.: Scheduling of Flow-Shop, Job-Shop, and Combined Scheduling Problems using MOEAs with Fixed and Variable Length Chromosomes. In: Dahal, K.P., Tan, K.C., Cowling, P.I. (eds.) Evolutionary Scheduling. SCI, vol. 49, pp. 49–99. Springer, Heidelberg (2007)

Kobayashi, S., Ono, I., Yamamura, M.: An efficient genetic algorithm for job shop scheduling problems. In: Proceedings of the 6th International Conference on Genetic Algorithms, pp. 506–511. Morgan Kaufmann Publishers Inc., San Francisco (1995)

Mattfeld, D.C., Bierwirth, C.: An efficient genetic algorithm for job shop scheduling with tardiness objectives. European Journal of Operational Research 155, 616–630 (2004)

Mesghouni, K., Hammadi, S., Borne, P.: Evolution programs for job-shop scheduling. In: 1997 IEEE International Conference on Systems, Man, and Cybernetics, Computational Cybernetics and Simulation, vol. 1, pp. 720–725 (1997)

Mesghouni, K., Pesin, P., Trentesaux, D., Hammadi, S., Tahon, C., Borne, P.: Hybrid approach to decision making for job-shop scheduling. Prod. Plann. Contr. J. 10(7), 690–706 (1999)

Nakano, R., Yamada, T.: Conventional genetic algorithm for job shop problems. In: International Conference on Genetic Algorithms, ICGA 1991, pp. 474–479 (1991)

Norman, B., Bean, J.: Random keys genetic algorithm for scheduling:unabridged version, technical report 95-10. Technical report, Dept. of Industrial and Operations Engineering, University of Michigan (1995)

Phan, H.T.: Constraint Propagation in Flexible Manufacturing. Springer-Verlag New York, Inc. (2000)

Ponnambalam, S.G., Aravindan, P., Rao, P.S.: Comparative evaluation of genetic algorithms for job-shop scheduling. Production Planning and Control 12(6), 560–574 (2001)

Rothlauf, F.: Representations for evolutionary algorithms. In: Proceedings of the 2008 GECCO Conference Companion on Genetic and Evolutionary Computation, GECCO 2008, pp. 2613–2638 (2008)

Roy, B., Sussmann, B.: Note ds no 9 bis: Les probl'emes d'ordonnancement avec contraintes disjonctives. Technical report, SEMA, Paris (1964)

Syswerda, G.: Uniform crossover in genetic algorithms. In: Proceedings of the 3rd International Conference on Genetic Algorithms, pp. 2–9. Morgan Kaufmann Publishers Inc., San Francisco (1989)

Syswerda, G.: Schedule Optimization Using Genetic Algorithms. In: Handbook of Genetic Algorithms, pp. 332–349. Van Nostrand Reinhold, New York (1991)

Wang, Y., Yin, H., Wang, J.: Genetic algorithm with new encoding scheme for job shop scheduling. The International Journal of Advanced Manufacturing Technology 44, 977–984 (2009)

Widmer, M., Hertz, A., Costa, D.: Metaheuristics and Scheduling. In: Production Scheduling, pp. 33–68. Wiley (2008)

Wu, Y., Li, B.: Job-shop scheduling using genetic algorithms. In: Proc. IEEE Int'l Conf. on System, Man and Cybernetics. IEEE SMC 1996, vol. 3, pp. 1994–1999 (1996)

Xhafa, F.: A hybrid evolutionary heuristic for job scheduling on computational grids. In: Abraham, A., Grosan, C., Ishibuchi, H. (eds.) Hybrid Evolutionary Algorithms. SCI, vol. 75, pp. 269–311. Springer, Heidelberg (2007)

Yamada, T.: Studies on Metaheuristics for Jobshop and Flowshop Scheduling Problems. PhD thesis, Kyoto University (2003)

Yamada, T., Nakano, R.: A genetic algorithm applicable to large-scale job-shop problems. In: Parallel Problem Solving from Nature: PPSN II, pp. 281–290. North-Holland, Elsevier Science Publishers (1992)

Yan, Z., Hongze, Q.: A symbiotic evolutionary algorithm for flexible job scheduling problem. In: Second International Workshop on Computer Science and Engineering, WCSE 2009, vol. 1, pp. 79–83 (2009)

Zhang, C.-Y., Li, P., Rao, Y., Li, S.: A New Hybrid GA/SA Algorithm for the Job Shop Scheduling Problem. In: Raidl, G.R., Gottlieb, J. (eds.) EvoCOP 2005. LNCS, vol. 3448, pp. 246–259. Springer, Heidelberg (2005)

Multi-objective Grid Scheduling

María Arsuaga-Ríos and Miguel A. Vega-Rodríguez

Abstract. Grid computing is a distributed paradigm that coordinates heterogeneous resources using decentralized control. Grid computing is commonly used by scientists for executing experiments. Scheduling jobs within Grid environments is a challenging task. Scientists often need to ensure not only a successful execution for their experiments but also they have to satisfy constraints such as deadlines or budgets. Both of these constraints, execution time and cost, are not trivial to satisfy, as they are conflict with each other, eg cheaper resources are usually slower than expensive ones. Hence, a multi-objective scheduling optimization is a more challenging task in Grid infrastructures. This chapter presents a new multi-objective approach, MOGSA (Multi-Objective Gravitational Search Algorithm), based on the gravitational search behaviour in order to optimize both objectives, execution time and cost, with the same importance and also at the same time. Two studies are carried out in order to evaluate the quality of this new approach for grid scheduling. Firstly, MOGSA is compared with the multiobjective standard and well-known NSGA-II (Non-Dominated Sorting Genetic Algorithm II) to prove the multi-objective optimization suitability of the proposed algorithm. Secondly two real grid schedulers (WMS and DBC) are also compared with MOGSA. The WMS (Workload Management System) is considered because of it is part of the most used European grid middleware - gLite - and also the DBC (Deadline Budget Constraint) algorithm from Nimrod-G participates in this evaluation due to its good performance keeping the deadline and budget per job. Results point out the superiority of MOGSA in all the studies carried out. MOGSA offers more quality solutions than NSGA-II and also better performance than current real schedulers.

María Arsuaga-Ríos
Beams Department, European Organization for Nuclear Research,
CERN, CH-1211, Geneva 23, Switzerland
e-mail: maria.arsuaga.rios@cern.ch

Miguel A. Vega-Rodríguez
ARCO Research Group, University of Extremadura,
Dept. Technologies of Computers and Communications,
Escuela Politécnica, Cáceres, Spain
e-mail: mavega@unex.es

A.Ş. Etaner-Uyar et al. (eds.), *Automated Scheduling and Planning*,
Studies in Computational Intelligence 505,
DOI: 10.1007/978-3-642-39304-4_9,

1 Introduction

Traditionally many real world optimization problems are solved by techniques that minimize or maximize a single objective, but many of these problems have more than one objective to satisfy. Recognising time constraints is one of the most important aspects within scheduling problems. Other significant oconstraints could include the length of the schedule, the availability of resources or their cost (machine scheduling), preferences of human resources (personnel scheduling), or compliance with regulations (educational timetabling). Traditional techniques try to combine multiple objectives into a single scalar value by using weights according to the importance suggested by the experts. However, normally, these objectives have the same importance and also they are conflictive each other. Currently multi-objective optimization techniques are emerging in scheduling problems ([22], [20]) giving the possibility to optimize more than one objective with the same importance and also providing decision support for the end users.

In this chapter, a multi-objective approach is presented to optimize the scheduling of tasks within grid environments. Grid scheduling [9] is a challenging task that manages the job submission on distributed and heterogeneous resources. Grid users have restrictions in terms of timing for their applications but also they have to operate within their budgets. Multi-objective optimization techniques may be utilised to accomplish both objectives (deadline and budget). Nowadays, multi-objective approaches are emerging to solve the grid scheduling problem ([30], [31], [32], [25]). These studies are often based on genetic algorithms to optimize the execution time and cost for a group experiments to be scheduled on the grid. However, their test environments lack of specific topologies that consider network features such as baud rate, delay, MTU (Maximum Transfer Unit) etc. Detailed resource data (operating system, number of machines, CPUs, speed, cost, etc.) is not taken into account. Moreover, most of these studies are based on the execution of simple experiments comprising independent jobs. Experiments based upon workflows that follow DAG (Directed Acyclic Graph) models are more challenging for grid scheduling due to its complexity and critical behaviour from the dependencies among jobs.

The research and the study of the proposed multi-objective algorithm - MOGSA (Multi-objective Gravitational Search Algorithm) addresses the weaknesses mentioned above. The MOGSA is compared with two meta-schedulers to evaluate the goodness of this contribution. The first meta-scheduler to be considered is the Workload Management System (WMS)[1] which belongs to the most extended European middleware gLite - Lightweight Middleware for Grid Computing -. The second meta-scheduler selected for this study is the Deadline Budget Constraint Algorithm (DBC)[3] from Nimrod-G. The DBC algorithm uses a greedy algorithm to attain the budget and deadline for an experiment.

MOGSA is the multi-objective version of the novel Gravitational Search Algorithm (GSA) [21], based on a swarm behaviour that comes from the gravitational forces among planets. MOGSA is also compared with the multi-objective standard

[1] `http://web.infn.it/gLiteWMS/`

and genetic algorithm Non-dominated Sorting Genetic Algorithm II (NSGA II) [8] to give more reliability to this multi-objective study.

The rest of this chapter is organized as follows, section 2 defines basic multi-objective optimization concepts, section 3 contains a survey of multi-objective scheduling optimization applied in several fields. In Section 4, a case study of multi-objective grid scheduling optimization using MOGSA compared with NSGA II is described. The conclusions are presented in Section 5.

2 Multi-objective Optimization

Diverse real world problems from different fields may be solved by multi-objective optimization techniques. In this section, a background and some basic multi-objective concepts are described.

A Multi-objective Optimization Problem (MOP) is defined as the task of finding a decision variables vector that satisfies the problem constraints and also optimizes an objective functions vector [13]. Those functions are usually in conflict each other, it being impossible to improve one objective function without worsening another objective function. The optimization goal is to find a solution vector with acceptable values for all the objective functions. Multi-objective problems must optimize `k` objective functions at the same time, dealing with the minimization/maximization of all the objective functions. For more information about multi-objective optimization the reader is directed to [33], [7] and [5]. Within thsi chapter we employ the following definitions:

Definition 1: ***Multi-objective Optimization Problem (MOP).*** A Multi-objective Optimization Problem includes a set of `n` parameters (decision variables), a set of `k` objective functions, and a set of `m` constraints. Objective function and constraints are function of the decision variables. The mathematical definition is shown as follows (Equation 1):

$$\begin{aligned} Optimize \quad & \mathbf{y} = \mathbf{f}(x) = (f_1(x), f_2(x), \ldots, f_m(x)) \\ subject\ to \quad & \mathbf{e}(x) = (e_1(x), e_2(x), \ldots, e_k(x)) \leq 0 \\ where \quad & \mathbf{x} = (x_1, x_2, \ldots, x_n) \in X \\ & \mathbf{y} = (y_1, y_2, \ldots, y_m) \in Y \end{aligned} \tag{1}$$

where `x` is the decision vector that belongs to the decision space `X` and `y` is the objective vector that is represented in the objective space `Y`. The decision variables vector can be discrete or continuous while the objective functions may be, lineal and discrete or continuous. The function `f : X → Y` is a transformation of the decision variables vector `x` on a response vector `y` (see Figure 1).

In mono-objective optimization problems, the optimization process finds one optimum solution that minimizes or maximizes a single objective function. When the problem is multi-objective, the optimum meaning must be redefined because of the presence of multiple objectives that are in conflict each other. The conflict among

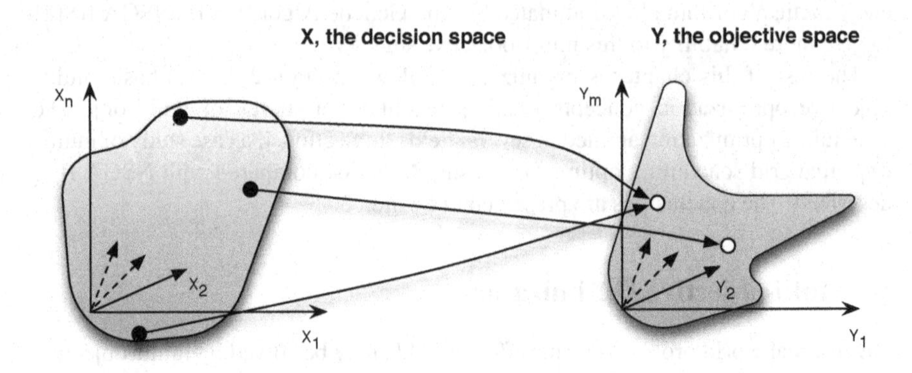

Fig. 1 Multi-objective optimization function

objectives may not allow the improvement of one of them without worsening the others. Thus, a multi-objective optimization consists in finding the best compromise among those objectives. The best compromise is called Pareto optimum.

Definition 2: ***Pareto Optimality.*** A decision vector x $\in X_f$ is a Pareto Optimum regarding a set A $\subseteq X_f$ if and only if:

$$\nexists a \in A : a \prec x \tag{2}$$

where X_f is the set of feasible solutions that satisfy the constraints of the problem (**e**(x)). This definition specifies x as a Pareto Optimum if does not exist other feasible vector a that decreases any of the objective functions without increasing simultaneously other one (assuming the minimization in all the objective functions). This comparison is also known as dominance ($\prec$).

Definition 3: ***Pareto Dominance.*** For any two decision vectors a and b (assuming a minimization problem):

$$\begin{aligned} a \prec b \; (a\, dominates\, b) &\Longleftrightarrow \mathbf{f}(a) \; < \; \mathbf{f}(b) \\ a \preceq b \; (a\, weakly\, dominates\, b) &\Longleftrightarrow \mathbf{f}(a) \; \leq \; \mathbf{f}(b) \\ a \sim b \; (a\, is\, indifferent\, to\, b) &\Longleftrightarrow \mathbf{f}(a) \; \nleq \; \mathbf{f}(b) \wedge \mathbf{f}(b) \; \nleq \; \mathbf{f}(a) \end{aligned} \tag{3}$$

Figure 2 shows a graphical representation of dominance regions regarding the solution F in a minimization problem. The solutions that are in the green region (A, C and D) dominate the solution F. In case of A and D, both values for the objective functions are better than the obtained by F. Also, despite of C and F obtain the same value for the objective function f_2, the solution C has better value for the objective function f_1. On the other hand, F is not dominated by the solutions B, E, G and I, being equally good than them. Finally, the solutions in the blue region, H and J, are dominated by F.

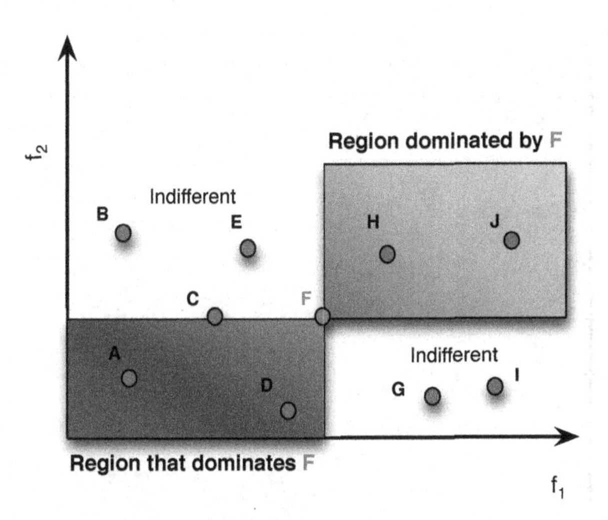

Fig. 2 Graphical representation of dominance regions regarding the solution F

Definition 4: ***Optimal Pareto Set.*** The Optimal Pareto set for a multi-objective problem is denoted as P^* and it is defined as (Equation 4):

$$P^* := \{x \in X \mid \nexists x' \in X, \quad \mathbf{f}(x') \preceq \mathbf{f}(x)\} \tag{4}$$

The optimal Pareto set comprises all the solutions from the decision space whose objective vectors cannot be improved at the same time. The objective vectors from the optimal Pareto set are called non-dominated solutions and compose the Pareto Front (PF^*).

Definition 5: ***Pareto Front.*** The Pareto Front PF^* from an Optimal Pareto set P^* is defined as:

$$PF^* := \{a = \mathbf{f}(x) \mid x \in P^*\} \tag{5}$$

The solutions found by multi-objective optimization may be presented as Pareto Front plots in order to evaluate them. In Figure 3, a Pareto front example is shown. Usually, the solutions that represent the best possible trade-offs among the objectives are the aim of the search (in case of Figure 3, solutions lying on the "knee" of the Pareto curve).

3 Multi-objective Optimization Applied on Scheduling Problems

Scheduling problems exist in many diverse real-world situations. The entities within a scheduling problem (people, tasks, vehicles, meetings, etc.) usually follow a space-time pattern in which some constraints must be satisfied and certain objectives have to be achieved. Such scheduling problems comprise large search spaces, the

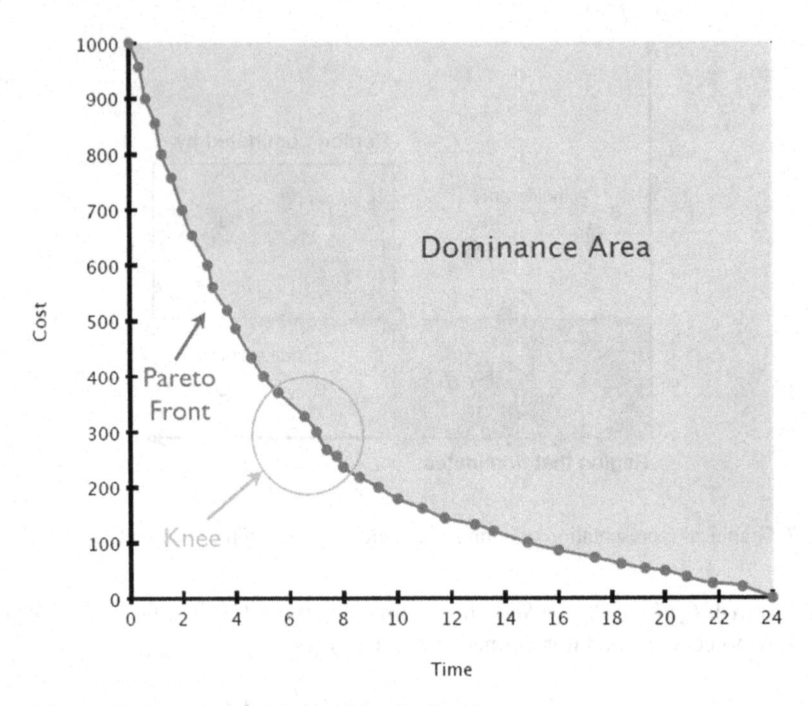

Fig. 3 Pareto Front example with two objective functions

objective being to find schedules that satisfy the user needs. Many scheduling problems are multi-objective by nature, because the users have more than one goal, such as minimizing the length of the schedule, satisfying preferences among human resources (personal scheduling), maximizing the compliance with regulations (educational timetabling), minimizing the tardiness of orders (production scheduling), optimizing job scheduling (machine scheduling) etc.

As mentioned in Section 1 some previous multi-objective optimisers have utilised an evaluation function comprising a weighted combination of their objectives in order to produce only one objective function value. In many real multi-objective scheduling problems, it is more desirable to consider the different objectives separately in order to obtain a better trade-off among all the conflictive objectives. This approach is supported by the Pareto optimization techniques. Currently multi-objective algorithms are utilising Pareto optimality to support decision makers, by allowing the user to choose between the solutions presented within the Pareto Front. For further information, the reader is directed to [23] and [20].

3.1 Multi-objective Personnel Scheduling

Personnel scheduling attempts to satisfy constraints and objectives arising from the interests of employees and employers taking into account the working regulations.

Mobasher [18] explored an example of a multi-objective approach in a medical context. The problem consists of a multi-objective nurse scheduling problem where both shift preferences, as a proxy for job satisfaction, and patient workload, as a proxy for patient dissatisfaction, are considered. A more general approach also applicable to nurse scheduling is presented by Li et al. [15]. They introduce a hybrid algorithm combining goal programming and meta-heuristic search to deal with compromise solutions in difficult employee scheduling problems. Moudani et al. [19] use a genetic algorithm and a greedy algorithm to manage a multi-objective airline crew optimization problem. Two objectives are optimized, minimizing operations cost and maximizing the crew staff overall degree of satisfaction. Yannibelli and Amandi [29] propose another multi-objective evolutionary algorithm for Project scheduling incorporating human resources. This approach tries to minimize the makespan for the Project and also assign the most effective set of human resources to each Project activity.

3.2 Multi-objective Educational Timetabling

Educational timetabling can be challenging due to inaccurate prediction of student enrollment, mistakes in the event list or resources availability and inadequate selection of hard and soft constraints [23]. A multi-objective linear programming model was proposed by Ismayilova et al. [11] to consider the administrations and instructors preferences by using a weight priority to schedule the class-course timetable. Educational timetabling also considers the exams timetabling, Côté et al. [6] present a multi-objective evolutionary algorithm that optimizes the maximum free time for the students while satisfying the clashing constraint (exam conflicts). Moreover, this multi-objective approach takes into account the timetable length as an optimization objective. The Balanced Academic Curriculum Problem requires the assigning of courses to teaching terms, satisfying prerequisites and balancing the credit course load within each term. Castro et al. [4] presented a multi-objective genetic algorithm to deal with this problem.

3.3 Multi-objective Production Scheduling

In manufacturing, the purpose of scheduling is to minimize the production time and costs, by instructing a production facility on timming and the utilisation of staff and equipment utilisation. Production scheduling aims to maximize the efficiency of the operation and reduce costs. Multi-objective production scheduling problems are widely studied in several domains [14]. Possible objectives to be considered in production scheduling problems include the makespan (response time), the mean completion time (the mean of the slower time activity during the production), the maximal tardiness or the mean tardiness (the maximum and minimum mean times obtained after several productions). In the study carried out by Loukil et al. [16], these objectives are optimized using a multi-objective simulated annealing approach. Multi-objective algorithms offer decision support, Mansouri et al.

[17] aim to identify the gaps in decision-making support based on multi-objective optimization (MOO) for build-to-order supply chain management. Although, these scheduling problems are based on the economy, multi-objective approaches allow considering also other aspects as the intangible value of freshness in their products [1].

3.4 *Multi-objective Machine Scheduling*

Machine scheduling refers to problems where a set of jobs or tasks has to be scheduled for processing by a sequence of one or more machines [20]. Each job or task consists of one or more operations (sub-tasks), usually a number of additional constraints must be also satisfied. Examples of such constraints are precedence relations between the jobs and limited availability of resources. Machine scheduling is widely studied and multiple multi-objective approaches are applicable to it [20]. Xiong et al. [28] address a robust scheduling for a flexible job-shop scheduling problem with random machine breakdowns. Makespan and robustness objectives are simultaneously optimized by using a multi-objective evolutionary algorithm. There exists other approaches, such as that adopted by Hamta et al. study [10]. The approach taken by Hamta et al. optimizes more objectives, such as cycle time, total equipment cost and the smoothness index. The problem being considered is a single-model assembly line balancing problem, where the operation times of tasks are unknown variables and the only known information is the lower and upper bounds for operation time of each task.

4 A Case Study: Multi-objective Machine Scheduling in Grid Environments

Grid computing [9] is a distributed computing paradigm in which all the resources of an unknown number of computers are subsumed to be treated as a single supercomputer in a transparent way. This innovative infrastructure allows the coordination of these heterogeneous resources (storage, computing and specific applications) using a decentralized control.

One of the emerging problems within in grid computing is that of job scheduling. Job scheduling consists in allocating jobs on grid resources whilst satisfying the end user needs. In this chapter we will consider the allocation of jobs that used to carry out software based experiments. The two most critical requirements for the grid based experiments as discussed in this chapter are deadline and budget. These parameters are conflictive each other because slower resource (in terms of processing time) is usually cheaper than resources with greater processing power. Therefore, a multi-objective approach is required to deal with this problem. Our approach takes into account two objectives (time and cost).

Given a set of jobs $J = \{J_i\}$, $i = 1,..,m$ and a set of resources $R = \{R_j\}$, $j = 1,..,n$ the fitness functions are described as follows:

$$Min\ F = (F_1, F_2) \tag{6}$$

$$F_1 = max\,time\,(J_i, f_j(J_i)) \tag{7}$$

$$F_2 = \sum cost\,(J_i, f_j(J_i)) \tag{8}$$

where $f_j(J_i)$ is a mapping function that assigns J_i onto resource R_j. Function *time* $(J_i, f_j(J_i))$ denotes the completion time and *cost* $(J_i, f_j(J_i))$ is the input data transmission cost and resource cost for processing the solution.

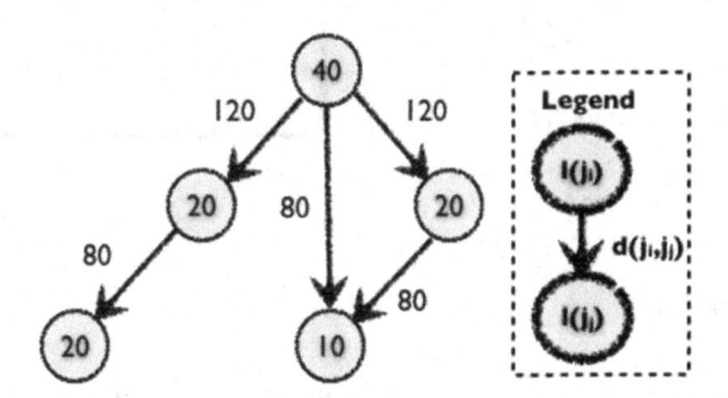

Fig. 4 A simple workflow graph

Many experiments are composed of jobs that depend on each other, this type of workflow influences both the running time and the cost of the experiment. This is because each workflow (or experiment) is modelled by a weighted directed acyclic graph (DAG) $JG = (V, E, l, d)$, where V is a set of nodes and E is a set of edges. Jobs are represented by the nodes and they have assigned a length $l(j)$, which denotes its length in terms of thousands of MI (Million of instructions). The precedence constraint is indicated by an edge $\langle j, j' \rangle$ in E $(j, j' \in V)$ from j to j'. This means that job j' cannot be executed until the job j has been completed successfully and j' receives all necessary data from j. The length (bytes) of the data transference $d(j \rightarrow j')$ between jobs is expressed by a label in the edge $j \rightarrow j'$. An example is shown in Figure 4. This problem formulation also takes into consideration dependent jobs as a constraint to calculate the fitness of its objective functions.

Each agent used in MOGSA as mass or NSGA-II as individual, represents a candidate solution. Candidate solutions have to take into account the scheduling of the jobs by allocating them to the resource where they will be executed. The order of job execution has to be considered. To build on the candidate solutions two vectors are created, the allocation and order vectors, based on Talukder et al [25] and [26]. The length of both vectors is the total of jobs that comprise the experiment to be executed $|J|$.

- Allocation Vector: Represents the job assigments within the available grid resources. Such that $a(i) = j$, where $0 \leq i < |J|$ and $0 \leq j < |R|$, i.e. job J_i is assigned to resource R_j, being $|R|$ the total number of available grid resources.

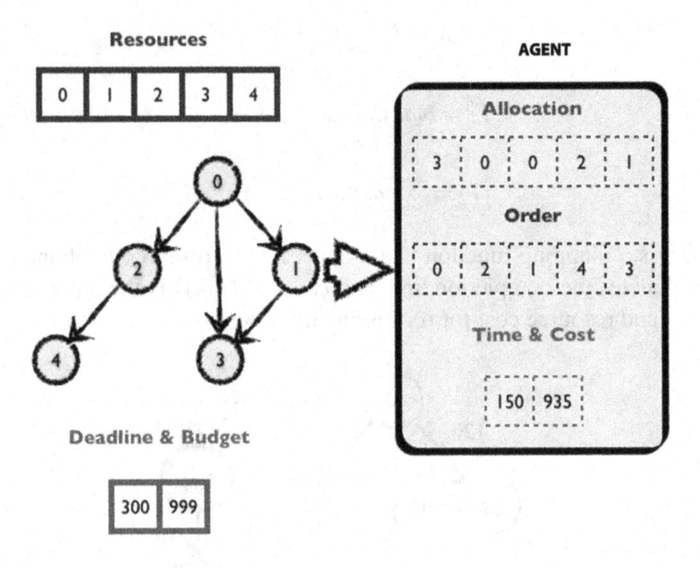

Fig. 5 Agent representation

- Order Vector: Indicates the order of the jobs execution following the precedence constraints of the DAG model from the workflow to run on the grid. Such that $o(k) = i$, where $0 \leq i,k < |J|$, and each J_i just appears once in the vector.

According to these vectors, candidate solutions provide a feasible solution with its total execution time and cost. The GridSim[2] [2] tool supplies these values and simulates the execution of the workflow, considering the mass vectors. In Figure 5, it is shown an example of a candidate solution with $|J| = 5$ and $|R| = 5$.

4.1 MOGSA: Multi-objective Gravitational Search Algorithm

MOGSA is a new version of GSA [21] with multi-objective properties. GSA is an algorithm that is mainly used to solve combinatorial optimization problems, or unconstrained numerical optimization, following the Newtonian gravity laws as a meta-heuristics among masses (solutions). Masses are represented as candidate solutions and their size represents their goodness, the attraction among them depends directly on their size. Big masses have major attraction force than others with smaller size, hence bigger masses are considered better solutions with best values of fitness functions than the smaller ones. Although, GSA is similar to the well-known Particle Swarm Optimization (PSO) [12], a comparative study is carried out in [21] . MOGSA adapts the behaviour of these masses with a multi-objective perspective. The steps of the algorithm are shown in Algorithm 1.

This new algorithm needs the parameters from GSA [21]:

[2] `http://www.buyya.com/gridsim/`

Algorithm 1. MOGSA pseudocode

INPUT: Population Size, Termination criteria, Initial Gravity (G_0), Min_{Kbest}, α, ε
OUTPUT: Set of Solutions

```
 1: Initialize population;
 2: Evaluate population (Time and Cost);
 3: Initialize Pareto set to empty;
 4: while termination criteria: maximum time of execution is 2 min do
 5:     Update Gravity using α , best and worst of population;
 6:     Calculate size of masses;
 7:     Calculate Force based on ε and Acceleration between masses;
 8:     Update Velocity and Position per each mass;
 9:     Update Pareto set with the best solutions;
10: end while
```

- Population size: Indicates the number of masses that are going to be in each algorithm iteration.
- Termination criteria: Corresponds to the maximum time that the algorithm is going to run, it is also called the stop condition.
- Initial Gravity (G_0): Represents the initial gravity force among masses, which is decreasing per algorithm iteration.
- Min_{Kbest}: Indicates the minimum number of masses that are going to exert their forces over the others. When the algorithm starts, all the masses are considered to exert their forces, but as the time progresses, this number of masses decreases until they achieve the number indicated by this parameter.
- α: Represents the reduction coefficient of the gravity.
- ε: A small constant used to calculate the force per each mass.

As multi-objective algorithm, the output is a set of solutions that belong to the first Pareto front (non-dominated solutions). These solutions will give decision support to the final user. The steps of MOGSA adapted to the grid job scheduling problem are detailed in the following sections.

4.1.1 Initial Population

A combination of both allocation and order vectors is considered as one vector called AO (allocation + order vector) in order to represent the location of the masses. The algorithm starts with a random initialization of the population and it is evaluated by ranking the masses in Pareto fronts with a rank value r_i for each mass M_i. This ranking is carried out by the *Pareto front ranking* operator, which classifies all the candidate solutions by applying the dominance concept. The value cd_i of the *Crowding distance* operator is stored per each mass M_i with the aim of calculating the MOFitness (equation 9). Crowding distance is a measure of how close a mass is respect to other masses. Large average crowding distance will result in better diversity in the population.

$$MOFitness_i = (2^{(r_i)} + \frac{1}{1+cd_i})^{-1} \tag{9}$$

4.1.2 Update Gravity Using α, Best and Worst of Population

The gravity G is updated through equation 10, where t is the current time and T indicates the total time.

$$G = G_0 \exp^{(-\alpha)\frac{t}{T}} \tag{10}$$

At the same time, best and worst MOFitness masses from the population are selected by using the operator before mentioned.

4.1.3 Calculate Size of Masses

The highest and lowest values obtained by selecting the best and worst masses are used to calculate the size of the masses (equation 11 and equation 12). A heavier mass is interpreted as a better candidate solution.

$$q_i = \frac{MOFitness_i - MOFitness_{worst}}{MOFitness_{best} - MOFitness_{worst}} \tag{11}$$

$$s_i = \frac{q_i}{\sum_1^N q_j} \tag{12}$$

4.1.4 Calculate Force based on ε and Acceleration, Update Velocity and Position per Each Mass

The exerting forces on each mass in every dimension are calculated, but only the K_{best} masses can exert their forces over the others. In equations 13 and 14, an Euclidean distance between each pair of agents is calculated for being used in the corresponding force. The calculation of the Euclidean distance is carried out for each dimension, corresponding to the AO vectors of the pair of masses. To encourage exploration by the MOGSA, the total force that acts on M_i uses a random weight in every dimension (equation 15).

$$R_{i,j} = \|AO_i, AO_j\|, \forall i, 1 \le i \le N; \forall j, 1 \le j \le K_{best} \tag{13}$$

$$F_{ij}^d = G \times \frac{s_i \times s_j}{R_{ij} + \varepsilon} \times (AO_j^d - AO_i^d) \tag{14}$$

$$f_i^d = \sum_{j \in K_{best}, j \neq i}^{N} rand\,[0,1] \times F_{ij}^d \tag{15}$$

The acceleration per mass is calculated per dimension by using the force in each dimension d and also considering its size (equation 16).

$$a_i^d = \frac{f_i^d}{s_i} \tag{16}$$

The velocity (equation 17) and the position of the masses (equation 18) are updated. The updating of the position consists of increasing the resource identifiers in the allocation vector and the order positions in the order vector, without forgetting their precedence constraints.

$$v_i^d = rand\,[0,1] \times v_i^d + a_i^d \tag{17}$$

$$AO_i^d = AO_j^d + v_i^d \tag{18}$$

4.1.5 Best Solutions and New Generation

Once all the positions are updated, MOGSA applies an improvement in case a stagnation stage occurs. The stagnation process avoids this problem by using heuristics specific to the grid scheduling problem. The worst node of the population is substituted by another created using heuristic methods applied to each vector (allocation and order vector). The order vector is updated by comparing itself with another order vector generated by a greedy algorithm. The greedy algorithm creates an order vector allocating the jobs with dependent jobs to the first positions without neglecting the precedence constraint. The heuristic method for the allocation vector considers the information within the new order vector when calculating the total execution time. Assuming that jobs with no dependencies can be executed in parallel. Each job is assigned to the resource that best reduces the total execution time, having preferences for the resources with best value of *processing speed/cost* and also the possible overhead when independent jobs are executed sequentially in the same resource. The resulting Pareto front is updated according to these changes and it is compared with Pareto front calculated in the last iteration.

4.2 NSGA II: Non-dominated Sorting Genetic Algorithm II

The NSGA-II (Non-Dominated Sorting Genetic Algorithm) was proposed by [8]. The main steps of NSGA-II algorithm are described in Algorithm 2.

The parameters of NSGA II are the essential parameters for a general genetic algorithm.

- Population size: Indicates the number of individuals per each generation.
- Termination criteria: Indicates the stop condition for the algorithm.
- Crossover probability (p_c): Represents the probability of crossover.
- Mutation Probability (p_m): Represents the probability of mutation.

Algorithm 2. NSGA II pseudocode.

INPUT: Population size, Termination criteria, Crossover (p_c) and Mutation (p_m) Probability
OUTPUT: Set of Solutions

1: Initialize population;
2: Evaluate population (Time and Cost);
3: **while** termination criteria: maximum time of execution is 2 min **do**
4: Multi-objective operators;
5: Binary Tournament Selection;
6: Crossover with p_c;
7: Mutation with p_m;
8: Best solutions will pass to the next generation population;
9: **end while**

4.2.1 Initial Population

In genetic algorithms, each individual represents a candidate solution to the problem. In this problem, each individual are encods both allocation and order vectors, where each cell represents a gene of this genetic algorithm. Allocation and Order vectors are created for each individual and they are simulated in GridSim to obtain the execution time and cost from a specific workflow. Both values are considered as the fitness values as they indicate the goodness of the generated candidate solution. The initial population is generated randomly according to the dependencies among jobs of the workflow to execute and the size of the population N. The population is ranked into Pareto fronts by applying the dominance concept among the fitness values supplied by GridSim.

4.2.2 Multi-objective Operators

Two multi-objective operators are applied during the execution of this algorithm: *Classification of Pareto fronts* and *Crowding distance*. These operators allow the comparison of individuals and the obtaining of the best solutions from a set of solutions. The population is sorted according to levels of non-dominance (ranking of Pareto fronts, *F1*, *F2*, ...) using the classification of Pareto fronts operator . Each solution has a rank that indicates the allocated front. Cowding distance is calculated per front to compare solutions that belong to the same front. The tournament selection, crossover and mutation operators are used to create the offspring population which is the same size as the initial population.

4.2.3 Binary Tournament Selection

Binary Tournament Selection is used to select the parents, using two tournaments (one per parent). These tournaments are conducted using the logic of the NSGA-II, i.e., every tournament selects an individual if both individuals are in the same front then crowding distance is used to determine the winner individual.

4.2.4 Crossover

Two types of crossover are applied per individual. These crossovers are based on Talukder et al. work ([25] and [26]). They consist ofm odifying the allocation and order vectors:

- Allocation crossover: This crossover selects random a position from allocation vector. The parent vectors swap their vectors from the random position, creating two new individuals (see Figure 6).

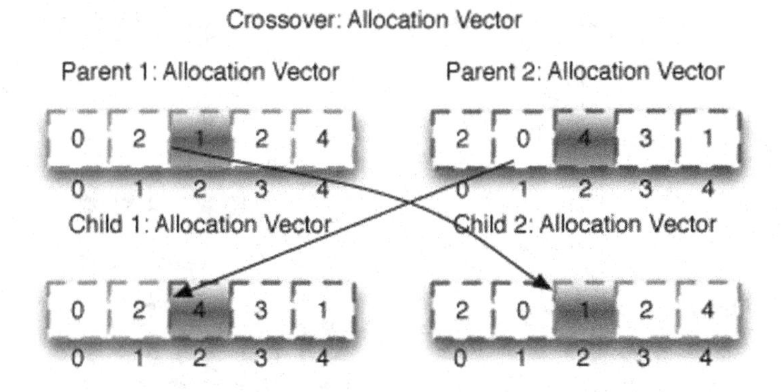

Fig. 6 NSGA-II: Allocation Crossover. The number indicated in each cell is the identification number of each grid resource and the index of each cell represents the job that is to be executed. For example, job 0 (the first job) is executed on grid resource 0 in case of Parent 1 or on grid resource 2 in case of Parent 2.

- Order crossover: The order crossover operator is similar to allocation crossover, but it has to keep the precedence constraints among jobs. After the swapping process, this method checks the constraints and for repeated order positions. To avoid this, it first checks if there is not the order position in the order vector. If it occurs, this position is stored and when there is a repeated position, one position stored before is selected randomly (see Figure 7).

These new individuals are simulated, obtaining their execution time and cost, and they are added to the population, the population size being *2N*.

4.2.5 Mutation

The mutation process is composed of two types of mutation operaror, Replacing mutation and Reordering mutation.

- Replacing Mutation: The allocation vector follows this mutation process, each job from the employed allocation vector selects a random number and, if the number is less than the mutation probability, a new grid resource is assigned to

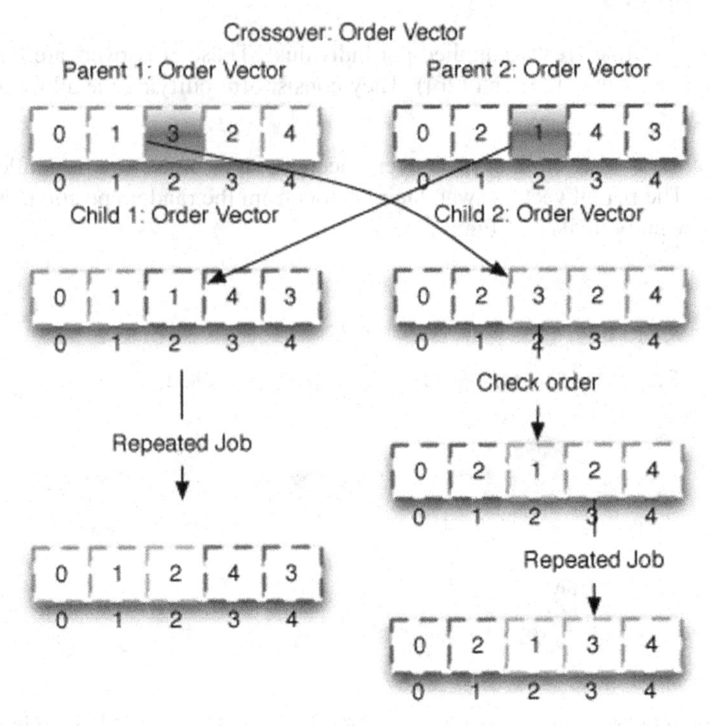

Fig. 7 NSGA-II: Order Crossover. The number indicated in each cell is the identification number of each job and the index of each cell represents the order of execution. For example the job 0 is executed in first place in all the cases. This example assumes the workflow shown in Figure 5.

this job. This new grid resource is selected randomly from the list of available resources (see Figure 8).

- Reordering Mutation: The Order vector is mutated by this process. This mutation method differs from the replacing mutation as reordering mutation considers the constraints of the DAG model. The process begins with the selection of the jobs to mutate in the same way as the replacing mutation, taking into account the mutation probability. When a job is selected to be mutated, the process identifies the last position of the vector in where a parent job (according to the DAG model) of the current job is placed. Then, the positions of the child jobs (of the job to mutate) are searched across the order vector and the process selects the first position of them. Finally, a new position for the job to mutate is chosen randomly between the last parent position and the first child position (see Figure 9).

All new individuals are executed in GridSim to calculate their cost and time.

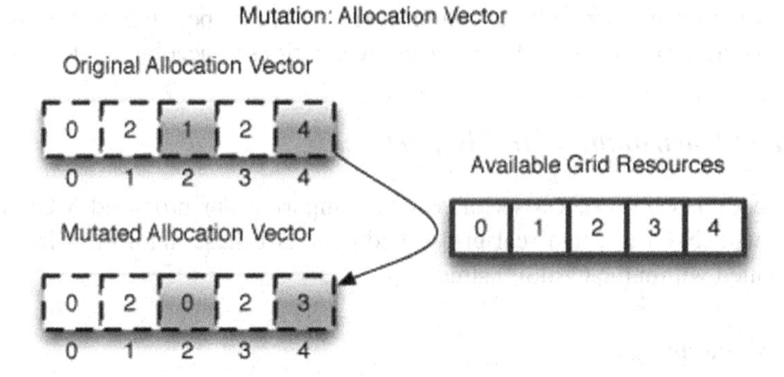

Fig. 8 NSGA-II: Allocation Mutation. The number indicated in each cell is the identification number of each grid resource and the index of each cell represents the job that is going to be executed. In this example, two mutations are performed (positions 2 and 4 of the allocation vector).

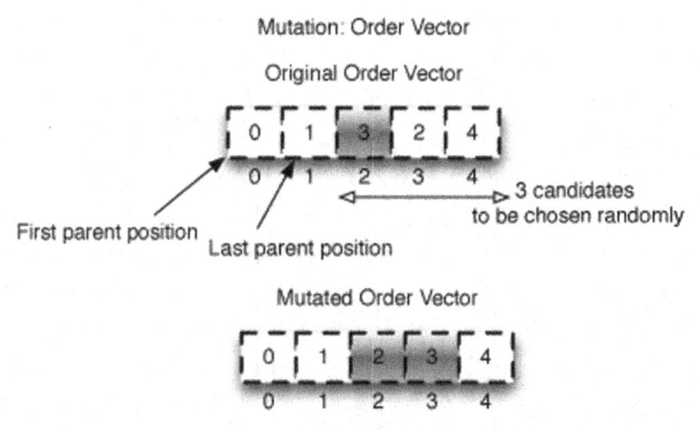

Fig. 9 NSGA-II: Order Mutation. The number indicated in each cell is the identification number of each job and the index of each cell represents the order of execution. As this example assumes the workflow shown in Figure 5, job 0 and job 1 are the parents of job 3 and job 3 does not have any child job. Therefore, job 3 has three candidate order positions. Finally, job 3 changes its order position from 2 to 3.

4.2.6 Best Solutions and New Generation

Once the non-dominated sorting process is complete, the new population is generated from the solutions of non-dominated fronts. This new population is first built with the best non-dominated front (*F1*), it continues with the solutions of the second front (*F2*), third (*F3*) and so on. As the population size after the genetic operators is *2N*, and there are only *N* solutions that make the descendants, not all the solutions belonging to these population fronts can be accommodated in the new population.

Those solutions that cannot be accommodated are deleted. In case not all the solutions of a particular front can be accommodated in the new population, the crowding distance operator is used in order to select the solutions to accommodate.

4.3 *Test Environment and Experiments*

In this section, we present a set of results comparing our proposed MOGSA algorithm with NSGA II and real grid schedulers. All these algorithms have been implemented within the GridSim simulator.

4.3.1 Methodology

Within GridSim a grid topology has been implemented taking into account network and resource characteristics. Figure 10 shows the topology implemented. This topology is based on the EU DataGrid test-bed [24] and it incorporates grid resource characteristics from WWG test-bed from the work studied in [3]. These characteristics are presented in Table 1.

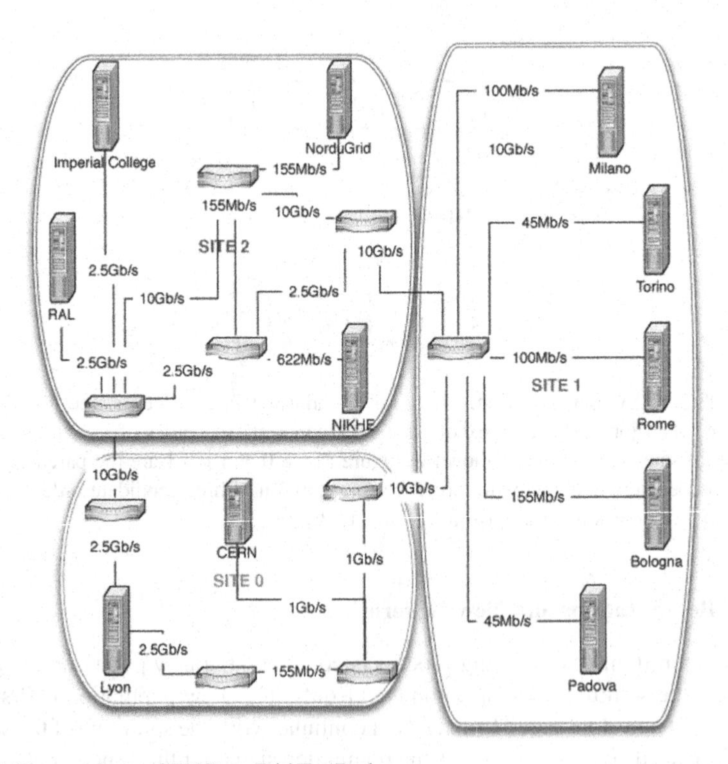

Fig. 10 The topology of EU DataGrid TestBed

Table 1 Resource Characteristics (G$ means Grid dollars).

Resource Name	Features (Vendor, Type, OS, CPUs/WN)	Resource Manager Type	MIPS per G$	Price (G$/ CPU time)
LYON	Compaq, AlphaServer, OSF1, 4	Time-shared	515	8
CERN	Sun, Ultra, Solaris, 4	Time-shared	377	4
RAL	Sun, Ultra, Solaris, 4	Time-shared	377	3
IMPERIAL	Sun, Ultra, Solaris, 2	Time-shared	377	3
NORDUGRID	Intel, Pentium/VC820, Linux, 2	Time-shared	380	2
NIKHEF	SGI, Origin 3200, IRIX, 6	Time-shared	410	5
PADOVA	SGI, Origin 3200, IRIX, 16	Time-shared	410	5
BOLOGNA	SGI, Origin 3200, IRIX, 6	Space-shared	410	4
ROME	Intel, Pentium/VC820, Linux, 2	Time-shared	380	1
TORINO	SGI, Origin 3200, IRIX, 4	Time-shared	410	6
MILANO	Sun, Ultra, Solaris, 8	Time-shared	377	3

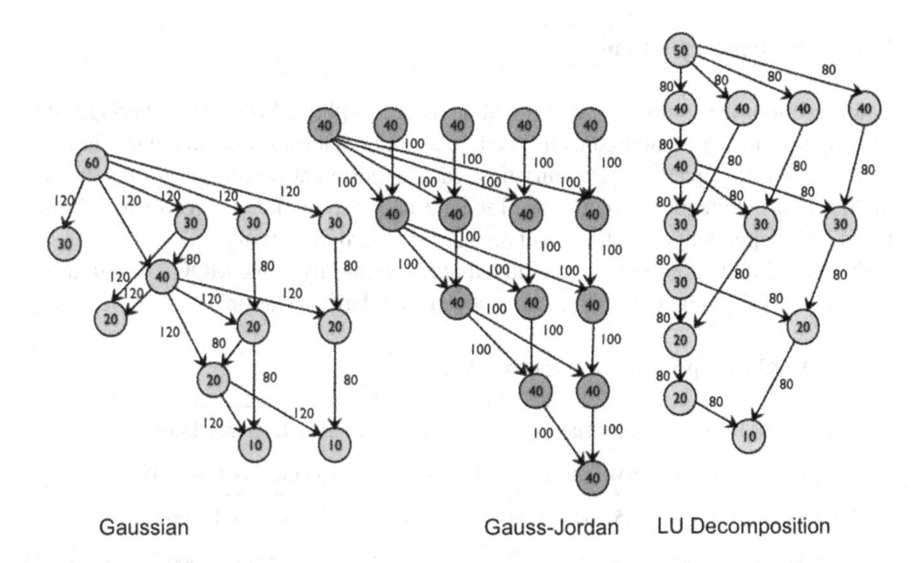

Fig. 11 Workflows: Gaussian, Gauss-Jordan and LU Decompostion

In order to study the different behaviours of the schedulers implemented, we utilise specific parallel numerical computation techniques such as the Parallel Gaussian Algorithm and the Parallel Gauss-Jordan Algorithm to solve systems of equations and Parallel LU decompositions [27]. These workflows are shown in Figure 11.

4.3.2 Parameterization

Previously studies (each single experiment was repeated 30 times independently) have found the best parameter configuration for each algorithm:

- MOGSA: Population size = 25, maximum time of execution = 2 minutes, G0 = 10000, $Min_{Kbest} = 5$, $\alpha = 2$, $\varepsilon = 1$.
- NSGA II: Population size = 100, maximum time of execution = 2 minutes, crossover probability = 0.9, mutation probability = 0.1, selection with binary tournament.

4.3.3 Analysis

The analysis may be divided into two sections, the first studies the multi-objective features of MOGSA and compares it to NSGA II. MOGSA is compared with real grid schedulers in the second section.

Multi-objective Algorithms

In this section, we compare two multi-objective algorithms, MOGSA and NSGA II. The algorithms are compared using the hypervolume metrics. Due to the stochastic nature of multi-objective metaheuristics, each experiment performed in this study includes 30 independent executions. Tables 2 and 3 show the average results from these 30 independent executions and also the standard deviation.

In Table 2 and Table 3, we appreciate that the reliability in the MOGSA is greater than in the NSGA II (MOGSA has a lower standard deviation in all the cases).

Table 2 MOGSA hypervolume per each workflow

Workflows	Average Hypervolume (%)	Standard Deviation of Hypervolume	Reference Point (Time (s), Cost (G$))
Gaussian	55.06	0.28	(1000, 10000)
Gauss-Jordan	55.53	0.22	(1200, 22000)
LU	54.48	0.47	(1200, 22000)

Table 3 NSGA II hypervolume per each workflow

Workflows	Average Hypervolume (%)	Standard Deviation of Hypervolume	Reference Point (Time (s), Cost (G$))
Gaussian	45.89	1.08	(1000, 10000)
Gauss-Jordan	47.13	0.66	(1200, 22000)
LU	48.02	0.70	(1200, 22000)

Table 4 Set Coverage comparison of NSGA II and MOGSA per each workflow

Coverage A ≥ B					
Algorithm		Workflows			Average
A	B	Gaussian	Gauss-Jordan	LU	
MOGSA	NSGA II	100%	100%	100%	100%
NSGA II	MOGSA	0%	0%	0%	0%

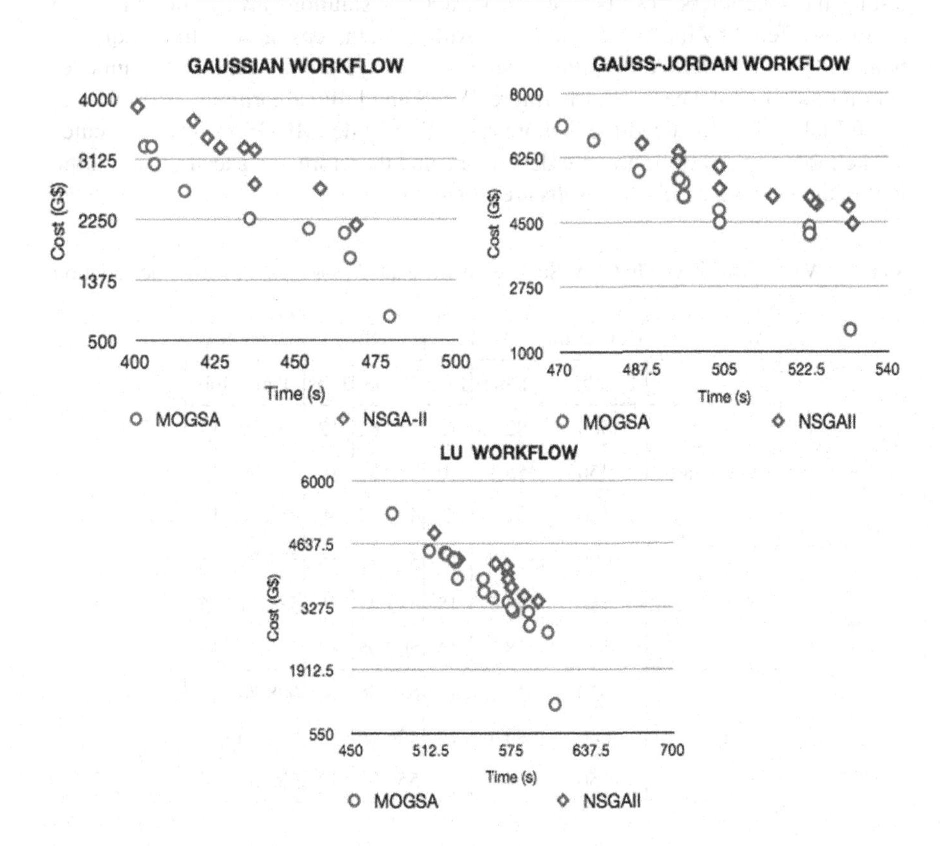

Fig. 12 NSGA II vs MOGSA: Pareto fronts per workflow

Moreover, the average hypervolume of MOGSA is around a 55 percent, being better than the 47 percent from NSGA II algorithm. This means that the solutions from MOGSA are better than the solutions offered by NSGA II. Furthermore, to study in more detail the resulting solutions, we have compared the Pareto fronts per each workflow. In Figure 12 it is shown that the Pareto fronts obtained by the NSGA II algorithm per wokflow are always dominated by the Pareto fronts of the MOGSA. Finally, the direct comparison of both algorithms regard to set coverage metrics is given in Table 4. Each cell gives the fraction of non-dominated solutions evolved by algorithm B, which are covered by the non-dominated points achieved by algorithm A [33]. Again, MOGSA covers all the results obtained by NSGA II.

Real Grid Schedulers

The Workload Management System (WMS)[3] based on the European middleware gLite[4] and the Deadline Budget Constraint (DBC) [3] from Nimrod-G are implemented in GridSim to compare and evaluate their results with our MOGSA. After testing the schedulers, results prove that MOGSA solutions always dominate the solutions offered by the WMS and DBC. Although, the cost is sometimes equal in both. DBC and MOGSA algorithms, MOGSA always provides the minimum execution time in all the cases. Furthermore, WMS and DBC algorithms report unsuccessful jobs when the deadline is more restrictive while, MOGSA always executes all the jobs required (jobs are interdependent, and therefore, the execution of all the jobs could be a must). These results are shown in Table 5.

Table 5 WMS vs DBC vs MOGSA: Successfully executed jobs regard to deadline variation

Workflows	Constraint	WMS		DBC		MOGSA	
	Deadline	Time	Jobs	Time	Jobs	Time	Jobs
Gaussian	500	482.68	12	480.82	12	479.12	12
	450	455.11	10	450.58	10	435.56	12
	400	401.01	7	400.77	9	391.51	12
Gauss-Jordan	600	534.70	15	533.41	15	531.71	15
	550	534.70	15	533.41	15	531.71	15
	500	428.57	14	500.08	14	496.25	15
LU	650	612.29	14	610.46	14	608.76	14
	600	585.78	13	596.66	14	594.96	14
	550	504.12	10	550.00	12	532.52	14

5 Conclusions

Multi-objective scheduling problems cover many activities in day-to-day life. Therefore, diverse techniques are emerging to consider the conflicting objectives and to give decision support to the users. In this chapter, we have explained the main concepts of multi-objective optimization and its application in scheduling problems.

[3] `http://web.infn.it/gLiteWMS/`
[4] `http://glite.cern.ch/`

A case study is presented to deal with the multi-objective optimization in grid environments. We have compared the novel Multi-Objective Gravitational Algorithm (MOGSA) with the standard and well-known multi-objective algorithm NSGA II, in order to know the performance of MOGSA. These algorithms allow the optimization of two conflicting objectives, execution time and cost, by using workflows with dependencies among jobs. MOGSA and NSGA II have been implemented using the simulator GridSim and modifying it to simulate a realistic approach of a grid environment with all its features.

In our analysis, NSGA II obtains worse results than MOGSA in terms of hypervolume and coverage metrics. Moreover, we have implemented real schedulers as WMS and DBC in GridSim to compare with MOGSA, and MOGSA has demonstrated in all the cases that its results are better than the offered by the other schedulers. In addition, MOGSA gives a good range of solutions for decision support.

In future work, MOGSA will be compared with other multi-objective techniques and also over other grid infrastructures.

References

[1] Amorim, P., Günther, H.O., Almada-Lobo, B.: Multi-objective integrated production and distribution planning of perishable products. International Journal of Production Economics 138(1), 89–101 (2012)

[2] Buyya, R., Murshed, M.: Gridsim: a toolkit for the modeling and simulation of distributed resource management and scheduling for grid computing. Concurrency and Computation: Practice and Experience 14(13), 1175–1220 (2002)

[3] Buyya, R., Murshed, M., Abramson, D.: A deadline and budget constrained cost-time optimisation algorithm for scheduling task farming applications on global grids. In: Int. Conf. on Parallel and Distributed Processing Techniques and Applications, Las Vegas, Nevada, USA, pp. 2183–2189 (2002)

[4] Castro, C., Crawford, B., Monfroy, E.: A genetic local search algorithm for the multiple optimisation of the balanced academic curriculum problem. In: Shi, Y., Wang, S., Peng, Y., Li, J., Zeng, Y. (eds.) MCDM 2009. CCIS, vol. 35, pp. 824–832. Springer, Heidelberg (2009)

[5] Coello, C., Van Veldhuizen, D., Lamont, G.: Evolutionary Algorithms for Solving Multi-Objective Problems. In: Genetic Algorithms and Evolutionary Computation. Kluwer (2002)

[6] Côté, P., Wong, T., Sabourin, R.: Application of a hybrid multi-objective evolutionary algorithm to the uncapacitated exam proximity problem. In: Proceedings of the 5th International Conference on Practice and Theory of Automated Timetabling, pp. 151–167 (2004)

[7] Deb, K.: Multi-Objective Optimization using Evolutionary Algorithms. John Wiley & Sons (2001)

[8] Deb, K., Pratap, A., Agarwal, S., Meyarivan, T.: A fast elitist multi-objective genetic algorithm: Nsga-ii. IEEE Transactions on Evolutionary Computation 6(2), 182–197 (2002)

[9] Foster, I., Kesselman, C., Tuecke, S.: The anatomy of the grid. In: Grid Computing-Making the Global Infrastructure a Reality. John Wiley Sons (2010)

[10] Hamta, N., Ghomi, S.F., Jolai, F., Shirazi, M.A.: A hybrid pso algorithm for a multi-objective assembly line balancing problem with flexible operation times, sequence-dependent setup times and learning effect. International Journal of Production Economics (2012)
[11] Ismayilova, N.A., Sagir, M., Gasimov, R.N.: A multiobjective faculty-course-time slot assignment problem with preferences. Mathematical and Computer Modelling 46(7-8), 1017–1029 (2007)
[12] Kennedy, J., Eberhart, R.: Particle swarm optimization. In: Proceedings of IEEE International Conference on Neural Networks, vol. 4, pp. 1942–1948 (1995)
[13] Fonseca, C.M., Fleming, P.J., Zitzler, E., Deb, K., Thiele, L. (eds.): EMO 2003. LNCS, vol. 2632. Springer, Heidelberg (2003)
[14] Lei, D.: Multi-objective production scheduling: a survey. The International Journal of Advanced Manufacturing Technology 43(9-10), 926–938 (2009)
[15] Li, J., Burke, E.K., Curtois, T., Petrovic, S., Qu, R.: The falling tide algorithm: A new multi-objective approach for complex workforce scheduling. Omega 40(3), 283–293 (2012)
[16] Loukil, T., Teghem, J., Fortemps, P.: A multi-objective production scheduling case study solved by simulated annealing. European Journal of Operational Research 179(3), 709–722 (2007)
[17] Mansouri, S.A., Gallear, D., Askariazad, M.H.: Decision support for build-to-order supply chain management through multiobjective optimization. International Journal of Production Economics 135(1), 24–36 (2012)
[18] Mobasher, A.: Nurse scheduling optimization in a general clinic and an operating suite. PhD thesis, University of Houston (2012)
[19] El Moudani, W., Cosenza, C.A.N., de Coligny, M., Mora-Camino, F.: A bi-criterion approach for the airlines crew rostering problem. In: Zitzler, E., Deb, K., Thiele, L., Coello Coello, C.A., Corne, D.W. (eds.) EMO 2001. LNCS, vol. 1993, pp. 486–500. Springer, Heidelberg (2001)
[20] Pinedo, M.L.: Scheduling: Theory, Algorithms, and Systems, 4th edn. Prentice-Hall (2012)
[21] Rashedi, E., Nezamabadi-Pour, H., Saryazdi, S.: Gsa: A gravitational search algorithm. Information Sciences 179(13), 2232–2248 (2009)
[22] Silva, A., Burke, E.K.: A tutorial on multiobjective metaheuristics for scheduling and timetabling. In: Multiple Objective Meta-Heuristics. LNEMS. Springer (2004)
[23] Silva, A., Burke, E.K., Petrovic, S.: An introduction to multiobjective metaheuristics for scheduling and timetabling. In: Grandibleux, X., Sevaux, M., Sörensen, K., T'Kindt, V. (eds.) Metaheuristic for Multiobjective Optimisation. LNEMS, vol. 535, pp. 91–129. Springer, Heidelberg (2004)
[24] Sulistio, A., Poduval, G., Buyya, R., Tham, C.: On incorporating differentiated levels of network service into gridsim. Future Gener. Comput. Syst. 23(4), 606–615 (2007)
[25] Talukder, A.K.A., Kirley, M., Buyya, R.: Multiobjective differential evolution for workflow execution on grids. In: MGC 2007: Proceedings of the 5th International Workshop on Middleware for Grid Computing, pp. 1–6. ACM, New York (2007)
[26] Talukder, A.K.A., Kirley, M., Buyya, R.: Multiobjective differential evolution for scheduling workflow applications on global grids. Concurr. Comput. Pract. Exper. 21(13), 1742–1756 (2009)
[27] Tsuchiya, T., Osada, T., Kikuno, T.: Genetics-based multiprocessor scheduling using task duplication. Microprocessors and Microsystems 22(3-4), 197–207 (1998)

[28] Xiong, J., Xing, L., Chen, Y.: Robust scheduling for multi-objective flexible job-shop problems with random machine breakdowns. International Journal of Production Economics (2012)
[29] Yannibelli, V., Amandi, A.: Project scheduling: A multi-objective evolutionary algorithm that optimizes the effectiveness of human resources and the project makespan. Engineering Optimization, 1–21 (2012)
[30] Ye, G., Rao, R., Li, M.: A multiobjective resources scheduling approach based on genetic algorithms in grid environment. In: International Conference on Grid and Cooperative Computing Workshops, pp. 504–509 (2006)
[31] Yu, J., Kirley, M., Buyya, R.: Multi-objective planning for workflow execution on grids. In: GRID 2007: Proceedings of the 8th IEEE/ACM International Conference on Grid Computing, pp. 10–17. IEEE Computer Society, Washington, DC (2007)
[32] Zeng, B., Wei, J., Wang, W., Wang, P.: Cooperative grid jobs scheduling with multi-objective genetic algorithm. In: Stojmenovic, I., Thulasiram, R.K., Yang, L.T., Jia, W., Guo, M., de Mello, R.F. (eds.) ISPA 2007. LNCS, vol. 4742, pp. 545–555. Springer, Heidelberg (2007)
[33] Zitzler, E., Thiele, L.: Multiobjective optimization using evolutionary algorithms - a comparative case study. In: Eiben, A.E., Bäck, T., Schoenauer, M., Schwefel, H.-P. (eds.) PPSN 1998. LNCS, vol. 1498, pp. 292–304. Springer, Heidelberg (1998)

Dynamic Multi-objective Job Shop Scheduling: A Genetic Programming Approach

Su Nguyen, Mengjie Zhang, Mark Johnston, and Kay Chen Tan

Abstract. Handling multiple conflicting objectives in dynamic job shop scheduling is challenging because many aspects of the problem need to be considered when designing dispatching rules. A multi-objective genetic programming based hyper-heuristic (MO-GPHH) method is investigated here to facilitate the designing task. The goal of this method is to evolve a Pareto front of non-dominated dispatching rules which can be used to support the decision makers by providing them with potential trade-offs among different objectives. The experimental results under different shop conditions suggest that the evolved Pareto front contains very effective rules. Some extensive analyses are also presented to help confirm the quality of the evolved rules. The Pareto front obtained can cover a much wider ranges of rules as compared to a large number of dispatching rules reported in the literature. Moreover, it is also shown that the evolved rules are robust across different shop conditions.

1 Introduction

Job Shop Scheduling (JSS) is a well-known problem in the scheduling literature. Given a set of machines and a set of jobs that need to be processed on those machines in pre-defined orders, the aim of JSS is to find the sequence in which the machines process the jobs to minimise an objective of interest such as mean flowtime or maximum tardiness. JSS problems are usually classified as *static* and *dynamic*. The focus of this work is on a dynamic JSS (DJSS) in which jobs can arrive at random over time and the attributes (processing time, routes through the machines, due date) of jobs is not known in advance.

Su Nguyen · Mengjie Zhang · Mark Johnston
Victoria University of Wellington, New Zealand
e-mail: {su.nguyen,mengjie.zhang,mark.johnston}@msor.vuw.ac.nz

Kay Chen Tan
National University of Singapore, Singapore
e-mail: eletankc@nus.edu.sg

A.Ş. Etaner-Uyar et al. (eds.), *Automated Scheduling and Planning*,
Studies in Computational Intelligence 505,
DOI: 10.1007/978-3-642-39304-4_10,

Dispatching rules have been a major research topic in the DJSS literature because of their practical advantages, e.g., low computational effort and ease to explain to users. In the last few decades, a large number of dispatching rules have been proposed to deal with different requirements in real world applications. Two critical issues with dispatching rules are (1) the ability to incorporate global information from the state of the system and the machines, and (2) the ability to cope with multiple conflicting objectives. Significant effort has been made in the literature to deal with these two issues and new sophisticated dispatching rules have been proposed. However, it is noted that the number of potential dispatching rules is very large, especially when multiple objectives are considered. With the advances in computing power, several machine learning methods [10, 12, 13, 18, 19, 23, 29] have been proposed to discover new effective dispatching rules for DJSS problems. Within these methods, Genetic Programming (GP) [17, 3] is the most popular because it is a straightforward method to evolve dispatching rules in the form of a priority function.

Similar to the human design process, GP also needs to deal with the two issues discussed above. The first issue is normally handled by including different system and machine attributes into the terminal and function sets to allow GP to synthesise those pieces of information into the evolved dispatching rules. Miyashita [19] and Jakobovic and Budin [13] proposed different ways to enhance the quality of the evolved rules by identifying the bottleneck machines. Meanwhile, the second issue has not received much attention in the previous studies of GP for DJSS. Tay and Ho [29] is the first work that focuses on multi-objective DJSS (MO-DJSS) problems. In their study, they converted the multi-objective problem to a single objective problem by optimising a linearly weighted sum of all the objectives. However, this approach is only effective when we have a good knowledge about the search space of the objectives considered, which is not available in most cases. Hildebrandt et al. [10] re-examined the GP system proposed by Tay and Ho [29] in different dynamic job shop scenarios and showed that rules evolved by Tay and Ho [29] are only slightly better than the earliest release date (ERD) rule and quite far away from the performance of the shortest processing time (SPT) rule with mean flowtime as the objective. This suggested that a linear combination of objectives may not be a suitable approach to deal with MO-DJSS. We proposed a cooperative coevolution GP method to evolve Pareto fronts of scheduling policies including dispatching rules and due date assignment rules to deal with MO-DJSS problems where due dates are internally assigned [21]. The results showed that the evolved scheduling policies can dominate the scheduling policies from combinations of existing rules, suggesting that Pareto-based approach is a promising approach for multi-objective DJSS problems.

This work focuses on using GP for evolving dispatching rules for MO-DJSS problems. The goal of this study is to examine the robustness of the evolved rules by testing them using different characteristics of the shop (due date tightness, utilisation, etc.) when multiple conflicting objectives are simultaneously considered. Different from [21], we do not consider the due date assignment decision in this work and the due date tightness is considered as a shop characteristic to assess the robustness of the evolved dispatching rules. The first part of this work gives an overview of DJSS and discusses different GP methods proposed in the literature to deal with DJSS. Then, we develop a multi-objective genetic programming based hyper-heuristic (MO-GPHH) method to design effective dispatching rules for the DJSS. Five scheduling objectives used here are mean flowtime, maximum flowtime (makespan), percentage of tardy jobs, mean tardiness and maximum tardiness. The HaD-MOEA [31] approach is used within in our proposed MO-GPHH to explore the Pareto front of non-dominated evolved dispatching rules. HaD-MOEA is used in this case because it was shown to work well when the number of objectives is large [31]. In this work, the evolved dispatching rules are be trained and tested on different simulated dynamic job shop scenarios. The evolved dispatching rules are also compared to a large number of existing dispatching rules in the DJSS literature and the insights from the evolved rules are discussed. Some selected evolved rules are also presented and analysed to show how they can effectively solve DJSS problems.

2 Background

This section gives a brief background about DJSS and a comprehensive literature review about genetic programming based hyper-heuristics (GPHH) for scheduling, particularly for JSS problems.

2.1 Dynamic Job Shop Scheduling

An example of a dynamic job shop is shown in Fig. 1. Different from static JSS problems when the number of jobs is known and no job will arrive during the timespan of the schedule, DJSS deals with situations in which the new job with unknown processing information will arrive dynamically. Stochastic simulation has been considered as a traditional approach to dealing with DJSS [2] and most of the research focuses on dispatching rules in order to identify the rules that can achieve good performance. Basically, a dispatching rule can be considered as a simple priority function to determine the priorities of jobs waiting in the queues of machines and the job with the highest (or lowest) priority will be processed next. For instance, idle machines select the job in its queue with smallest processing time to process first if shortest processing time (SPT) rule is employed. Since the concept and the implementation of dispatching rules are very straightforward, they have received a lot of attention from both researchers and practitioners.

Until now, there have been hundreds of dispatching rules proposed in the literature to deal with different types of manufacturing environments. Jones and Rabelo [16] categorised dispatching rules into three groups: (1) simple priority rules, which are mainly based on the information related to the jobs; (2) combinations of rules that are implemented depending on the situation that exists on the shop floor; and (3) weighted priority indices which employ more than one piece of information about each job to determine the schedule. Composite dispatching rules (CDR) [25, 15] can also be considered as a version of rules based on weighted priority indices, where

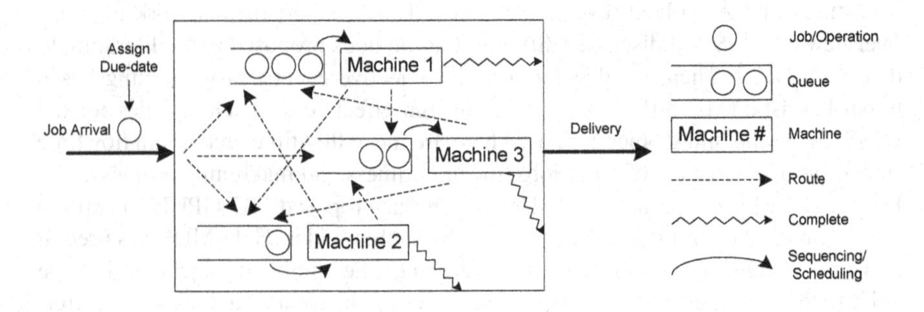

Fig. 1 Example of a dynamic job shop with 3 machines

scheduling information can be combined in more sophisticated ways instead of linear combinations. Panwalkar and Iskander [24] provided a very comprehensive survey on scheduling (dispatching) rules used in research and real world applications using a similar classification. Pinedo [25] also showed various ways to classify dispatching rules based on the characteristics of these rules. The dispatching rules in this case can be classified as *static* and *dynamic* rules, where dynamic rules are time dependent (e.g. minimum slack) and static rules are not (e.g. shortest processing time). Another way to categorise these rules is based on the information used by these rules (either local or global information) to make sequencing decisions. A *local* rule only uses the information available at the machine where the job is queued. A *global* rule, on the other hand, may use the information from other machines. The comparisons of different dispatching rules have been continuously done in many studies [28, 27, 15, 11, 10]. The comparison was usually performed under different characteristics of the shop because it is well-known that the characteristics of the shop can significantly influence the performance of the dispatching rules. Different objectives were also considered in these studies because they are the natural requirements in real world applications. Although many dispatching rules have been proposed, it is still a challenge for scheduling researchers to develop rules that can perform well on many objectives.

2.2 GPHH for Scheduling Problems

Genetic Programming based Hyper-Heuristics (GPHH) is a GP based method that has recently become popular [5]. Since the representation of GP is flexible, it can be easily used to represent heuristics in different forms for different computational problems. Many GPHH methods have also been proposed to automatically generate dispatching rules for different scheduling problems. Dimopoulos and Zalzala [7] employed a simple GP method to evolve dispatching rules for minimising the total tardiness in the single machine scheduling problem. Different scheduling statistics such as processing time and due date are included in the terminal set of GP and these terminals are combined by a function set of standard mathematical operators. The experimental results show that the rules evolved by GP are significantly better than the traditional rules even for some large and unseen instances. Jakobovic et al. [14] also applied GP for developing dispatching rules for the parallel machine scheduling problem in both static and dynamic environments and also showed very competitive results. Geiger et al. [9] presented a learning system that combines GP with a simulation model of an industrial facility. Both static and dynamic environments are also investigated in this study and they showed that the evolved rules are very promising. The paper also proposed a method to learn dispatching rules for multiple machine problems in which GP will evolve multiple trees simultaneously with modified crossover and mutation operators. Comparison with the optimal rule in a simple two-machine environment showed that the evolved rules are rather competitive. Geiger and Uzsoy [8] applied this system to learn dispatching rules for batch processor scheduling and obtained good results. For a stochastic single machine scheduling problem, Yin et al. [32] proposed a GP system employing a bi-tree structured representation scheme to deal with machine breakdowns. The empirical results under different stochastic environments showed that the GP can evolve high-quality predictive scheduling heuristics.

Several GPHH methods have also been proposed for JSS problems. Atlan et al. [1] applied GP for JSS problems. However, the focus of their paper is on finding the solution for a particular problem instance. Miyashita [19] made the first attempt to develop an automatic method using GP to design customised dispatching rules for job shops. In his study, he examined three potential multi-agent models to evolve dispatching rules in multiple machine environments: (1) a homogeneous model where all machines share the same dispatching rule, (2) a distinct agent model where each machine employs its own evolved rule, and (3) a mixed agent model where two rules can be selected to prioritise jobs depending on whether the machine is a bottleneck. The experiments showed that the distinct agent model provided better results in the training stage compared to the homogeneous model but had some over-fitting problems. The mixed agent model was the most robust in all the experiments. However, the use of the mixed agent model depends on the prior-knowledge about the bottleneck machine, which can change in dynamic situations. To handle this issue, Jakobovic and Budin [13] proposed a new GP method called GP-3 to provide some adaptive behaviour for the evolved rules. In their method, GP is used to evolve three components of the rules including a decision tree and two dispatching rules for bottleneck and non-bottleneck machines. The purpose of the decision tree is to identify whether a considered machine is a bottleneck and decide which one of the two evolved rules should be applied. The experiments showed that this method can provide better rules than a simple GP method. However, it is noted that the superior performance of GP-3 will depend the bottleneck machines. If the load levels between machines in the shops are rather similar (existence of multiple bottleneck machines), the information/output from the decision tree in GP-3 may not be very helpful.

Tay and Ho [29] performed a study on using GP for multi-objective JSS problems. In their method, three objectives are linearly combined (with the same weights) into an aggregate objective, which is used as the fitness function in the GP method. The experiments showed that the evolved rules are quite competitive as compared to simple rules but still have trouble dominating the best rule for each single objective. In another study, Hildebrandt et al. [10] explained that the poor performance of the rules evolved by Tay and Ho [29] is caused by the use of a *linear* combination of different objectives and the fact that the randomly generated instances cannot effectively represent the situations that happen in a long term simulation. For that reason, Hildebrandt et al. [10] evolved dispatching rules by training them on four simulation scenarios (10 machines with two utilisation levels and two job types) and only minimised the mean flow time. Some aspects of the simulation models were also discussed in their study. The experimental results showed that the evolved rules were quite complicated but effective as compared to other existing rules. Moreover, these evolved rules are also robust when tested with another environment (50 machines and different processing time distributions). However, their work did not consider how to handle multiple conflicting objectives. We proposed a cooperative coevolution MO-GPHH for multi-objective DJSS problems [21]. In that work, the due dates of new jobs are assumed to be assigned internally and two scheduling rules (dispatching rule and due date assignment rule) are simultaneously considered in order to develop effective scheduling policies. While the representation of the dispatching rules is similar to those in other GP methods, the operation-based representation [22] is used to represent the due date assignment rules. The results showed that the evolved scheduling policies can outperform scheduling policies from different combinations of existing dispatching rules and due-date assignment rules in different simulation scenarios.

Designing an effective dispatching rule is important task to achieve a good scheduling performance in DJSS. Since this is a very complicated process, manual design of effective rules is very challenging. GP has been shown to be a suitable method to facilitate this process. However, most previous studies mainly focus on a single objective while handling multiple objectives is an important issue for real world scheduling applications. It is clear that there have been only a very limited number of studies on MO-DJSS in general and GP for MO-DJSS in particular, mainly focusing on the application aspect. In this work, we provide a comprehensive study on GP for MO-DJSS to point out some crucial problems in this research direction.

3 MO-GPHH for DJSS

This section will show how the proposed MO-GPHH method is used to solve DJSS problems. The first part will show how dispatching rules are represented by GP and how they can be evaluated. Then, the proposed MO-GPHH algorithm is presented. Finally, we describe the simulation model of DJSS which is used for training/testing purposes and the statistical procedure to analyse the results.

3.1 Representation and Evaluation

Similar to previous applications of GP for JSS problems [10, 12, 13, 18, 19, 23, 29], the dispatching rules (DR) here are also represented by GP trees [17]. A GP tree in this case will play the role of a priority function which will determine the priorities of jobs waiting in the queue. As mentioned, more sophisticated representations of the dispatching rules are also proposed in the literature to enhance the quality of the evolved dispatching rules by taking into account bottleneck machines [13, 19]. However, the use of bottleneck machines may not be very useful if multiple bottleneck machines simultaneously exist and may also significantly increase the search space of GP. Therefore, in this work, we only consider the simple GP tree representation of dispatching rules. Moreover, we believe that GP is totally capable of evolving effective priority functions that can incorporate the global/local information of the shop for making sequencing decisions. The terminal and function sets of evolved dispatching rules are presented in Table 1. In this table, the upper part shows a number of terms that usually appear in the dispatching rules in the literature. The next part in this table shows the three terms that reflect the status of the current and downstream machines. It is noted that more global terms can also be used in this case. However, since dispatching rules will need to work in a dynamic environment where the global information can change rapidly, the use of global information may be outdated very soon after the sequencing decisions are made.

An example of the evolved rule is shown in Fig. 2. In the job shop, when a machine is idle and a new job arrives at that machine, that job will be processed immediately. In the case that a machine has just completed a job and there are still jobs waiting in the queue to be processed at that machine, the dispatching rule will be applied. To assign a priority to a waiting job, the information about that job will

Table 1 Terminal and function sets for DR

Symbol	Description
rJ	job release time (arrival time)
RJ	operation ready time
RO	number of remaining operation within the job j.
RT	work remaining of the job
PR	operation processing time
DD	due date d_j
RM	machine ready time
SL	slack of the job $j = \mathtt{DD} - (\mathtt{t} + \mathtt{RT})$
WT	is the current waiting time of the job $= \mathtt{max}(0, \mathtt{t} - \mathtt{RJ})$
#	Random number from 0 to 1
NPR	processing time of the next operation
WINQ	work in the next queue
APR	average operation processing time of jobs in the queue
Function set	$+, -, \times$, and protected division %, min, max, abs, and If

*t is the time when the sequencing decision is made.

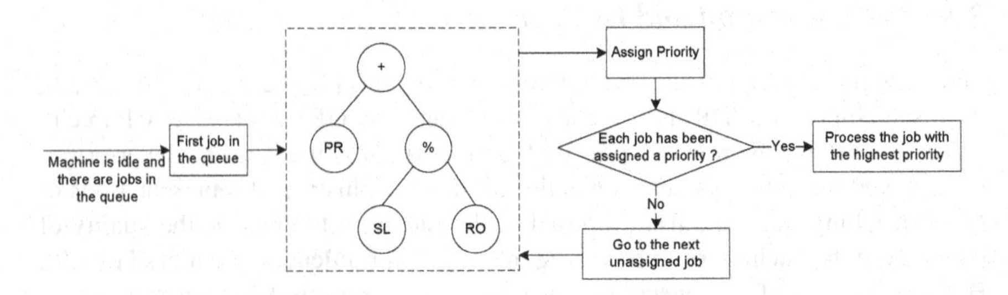

Fig. 2 Illustration of a dispatching rule

be extracted to be used in the terminals in Table 1. Then, the GP tree representing the dispatching rule will be evaluated and the output from this evaluation will be assigned to the considered job as its priority (refer to [17] for detailed discussion on how a GP tree is evaluated). This procedure will be applied until priorities are assigned to all waiting jobs and the job with highest priority will be processed next.

3.2 The Proposed MO-GPHH Algorithm

In this work, we want to evolve dispatching rules to minimise five popular objectives in the DJSS literature, which are the mean flowtime, maximum flowtime, percentage of tardy jobs, mean tardiness, and maximum tardiness [28, 11, 27]. The HaD-MOEA algorithm [31] is applied here to explore the Pareto front of non-dominated dispatching rules regarding the five objectives mentioned above. HaD-MOEA can be considered as an extension of NSGA-II [6] and it was shown to work well on the problems with many objectives. Algorithm 1 shows how the proposed MO-GPHH works. At first, a number of training simulation scenarios (more details will be shown in the next section) are loaded and the initial archive $\mathcal{P}^e$ (parent population) is empty. These scenarios will be used to evaluate the performance of an evolved dispatching rule. The initial GP population is created using the ramped-half-and-half method [17]. In each generation of MO-GPHH, all individuals in the population will be evaluated by applying them to each simulation scenario. The quality of the each individual in the population will be measured by the average value of the objectives across all simulation scenarios. After all individuals have been evaluated, we calculate the Harmonic distance [31] for each individual. Then, individuals in both archive $\mathcal{P}^e$ and population P are selected to update the archive $\mathcal{P}^e$ based on the Harmonic distance and the non-dominated rank [6]. The new population will generated by applying crossover and mutation to the current population. For crossover, GP uses the subtree crossover [17], which creates new individuals for the next generation by randomly recombining subtrees from two selected parents. Mutation is performed by subtree mutation [17], which randomly selects a node of a chosen individual in the population and replaces the subtree rooted at that node by a newly randomly-generated subtree. Binary tournament selection [6] is used to select the parents for the two genetic operations. The crossover rate and mutation rate used in

Algorithm 1. MO-GPHH to evolve dispatching rules for DJSS problems

load training simulation scenarios $\mathbb{S} \leftarrow \{S_1, S_2, \ldots, S_T\}$
randomly initialise the population $P \leftarrow \{\mathscr{R}_1, \mathscr{R}_2, \ldots, \mathscr{R}_{popsize}\}$
$\mathscr{P}^e \leftarrow \{\}$ and $generation \leftarrow 0$
while $generation \leq maxGeneration$ **do**
  foreach $\mathscr{R}_i \in P$ **do**
    $\mathscr{R}_i.objectives \leftarrow$ apply $\mathscr{R}_i$ to each scenario $S_k \in \mathbb{S}$
  end
  calculate the Harmonic distance [31] and the ranks for individuals in $P \bigcup \mathscr{P}^e$
  $\mathscr{P}^e \leftarrow$ select($P \bigcup \mathscr{P}^e$)
  $P \leftarrow$ apply crossover, mutation to $\mathscr{P}^e$
  $generation \leftarrow generation + 1$
end
return $\mathscr{P}^e$

the three methods are 90% and 10%, respectively. The maximum depth of GP trees is 8. A population size of 200 is used in this study and the results will be obtained after the proposed method runs for 200 generations.

3.3 Simulation Models for Dynamic Job Shop

Simulation is the most popular method to evaluate the performance of dispatching rules in the DJSS literature. Since our goal is to evolve robust dispatching rules, a general job shop would be more suitable than a specific shop. The following factors characterise a job shop:

- Distribution of processing times ($F1$)
- Utilisation ($F2$)
- Due date tightness ($F3$)

Utilisation indicates the congestion level of machines (and the shop). The performances of the scheduling decisions under different utilisation levels are of interest in most research in the DJSS literature. Meanwhile, the distribution of processing times and the due date tightness are also very important factors that can influence the performance of a dispatching rule. In this study, we employ a symmetrical (balanced) job shop model in which each operation of a job has equal probability to be processed at any machine in the shop. Therefore, machines in the shop expect to have the same level of congestion in long simulation runs. This model has been used very often in the DJSS literature [11, 27, 10]. The scenarios for training and testing of dispatching rules are shown in Table 2.

The simulation experiments have been conducted in a job shop with 10 machines. The triplet $\langle m, u, c \rangle$ represents the simulation scenario in which the average processing time is m (m is 25 or 50 when processing times follow discrete uniform distribution [1,49] or [1,99], respectively), the utilisation is u% and the allowance factor is c. In the training stage, two simulation scenarios (corresponding to the two utilisation

Table 2 Training and testing scenarios

Factor	Training	Testing
$F1$	Discrete Uniform[1,49]	Discrete Uniform[1,49] and [1,99]
$F2$	70%, 80%	85%, 95%
$F3$	c is randomly selected from $(3,5,7)$	$c=4$, $c=6$, $c=8$
Summary	$\langle 25,70,(3,5,7)\rangle$, $\langle 25,80,(3,5,7)\rangle$	$\langle 25,85,4\rangle$, $\langle 25,85,6\rangle$, $\langle 25,85,8\rangle$, $\langle 25,95,4\rangle$, $\langle 25,95,6\rangle$, $\langle 25,95,8\rangle$, $\langle 50,85,4\rangle$, $\langle 50,85,6\rangle$, $\langle 50,85,8\rangle$, $\langle 50,95,4\rangle$, $\langle 50,95,6\rangle$, $\langle 50,95,8\rangle$

Total Work Content (TWK) [2] with allowance factor c is used to set the due dates.

levels) and five replications will be performed for each scenarios. The average value for each objective from $2 \times 5 = 10$ replications will be used to measure the quality of the evolved rules (as described in the previous section). We use the shop with different characteristics here in order to evolve rules with good generality. The allowance factors, which decide the due date tightness, are selected randomly from the three values 3, 5, and 7 instead of a fixed allowance factor (for each scenario) in common simulation experiments for DJSS problems. If we train on scenarios with fixed allowance factors, the evolved rules will tend to focus more on the scenarios with small allowance factors to improve the due date performance (mean tardiness, maximum tardiness, etc.) because the values of the due date based objectives are higher in these cases. This may cause an overfitting problem for the evolved dispatching rules. Moreover, training on different scenarios with different fixed allowance factors will also increase the training time of our MO-GPHH method. Simulating multiple utilisation levels in a simulation scenario can be used to reduce the number of training scenarios but will increase significantly the running time of a replication to obtain the steady state performance of the rules, and indirectly increase the training time of the MO-GPHH method.

In the testing stage, 12 simulation scenarios with 50 replications for each scenario (or shop condition) resulting in $12 \times 50 = 600$ replications will be used to have a comprehensive assessment of the evolved rules. In each replication of a simulation scenario, we start with an empty shop and the interval from the beginning of the simulation until the arrival of the 500th job is considered as the warm-up time and the statistics from the next completed 2000 jobs [11] will be used to calculate the five objective values. The number of operations for each new job is randomly generated from the discrete uniform distribution [2,14] and the routing for each job is randomly generated, with each machine having equal probability to be selected (re-entry is allowed here but consecutive operations are not processed on the same machine). The arrival of jobs will follow a Poisson process with the arrival rate adjusted based on the utilisation level.

Table 3 gives formal definitions of the five objectives considered in this work. In this table, $\mathbb{C}$ is the collection of jobs recorded from a simulation run (2000 jobs) and $\mathbb{T} = \{j \in \mathbb{C} : C_j - d_j > 0\}$ is the collection of tardy jobs where C_j, f_j and d_j are

Table 3 Performance measures of dispatching rules

Mean Flowtime	$F = \frac{\sum_{j \in \mathbb{C}} f_j}{\vert\mathbb{C}\vert}$
Maximum Flowtime	$F_{max} = \max_{j \in \mathbb{C}}\{f_j\}$
Percentage of Tardy Jobs	$\%T = 100 \times \frac{\vert\mathbb{T}\vert}{\vert\mathbb{C}\vert}$
Mean Tardiness	$T = \frac{\sum_{j \in \mathbb{T}} C_j - d_j}{\vert\mathbb{T}\vert}$
Maximum Tardiness	$T_{max} = \max_{j \in \mathbb{T}}\{C_j - d_j\}$

the completion time, flowtime and due date of job j, respectively. The objectives are selected since they are very popular performance measures of dispatching rules for DJSS problems, which have been used regularly in previous studies [27, 15, 11].

3.4 *Benchmark Dispatching Rules*

Table 4 shows 31 dispatching rules that will be used to compare with the evolved rules in our work. The upper part of this table shows some original dispatching rules proposed and the lower part shows some extensions of the original rules that have been proposed in the literature. The parameters of ATC and COVERT are the same as those used in Vepsalainen and Morton [30] ($k = 3$ for ATC, $k = 2$ for COVERT, and the leadtime estimation parameter $b = 2$). More detailed discussion on these rules can be found in [30, 27, 24, 15, 11].

3.5 *Statistical Analysis*

Since DJSS is a stochastic problem, statistical analysis is required to compare the performance of dispatching rules obtained from simulation. In this work, we use the one-way ANOVA and Duncan's multiple range tests [20] to compare the performance of rules or a set of rules for each objective since this statistical analysis has been used in previous studies [27, 15, 11].

It is interesting to note that Pareto-dominance has not been considered before in the dispatching rule literature, even though it is an important concept in the multi-objective optimisation domain. Most studies on dispatching rules have been done mainly based on a single objective even when multiple objectives are investigated. The reason is that the focus of previous studies is on minimising a single objective and the performance on other objectives are not of interest. Also, since DJSS is a stochastic problem, statistical analysis is necessary to examine the Pareto-dominance of rules but there is no standard statistical procedure available for this task. In this work, we describe two procedures to check for the *statistical Pareto-dominance* of between two different rules.

Table 4 Benchmark dispatching rules

SPT	shortest processing time	LPT	longest processing time
EDD	earliest due date	FDD	earliest flow due date
FIFO	first in first out	LIFO	last in first out
LWKR	least work remaining	MWKR	most work remaining
NPT	next processing time	WINQ	work in next queue
CR	critical ratio	AVPRO	average processing time/operation
MOD	modified due date	MOPNR	most operations remaining
SL	negative slack	Slack	slack
PW	process waiting time	RR	Raghu and Rajendran
ATC	apparent tardiness cost	COVERT	cost over time
OPFSLK/PT	operational flow slack per processing time		
LWKR+SPT	least work remaining plus processing time		
CR+SPT	critical ratio plus processing time		
SPT+PW	processing time plus processing wating time		
SPT+PW+FDD	SPT+PW plus earliest flow due date		
Slack/OPN	slack per remaining operations		
Slack/RPT+SPT	slack per remaining processing time plus operation processing time		
PT+WINQ	procesting time plus work in next queue		
2PT+WINQ+NPT	double processing time plus WINQ and NPT		
PT+WINQ+SL	processing time plus WINQ and slack		
PT+WINQ+NPT+WSL	PT+WINQ plus next processing time and waiting slack		

3.5.1 Objective-Wise Procedure

In multi-objective optimisation, solution (or dispatching rule in this work) a is said to *Pareto-dominate* solution b if and only if $\forall i \in \{1,2,\ldots,n\} : f_i(a) \leq f_i(b) \wedge \exists j \in \{1,2,\ldots,n\} : f_j(a) < f_j(b)$ where n is the number of objective functions to be minimised. However, if $f_j(a)$ and $f_j(b)$ are random variables (i.e. solutions a and b produce different outputs in different runs/replications), we cannot use the above definitions to check for the Pareto-dominance. Therefore, we need to redefine the Pareto-dominance for this context. For the objective-wise procedure, solution a *statistically Pareto-dominates* solution b if and only if $\forall i \in \{1,2,\ldots,n\} : f_i(a) \leq_{\mathscr{T}} f_i(b) \wedge \exists j \in \{1,2,\ldots,n\} : f_j(a) <_{\mathscr{T}} f_j(b)$, where $f_i(a) \leq_{\mathscr{T}} f_i(b)$ means that a is significantly smaller (better) than or not significantly different from b based on the statistical test $\mathscr{T}$ (e.g. *z-test*); similarly, $f_j(a) <_{\mathscr{T}} f_j(b)$ means that a is significantly smaller than b based on $\mathscr{T}$. It should be noted that since multiple comparisons (n comparisons for n objectives) need to be done here, we have to adjust the value of the pre-set probability α of a type-1 error [20] in order to control the false positive rate. Many methods have been proposed for this problem such as Bonferroni method, Scheffe method, etc [26].

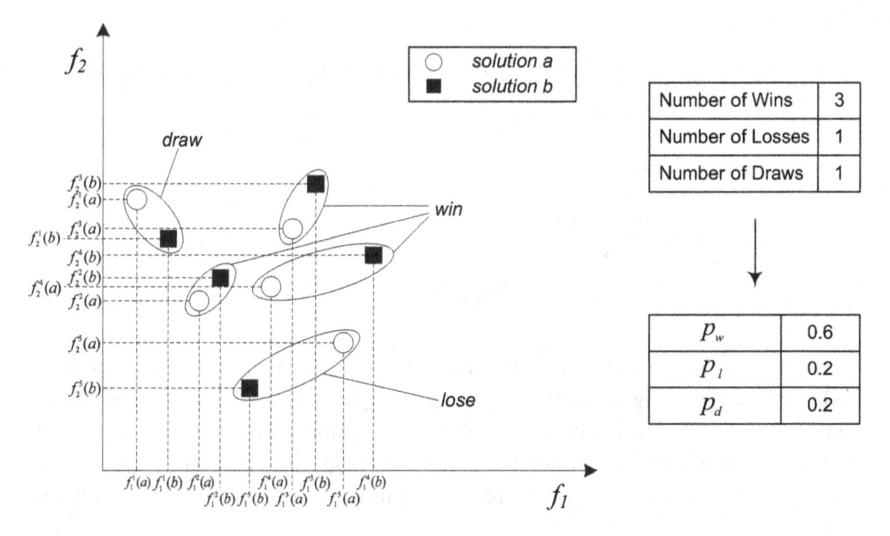

Fig. 3 Counting wins, loss and draws in replication-wise procedure

3.5.2 Replication-Wise Procedure

Different from the above method that examines the Pareto-dominance of two solutions based on the relative performance of each objective, the replication-wise procedure focuses on the Pareto-dominance in each replication/observation to detect the difference between two solutions. This procedure is adapted from the method proposed by Bhowan et al. [4] to compare the performance of different multi-objective GP methods on a run-by-run basis and determine whether a method significantly dominates another over all runs. In this procedure, the traditional Pareto-dominance is used to examine the dominance relation between two solutions in each replication. For instance, $f_j(a) = \{f_j^1(a), f_j^2(a), \ldots, f_j^N(a)\}$ and $f_j(b) = \{f_j^1(b), f_j^2(b), \ldots, f_j^N(b)\}$ are the values for objective j obtained by solutions a and b from N replications. In replication k, $\{f_1^k(a), f_2^k(a), \ldots, f_n^k(a)\}$ is compared to $\{f_1^k(b), f_2^k(b), \ldots, f_n^k(b)\}$ to determine the Pareto-dominance between a and b in this replication. Three possible outcomes from the comparison are (1) win for a if a dominates b, (2) loss for a if b dominates a, or (3) draw otherwise. The proportions of win (p_w), lose (p_l), and draw (p_d) over N replications is then recorded. Fig. 3 gives an example to show how p_w, p_l and p_d are calculated in the case with two objectives and $N = 5$. The outcomes here form a *multinominal* distribution since the proportions or probabilities for all outcomes always sum to one. In a multinominal distribution, the $(1-\alpha)\%$ confidence interval of the difference in the probability of win and lose ($p_w - p_l$) can be calculated as followed:

$$(p_w - p_l) \pm z_{\alpha/2}\sqrt{var(p_w - p_l)} \quad (1)$$

where

$$\begin{aligned}
var(p_w - p_l) &= var(p_w) + var(p_l) - (var(p_w + p_l) - var(p_w) - var(p_l)) \\
&= 2var(p_w) + 2var(p_l) - var(p_w + p_l) \\
var(p_w) &= \frac{p_w(1-p_w)}{N} \\
var(p_l) &= \frac{p_l(1-p_l)}{N} \\
var(p_w + p_l) &= \frac{(p_w + p_l)(1 - p_w - p_l)}{N}
\end{aligned}$$

The confidence interval obtained by the equation (1) can be used to determine whether one solution significantly dominates the other. Basically, if the lower bound of the confidence interval is positive, solution a significantly dominates solution b. If the upper bound of the confidence interval is negative, solution b significantly dominates solution a. Otherwise, there is no significant dominance between the two solutions.

There are some key differences between these two procedures. While the objective-wise procedure focuses more on the magnitude of the difference between average objectives obtained by the two methods, the replication-wise procedure only cares about the Pareto dominance regardless of the difference between the obtained objective values in each replication. If the variances of the objectives obtained from the simulation are high, the replication-wise procedure may not accurately determine the statistical Pareto dominance between two solutions. For example, when p_w and p_l are very close, it is very likely the replication-wise procedure will conclude that there is no dominance between the two solutions. However, it is intuitively not true if there are some *"big"* wins (there are large difference between pairs of objective values $f_j^k(a)$ and $f_j^k(b)$ for some $j \in \{1,2,\ldots,n\}$) for a solution in some replications. The advantage of the replication-wise procedure is that one statistical significance test needs to be performed as compared to multiple tests (which make the procedure more complicated) in the objective-wise procedure. In the later section, we will apply both procedures to determine the *statistical Pareto-dominance* between evolved dispatching rules and the dispatching rules reported in the literature.

4 Results

We perform 30 independent runs of the proposed MO-GPHH method and the non-dominated evolved rules from the evolved Pareto front $\mathscr{P}^e$ are recorded. We perform a post-processing step to extract the Pareto front $\mathscr{P}$ from $\mathscr{P}^e$ for each testing scenario based on the average values of five objectives in that scenario. The performance of the evolved rules in $\mathscr{P}$ will be presented in this section. We first examine the quality of these rules for each single objective. Then, we show the Pareto dominance of the evolved dispatching rules as compared to the dispatching rules reported in the literature.

4.1 Single Objective

Even though our target is to solve the MO-DJSS problems, it is important to know whether the evolved rules can provide satisfactory results for each single objective. This is also a good opportunity to make a proper comparison of the evolved dispatching rules from a multi-objective GP method and the existing rules which are usually designed for a specific objective. Fig. 4 and Fig. 5 show the performance of the evolved rules for each objective under different shop conditions. For each GP run, the evolved rule within $\mathscr{P}$ that performs best on the objective $\mathscr{O}$ ($\mathscr{O}$ can be F,F_{max}, $\%T$, T, or T_{max}) is denoted as $\mathscr{R}^*_{\mathscr{O}}$. The left box-plot in each plot in Fig. 4 and Fig. 5 represents the average values of the objective $\mathscr{O}$ obtained by $\mathscr{R}^*_{\mathscr{O}}$ from the 30 GP runs. The right box-plot shows the corresponding values obtained by the top five rules among the 31 existing rules shown in Table 4.

A quick observation of Fig. 4 and Fig. 5 shows that the proposed MO-GPHH method can effectively find rules that are better than, or as competitive as, the best existing dispatching rules for each objective under different shop conditions. The evolved rule $\mathscr{R}^*_{\mathscr{O}}$ can dominate the existing rules regarding F, F_{max}, $\%T$, and T. For T_{max}, the proposed MO-GPHH can find the rules that dominate the majority of the existing rules and the obtained $\mathscr{R}^*_{\mathscr{O}}$ from some GP runs can also dominate the best existing rule. This suggests that it is totally possible to evolve a superior rule for each single objective by the proposed MO-GPHH. However there are objectives that are more difficult to minimise, e.g., T_{max} in this case. Given that we try to evolve rules to minimise five objectives simultaneously in the general case, the results obtained here for single objective are very competitive.

Further statistical tests are also performed here to confirm the quality of the evolved rules. For a specific objective $\mathscr{O}$ and shop condition $\langle m,u,c\rangle$, we perform statistical analysis of the $\mathscr{R}^*_{\mathscr{O}}$ rule from each GP run and the best five dispatching rules in the literature (based on the average values of the corresponding objective) using the one-way ANOVA and Duncan's multiple range tests [20] with $\alpha = 0.01$. The summary of all statistical tests is shown in Table 5. For each shop condition, the first row shows the number of times the proposed MO-GPHH method is able is find the $\mathscr{R}^*_{\mathscr{O}}$ that is significantly better than the best existing rule for minimising $\mathscr{O}$, which is shown in the second row. In general, the results here are similar to those shown in Fig. 4 and Fig. 5. It is clear that the MO-GPHH method can always find a superior rule for minimising F while 2PT+WINQ+NPT is the best existing rule. These observations are consistent with those in [10] and [11]. Similar to [10] when using GP to evolve rules for minimising F, the evolved rules can easily beat 2PT+WINQ+NPT across different simulation scenarios. For F_{max}, the evolved rules also dominate the best rule, i.e., SPT+PW+FDD in this case, in the majority of GP runs. It is interesting to note that the 2PT+WINQ+NPT rule and SPT+PW+FDD rule are always the best existing rules for the two objectives (F and F_{max}) under all shop conditions. This suggests that the shop condition does not really have a big impact on the performance of the rules. However, the complexity of the objective may make the design of an effective rule more difficult.

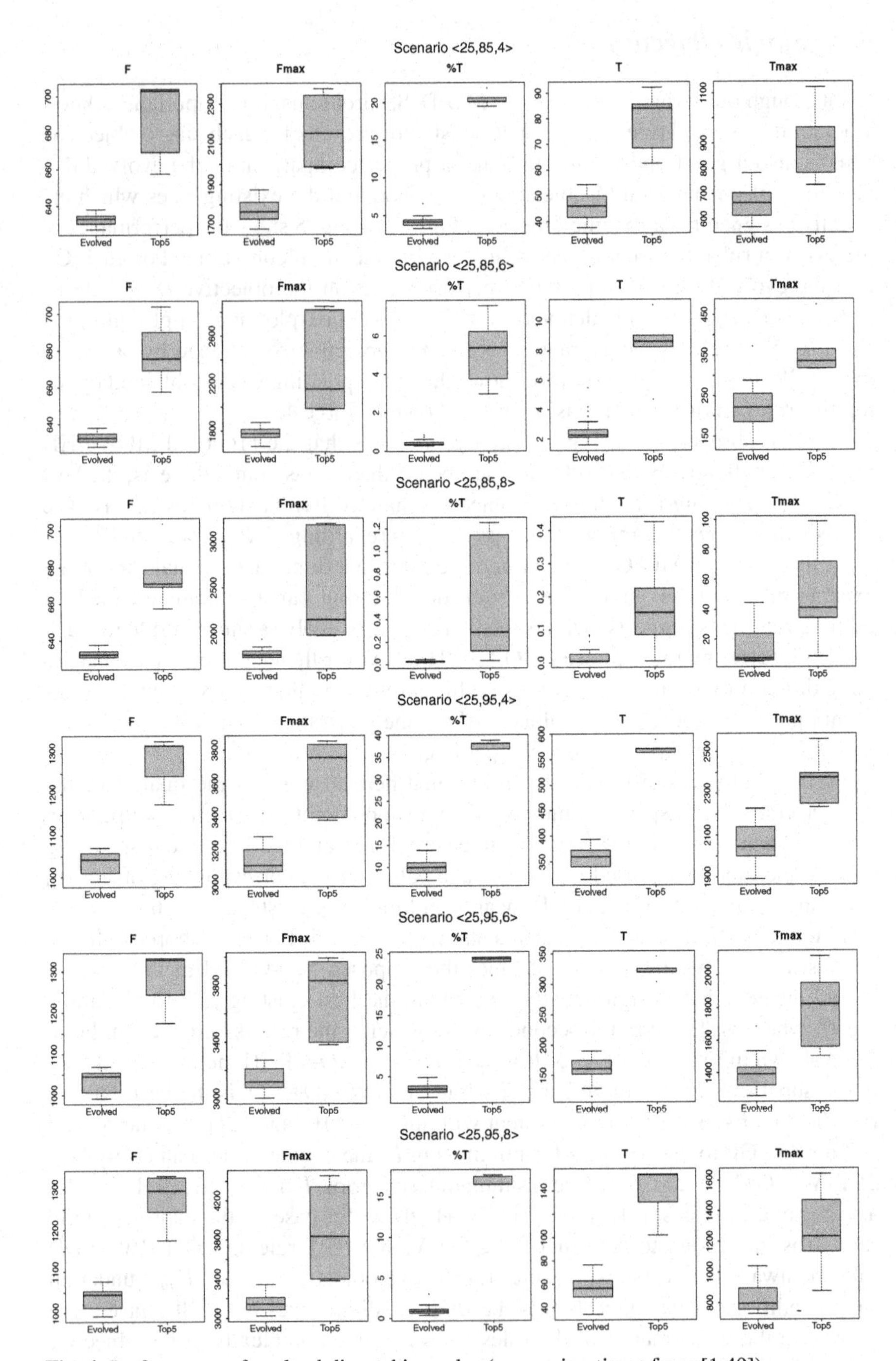

Fig. 4 Performance of evolved dispatching rules (processing times from [1,49])

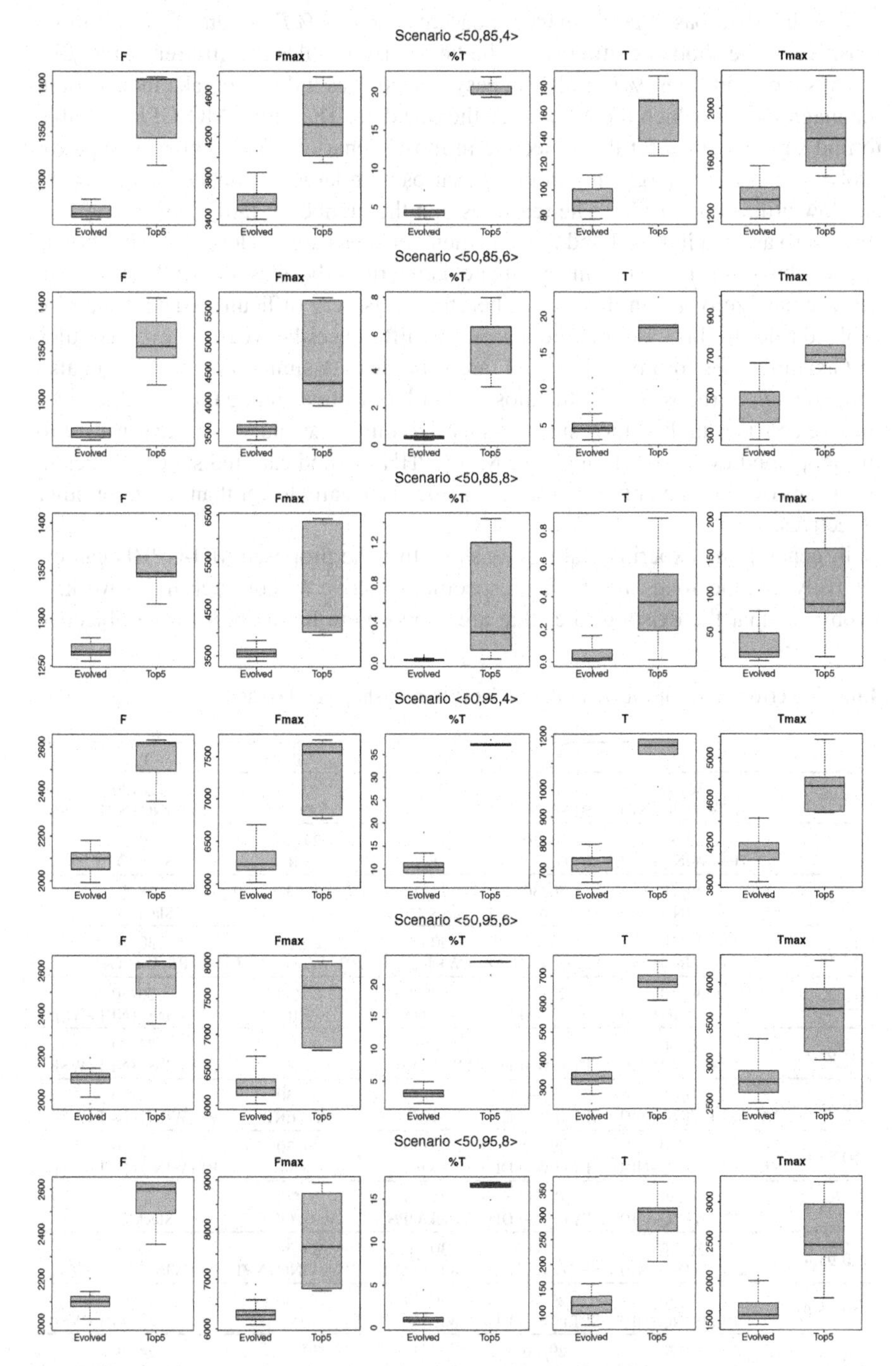

Fig. 5 Performance of evolved dispatching rules (processing times from [1,99])

The due date based performance measures such as $\%T$, T, and T_{max} are more sensitive to the shop condition since the best existing rules are different under different shop conditions. $\%T$ is also an easy objective as it does not take into account the magnitude in which the job misses the due date. Therefore, MO-GPHH is able to find superior rules for this objective in most scenarios. The number of superior evolved rules is not large only in the scenarios with large allowance factor ($c = 8$) and low utilisations (85%). The reason is that the number of tardy jobs is very low (near zero as seen in Fig. 4 and Fig. 5) when due dates are too loose and the shop is not very busy. It is noted that many other existing rules (besides Slack/OPN) can also achieve near zero $\%T$ in this case. Therefore, it is very difficult to detect superior evolved rules in this case. In other cases, the differences between the evolved rules and existing rules for minimising $\%T$ are very clear. A similar conclusion can also applied to T. Perhaps, T_{max} is the most difficult objective among the five objectives that we consider in this study since it is hard to minimise and also quite sensitive to the shop condition. Even though our MO-GPHH method can find superior rules in most runs overall, the number of superior rules is usually lower than those for other objectives.

In general, the experimental results show that the proposed MO-GPHH can effectively find the good rules for each specific objective we consider in this work. It is obvious that the existing rules that are supposed to be the best for an objective

Table 5 Performance of evolved rules under different shop conditions

		F	F_{max}	%T	T	T_{max}
⟨25,85,4⟩	*	30/30	30/30	30/30	30/30	26/30
	**	2PT+WINQ+NPT	SPT+PW+FDD	MOD	COVERT	PT+WINQ+NPT+WSL
⟨25,85,6⟩	*	30/30	30/30	30/30	29/30	29/30
	**	2PT+WINQ+NPT	SPT+PW+FDD	RR	RR	Slack/OPN
⟨25,85,8⟩	*	30/30	30/30	25/30	17/30	18/30
	**	2PT+WINQ+NPT	SPT+PW+FDD	Slack/OPN	Slack/OPN	Slack/OPN
⟨25,95,4⟩	*	30/30	29/30	30/30	30/30	30/30
	**	2PT+WINQ+NPT	SPT+PW+FDD	LWKR+SPT	2PT+WINQ+NPT	PT+WINQ+SL
⟨25,95,6⟩	*	30/30	29/30	30/30	30/30	26/30
	**	2PT+WINQ+NPT	SPT+PW+FDD	LWKR+SPT	RR	PT+WINQ+NPT+WSL
⟨25,95,8⟩	*	30/30	30/30	30/30	30/30	22/30
	**	2PT+WINQ+NPT	SPT+PW+FDD	LWKR+SPT	RR	PT+WINQ+NPT+WSL
⟨50,85,4⟩	*	30/30	29/30	30/30	30/30	28/30
	**	2PT+WINQ+NPT	SPT+PW+FDD	MOD	COVERT	PT+WINQ+NPT+WSL
⟨50,85,6⟩	*	30/30	29/30	30/30	29/30	29/30
	**	2PT+WINQ+NPT	SPT+PW+FDD	RR	RR	PT+WINQ+NPT+WSL
⟨50,85,8⟩	*	30/30	30/30	24/30	10/30	8/30
	**	2PT+WINQ+NPT	SPT+PW+FDD	Slack/OPN	Slack/OPN	Slack/OPN
⟨50,95,4⟩	*	30/30	30/30	30/30	30/30	30/30
	**	2PT+WINQ+NPT	SPT+PW+FDD	LWKR+SPT	2PT+WINQ+NPT	PT+WINQ+NPT+WSL
⟨50,95,6⟩	*	30/30	29/30	30/30	30/30	25/30
	**	2PT+WINQ+NPT	SPT+PW+FDD	LWKR+SPT	RR	PT+WINQ+NPT+WSL
⟨50,95,8⟩	*	30/30	29/30	30/30	30/30	24/30
	**	2PT+WINQ+NPT	SPT+PW+FDD	LWKR+SPT	RR	PT+WINQ+NPT+WSL

can also be outperformed by the evolved rules. Since we evolved the Pareto front of non-dominated rules for five objectives with a modest population of 200 individuals, the method may not always find the superior rules for some hard objectives. However, as shown in Table 5, because the shop condition can impact the performance of dispatching rules and their relative performance, the rules that are superior under one shop condition may not be the superior one under the other shop conditions. Therefore, evolving a set of non-dominated rules in our method is actually more beneficial than evolving a single rule (either for single objective in [10] or aggregate objective of multiple objective in [29]) since it can provide potential rules to deal with different shop conditions.

4.2 Multiple Objectives

The comparison above has shown that the proposed MO-GPHH method can simultaneously evolve superior rules for each specific objective. However, these superior performances may not come without any trade-off on other objectives. Previous studies have shown that there is no dispatching rule that can minimise all objectives. Therefore, dispatching rules in the literature are designed for minimising a specific objective only. Although it is true that these rules can effectively minimise the objective that it focuses on, it usually deteriorates other objectives significantly. For example, the 2PT+WINQ+NPT rule can successfully reduce the average flowtime but it performs badly on almost all other objectives. Since the existence of multiple conflicting objectives is a natural requirement in real world scheduling applications, it is crucial to include this issue into the design process of dispatching rules as well. In this part, we will examine the *Pareto-dominance* of the evolved rules against other dispatching rules in the literature.

For each MO-GPHH run, the evolved rules in the Pareto front $\mathscr{P}$ are compared to the set $\mathscr{D}$ of 31 benchmark dispatching rules. For each shop condition, we will employ the objective-wise (OBJW) and replications-wise (REPW) procedures discussed in Section 3.5 to determine the statistical Pareto dominance between each pair $(\mathscr{R}_i, \mathscr{B}_j)$ for all $\mathscr{R}_i \in \mathscr{P}$ and $\mathscr{B}_j \in \mathscr{D}$. Therefore, there are $|\mathscr{P}| \times |\mathscr{D}|$ comparisons in total for each MO-GPHH run and each statistical procedure. Both OBJW and REPW procedures will be performed with $\alpha = 0.01$. In the OBJW procedure, we use the Bonferroni method [20] to adjust the value of $\alpha^t = \alpha/n$ in each z-test (for each objective). From this point forward, we use *dominate* or *dominance* when mentioning about the statistical Pareto-dominance, unless otherwise indicated. After all the comparisons in each MO- GPHH run were done, an evolved dispatching rule $\mathscr{R}_i$ is classified into three categories:

1. *Non-dominated* if there is no dominance between $\mathscr{R}_i$ and $\mathscr{B}_j$ for $\forall \mathscr{B}_j \in \mathscr{D}$.
2. *Dominating* if $\mathscr{R}_i$ is not dominated by any $\mathscr{B}_j \in \mathscr{D}$ and $\exists \mathscr{B}_j \in \mathscr{D}$ such that $\mathscr{R}_i$ dominates $\mathscr{B}_j$.
3. *Dominated* if $\exists \mathscr{B}_j \in \mathscr{D}$ such that $\mathscr{R}_i$ is dominated by $\mathscr{B}_j$.

The proportions of evolved rules in the three categories for each $\mathscr{P}$ is determined and the average proportions from 30 MO-GPHH runs are shown in Fig. 6. The

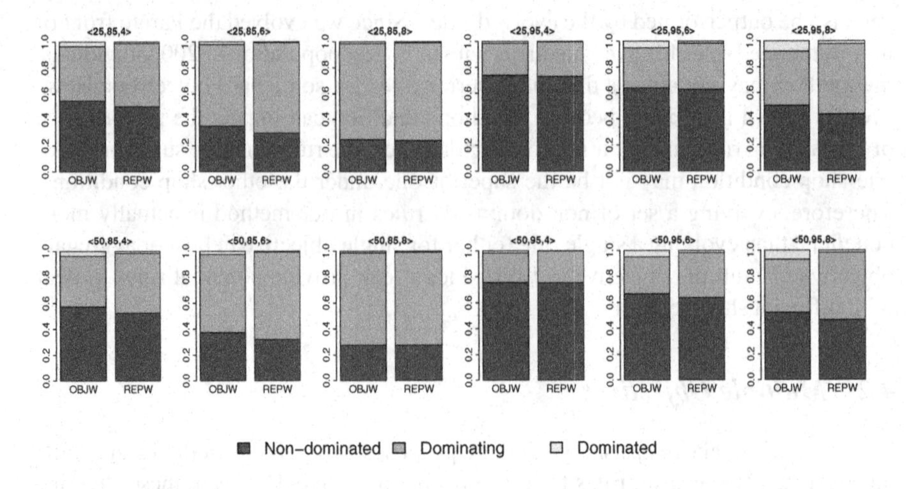

Fig. 6 Average Pareto dominance proportion of evolved dispatching rules

triplets in the figure indicate the shop conditions as explained in the previous section. It is clear that the proposed MO-GPHH method can always find rules that can dominate rules reported in the literature across all objectives. In the worst cases $\langle 25,95,4\rangle$ and $\langle 50,95,4\rangle$, there are still about 20% of the evolved rules are dominating rules. The number of dominated evolved rules are also very low and the highest proportions (about 10%) of dominated rules are in $\langle 25,95,8\rangle$ and $\langle 50,95,8\rangle$. There are also some interesting patterns in Fig. 6. Different from our comparison for single objective when there are fewer superior rules found when the allowance factor increases, it is easy to see that the number of non-dominated rules decreases and the number of dominating rules increases when the allowance factor increases from 4 to 8. This suggests that even when the MO-GPHH method cannot find a superior rule for a specific objective, it can easily find rules that can perform as good as the best existing rule on that objective while significantly improving other objectives. Another interesting pattern in Fig. 6 is that the number of dominated rules increases when the allowance factor increases with the shop utilisation of 85%. However, a reverse trend is found with the utilisation of 95% when the number of dominated rules decreases when the allowance factor increases. For the cases with utilisation of 85%, the higher allowance made the DJSS problems easier, at least for the due date based performance measure. Therefore, it is difficult for existing rules to dominate the evolved dispatching rules. In the case with utilisation of 95% and low allowance factor, it is very difficult to make a good sequencing decision to satisfy multiple objectives and to find a rule that is superior on all objectives. For that reason, the number of dominating and dominated rules are relatively small compared to the number of non-dominated rules. When the utilisation is 95% and the allowance factor is high, the number of dominated rules increases because these shop conditions (very busy shop and loose due dates) are quite different from the shop conditions used in the training stage.

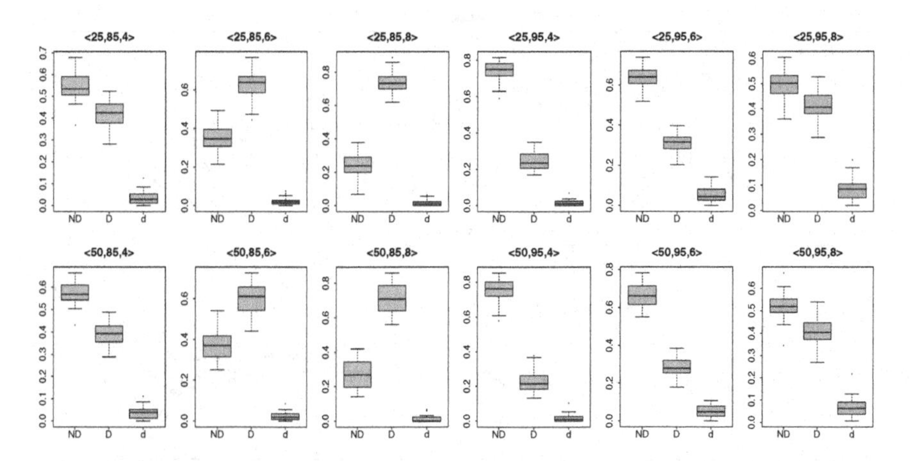

Fig. 7 Pareto dominance proportions of evolved rules

It is also noted that the results from OBJW and REPW in Fig. 6 are very consistent. The REPW procedure results in more dominating rules (fewer non-dominated rules) as compared to the OBJW procedure. Perhaps, this is because the OBJW procedure with the Bonferroni adjustment method is quite conservative, which makes the OBJW procedure more difficult to detect significant differences between two rules. However, the differences between the two procedure in our application is very small. Therefore, both OBJW and REPW are suitable procedures to analyse the results from our experiments. A more detailed Pareto dominance of evolved rules is shown in Fig. 7. In this figure, the box-plots represent the proportions from the OBJW procedure of non-dominated (ND), dominating (D) and dominated (d) from each MO-GPHH. This figure shows that the proposed MO-GPHH is quite stable since the obtained dominance proportions have low variances. Moreover, the proportions of non-dominated and dominating rules are always larger than that of dominated rules. In general, these results suggests that the evolved dispatching rules are significantly better or at least very competitive when compared to the existing dispatching rules.

Through all the comparisons, we also count the number of dominating evolved rules (NDER) in each MO-GPHH run that dominate a specific rule $\mathscr{B}_j$. These values can be used as an indicator for the competitiveness of the existing dispatching rules when multiple objectives are considered. The values of NDER for each rule $\mathscr{B}_j$ shown in Table 4 under different shop conditions from 30 MO-GPHH runs are shown in Fig. 8 and Fig. 9. In these figures, the rules are arranged from left to right in the order of decreasing values of the average NDER. It is quite obvious that the MO-GPHH method can easily evolve rules that dominate the simple rules such as LPT, MWKR, FIFO, etc. It is noted that most rules with low values of NDER are the ones which are designed for minimising due date based performance measures and the

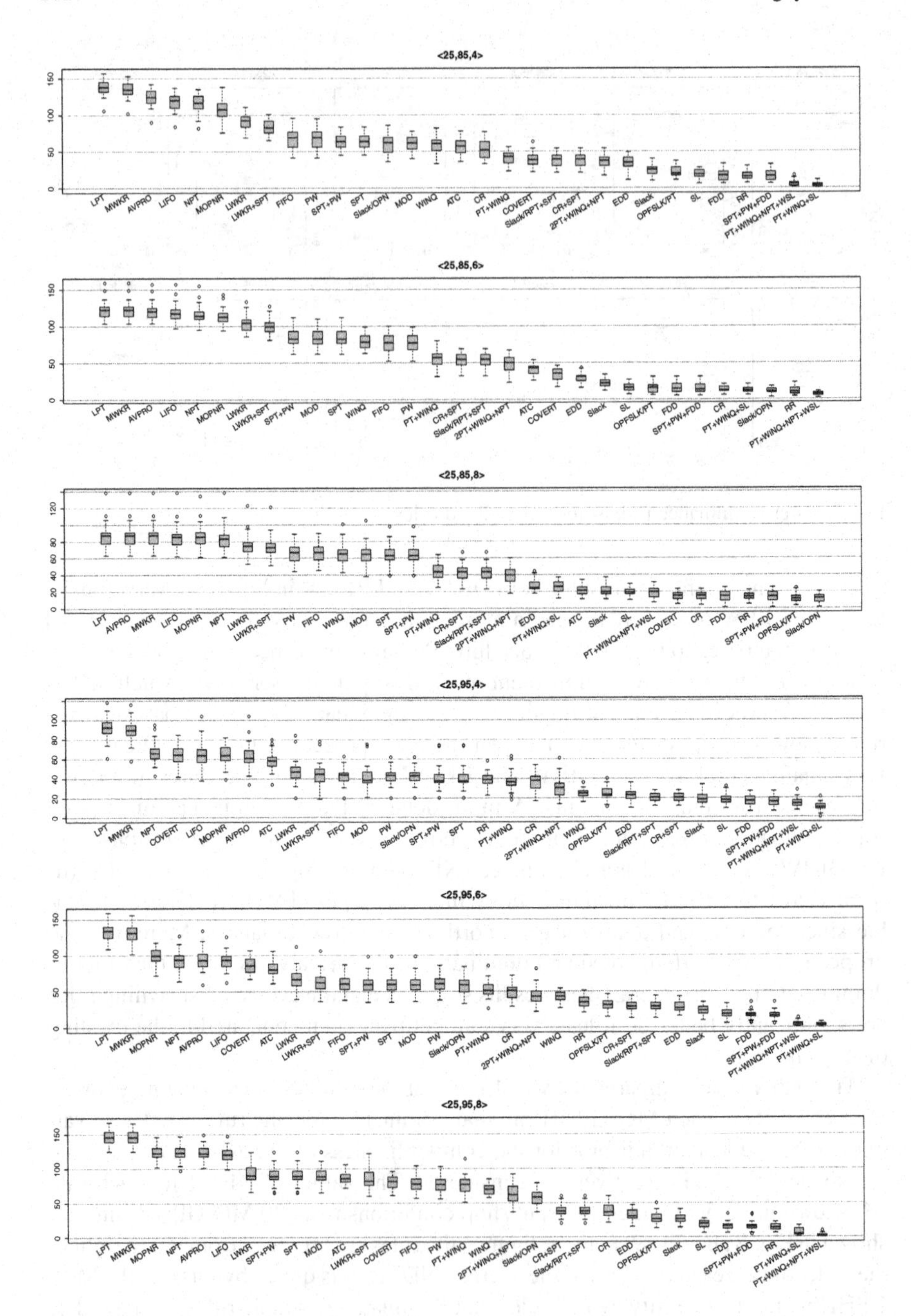

Fig. 8 NDER for each existing dispatching rules (processing times from [1,49])

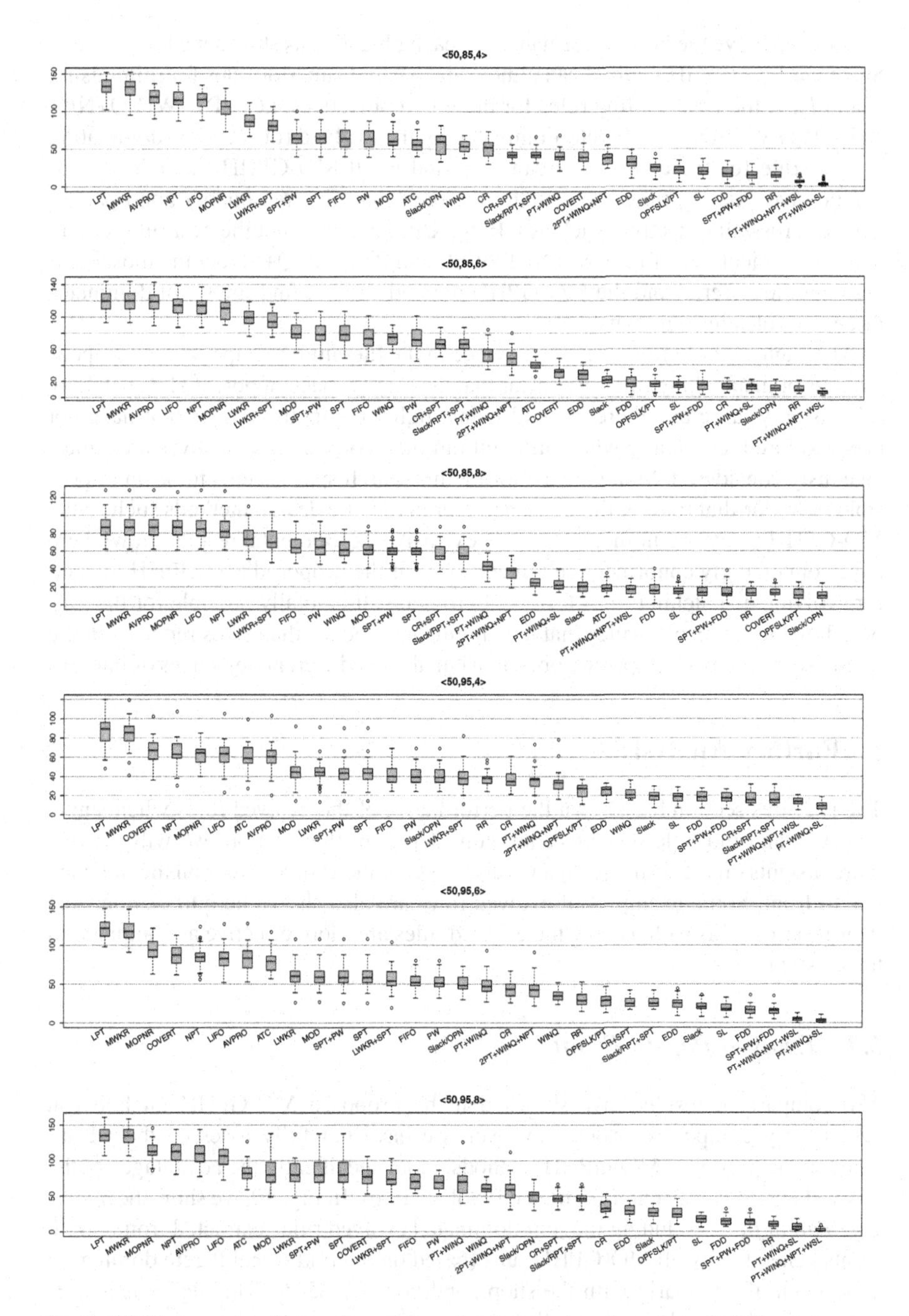

Fig. 9 NDER for each existing dispatching rules (processing times from [1,99])

ones that achieve the best performance for each objective as shown back in Table 5. Since the MO-GPHH method can almost always find superior rules for minimising F and F_{max}, the best existing rules for these two objectives, i.e., 2PT+WINQ+NPT and SPT+PW+FDD, are Pareto-dominated by the evolved rules easier (dominating evolved rule for these two rules can be found in all MO-GPHH runs). The most competitive existing rules are actually the ones that give reasonably good performance across all objectives such as OPFSLK/PT, which is not the best rule for any particular objective. PT+WINQ+NPT+WSL and PT+WINQ+SL are the most competitive rules overall and the MO-GPHH method can not find rules that dominates these two rules in some runs.

Although a lot of efforts have been made in the literature to improve the competitiveness of dispatching rules, it is clear that the search space of potential dispatching rules is very large and there are still many highly competitive rules that have not been explored, especially when different multiple conflicting objectives are simultaneously considered. Manually exploring this search space seems to be an impossible task. For that reason, there is a need for automatic design methods such as the MO-GPHH proposed in this work. The extensive experimental results shown here have convincingly confirmed the effectiveness of the proposed MO-GPHH method for evolving dispatching rules for DJSS problems. It is totally possible for the proposed method to evolve rules that are significantly better than rules reported in the literature, not only on a specific objective but also on different objectives of interest.

5 Further Analysis

The previous section has shown the performance of the evolved rules when single objective and multiple objectives are considered. In this section, we will provide more insights on the distribution and robustness of the evolved rules on the obtained Pareto front. Some examples of evolved rules are also shown here to demonstrate their robustness as well as how the evolved rules are more effective as compared to the existing rules.

5.1 Evolved Pareto Front

The comparison results have shown that the proposed MO-GPHH method can evolve very competitive rules. However, we have not fully assessed the advantages of the proposed MO-GPHH methods, more specifically the advantages of the evolved Pareto front of non-dominated evolved rules. In Fig. 10, we show the *aggregate Pareto front* including the non-dominated evolved rules extracted from Pareto fronts generated by all MO-GPHH runs (based on the traditional Pareto dominance concept) in the scenario with the shop condition $\langle 25, 85, 4\rangle$. This figure is a scatter plot matrix which contains all the pairwise scatter plots of the five objectives (the two scatter plots which are symmetric with respect to the diagonal are similar except that the two axes are interchanged). The objective values obtained by 31 existing rules are also plotted in this figure (as $+$).

The first observation is that the Pareto front can cover a much wider range of potential non-dominated rules compared to rules that have been discovered in the literature. The figure not only shows that the evolved rules can dominate the existing rules but the Pareto front of evolved rules also helps with understanding better about the possible trade-offs in this scenario. For example, it can be seen that the percentage of tardy jobs $\%T$ can be substantially reduced with only minor deterioration on other objectives. Obviously, this insight cannot be obtained with the available dispatching rules since these rules only suggest that other objectives will be deteriorated significantly when we try to reduce $\%T$ below 20%. However, we can see from the Pareto front that it is possible to reduce $\%T$ further to 10% without major deteriorations in other objectives. In fact, F and T will not be affected when we try to reduce $\%T$ to a level above 10%. When we try to reduce $\%T$ below 10%, F_{max} and T_{max} will be greatly deteriorated. In this scenario, we also see that there is a strong correlation between F_{max} and T_{max} when the values are high and the trade-offs between these two objectives are only obvious when they reach their lowest values. This makes sense since high values of F_{max} and T_{max} are caused by some extreme cases. Thus, as long as these extreme cases are handled well, both F_{max} and T_{max} can also be reduced. This observation also suggests that focusing on one of them should be enough if these two objectives are not very important.

This demonstration shows that decision makers can benefit greatly from the Pareto front found by the proposed MO-GPHH method. For DJSS problems, the ability to understand all possible trade-offs is very important since many aspects need to be considered when a decision needs to be made. Without the knowledge from these trade-offs, the decisions will be too extreme (only focus on a specific objective) and they can be practically unreasonable sometimes (e.g. double the maximum tardiness just for reducing $\%T$ by 1%). Moreover, the decision makers do not need to decide their preferences on the objectives before the design process, which could be quite subjective in most cases.

5.2 Robustness of the Evolved Dispatching Rules

It has been shown that the evolved Pareto fronts contain very competitive rules. In this section, we will investigate the robustness of the evolved rules, which is their ability to maintain their performance across different simulation scenarios. In the single objective problem, the robustness of the evolved rules can be easily examined by measuring and comparing the performance of the rules on different scenarios. However, the assessment of the robustness of the evolved rules are not trivial in the case of multi-objective problems because the robustness of rules depends not only on the values of all the objectives but also on the Pareto dominance relations of the rules. Unfortunately, there has been no standard method to measure the robustness of the evolved rules for the multi-objective problems. Therefore, we propose a method to help us roughly estimate the robustness of the evolved rules. In this work, the robustness of a rule $\mathscr{R}_i$ will be calculated as follows:

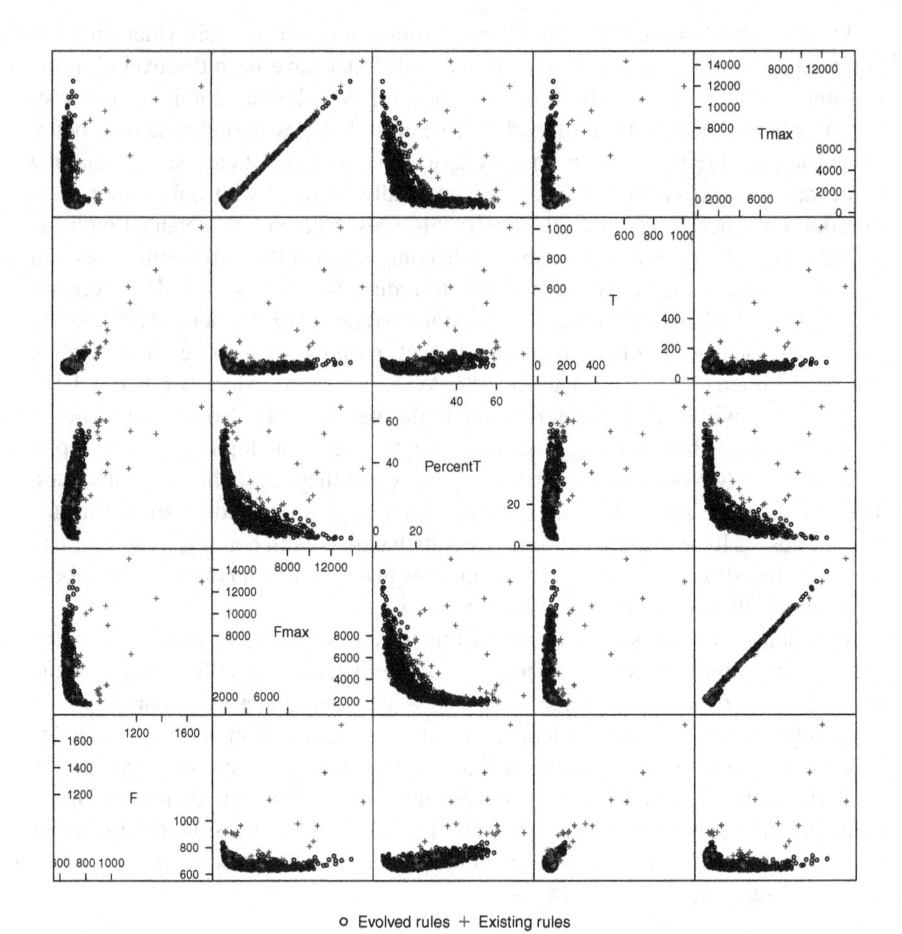

Fig. 10 Distribution of rules on the evolved Pareto front for the scenario $\langle 25, 85, 4\rangle$

$$robustness_i = 1 - \frac{\sum_{s\in\mathbb{S}} Hamming\_Distance(dom_{is}, dom^*_{is})}{|\mathbb{S}| \times |\mathbb{B}|} \tag{2}$$

where $dom_{is} = \{d_{is1}, \ldots, d_{isj}, \ldots, d_{is|\mathbb{B}|}\}$ is a binary array which stores the Pareto dominance between $\mathscr{R}_i$ and each rule $\mathscr{B}_j$ in the set $\mathbb{B}$ of reference rules. In a simulation scenario $s \in \mathbb{S}$ (12 test scenarios in our work), d_{isj} is assigned 1 when $\mathscr{R}_i$ statistically dominates $\mathscr{B}_j$, and 0 otherwise. Here, we include in $\mathbb{B}$ ten benchmark rules that are most competitive in Fig. 8 and Fig. 9 (FDD, Slack/OPN, OPFSLK/PT, SPT+PW+FDD, PT+WINQ+NPT+WSL, PT+WINQ+SL, SL, RR, 2PT+WINQ+NPT, and COVERT). Meanwhile, dom^*_{is} is also a binary array which contains the most frequent value of d_{isj} across all $s \in \mathbb{S}$. The second term in equation (2) measures the average Hamming distance per dimension between dom_{is} and dom^*_{is}. From this calculation, if the Pareto-dominance relations between $\mathscr{R}_i$ and each rule $\mathscr{B}_j$ are consistent across all $s \in \mathbb{S}$, this term will be zero and the robustness is one. In the worst case when the Pareto dominance relations are different greatly for each scenario s, the second term in equation (2) will approach 1 and the robustness will be near zero.

A histogram of robustness values of all evolved rules obtained by 30 MO-GPHH runs is shown in Fig. 11. It is clear that the distribution of robustness values is skewed to the right, which indicates that the evolved rules are reasonably robust. The majority of the rules have the robustness values from 0.8 to 0.95 and there is only a small proportion of evolved rules with small robustness. This result is consistent with our observation in Section 4.2 that a small number of evolved rules that do not perform well on unseen scenarios can be dominated by the benchmark rules, in which case their Pareto-dominance relations are changed.

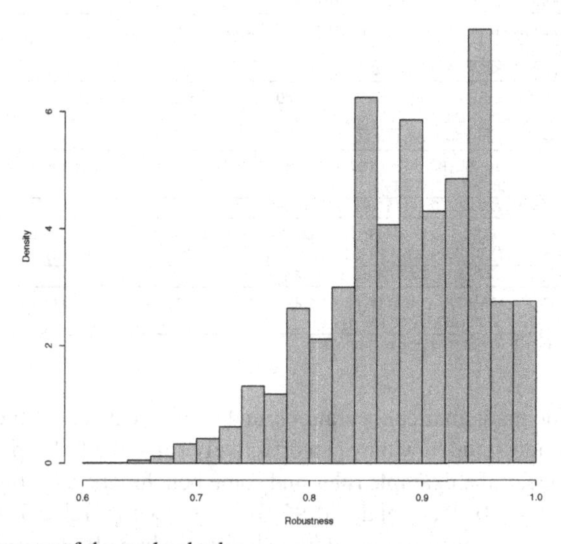

Fig. 11 Robustness of the evolved rules

5.3 *Examples of Evolved Dispatching Rules*

This section shows examples of the evolved dispatching rules. Since many rules have been evolved, it is impossible to list all of these rules. We only show here ten typical rules which can achieve balanced performance on all objectives that we considered in this work. These ten rules are shown in Table 6 along with their average objective values obtained from the training scenarios. In general, the example rules shown in the table are quite long and include different terminals from Table 1. This suggests that different information needs to be considered in order to make good sequencing decisions that can favour all objectives. Therefore, it seems to be infeasible to design such rules manually, especially when different trade-offs have to be taken into account. Although the rules here are quite long, they are mainly synthesised

Table 6 Some typical examples of evolved rules

Rule #1 – Objectives(757.16,3520.19,0.17,164.52,1811.77)
(((IF(SJ,RJ,max(PR,WT)) + (max(RO,RT) + (RJ/IF(SJ,PR,rJ)))) − WINQ) + (((max(RO,RT)+ IF(SJ,IF(SJ,PR,rJ),rJ)) + (−1 × (IF(SJ,PR,rJ)) + IF(SJ,DD/PR,rJ))) − min(SJ,(WINQ × min(PR,WINQ))))) −Abs((rJ − RT) + min(min(SJ,IF(SJ,PR,rJ)),(rJ − RT)))
Rule #2 – Objectives(828.45,2322.88,0.19,165.04,1931.40)
(−rJ − SJ + max(RO,RT) + (((((RJ/PR) + max(RO,RT)) + max(PR,max(RO,RT))) + (−PR − RT)) − 0.8968051)) −Abs(IF(min(SJ,WINQ),WINQ,DD/PR) + Abs(min(SJ,WINQ)))
Rule #3 – Objectives(720.28,4383.52,0.09,105.67,2401.59)
((max(RM,(Abs(min(WT,SJ)) × (RT × PR)))/PR)/Abs(PR + RO))/Abs(max(((PR × max(RT,max(APR,SJ))) × (PR +WINQ)),((Abs(PR) × (RT × PR)) × DD)%Abs(min(WT,(SJ/APR)))))
Rule #4 – Objectives(716.52,3842.36,0.11,82.67,1714.88)
((max((PR × APR),(Abs(min(WT,SJ)) × WINQ))/PR)%WINQ)/Abs(max(((PR × PR) × (max(Abs(RT),max(APR,SJ)) +min(WT,(SJ/APR)))),((PR × WINQ) × DD)%Abs(min(WT,(SJ/APR)))))
Rule #5 – Objectives(708.05,4141.63,0.13,109.08,1977.63)
(((((RT/rJ) + rJ)/max(min(DD,SJ),RT)) − min(−(IF(SJ,RJ,NPR)/(SJ + WINQ)),DD)) + (−WINQ + (−RO− min(min(SJ,WINQ),rJ)))) + ((max(SJ,rJ) + ((IF(SJ,RJ,−RO)/PR) − (rJ + max((WINQ + PR),0.371)))) − NPR)
Rule #6 – Objectives(687.85,5708.02,0.16,134.13,4046.06)
Abs((((RJ/SJ)/PR)/PR)/max(APR,WINQ)) × Abs(((((SJ/APR) − SJ)/min(RT,SJ)) × min(RT,SJ)) /min(((RJ/SJ) × (RJ/SJ)),RT))
Rule #7 – Objectives(798.58,3383.23,0.15,73.83,602.15)
((SJ/APR) + (−(min(min(RT,PR),rJ) − (min(PR,SJ) − min(WINQ,SJ))) − min(RT,SJ))) + ((RT + (((RJ/SJ)/PR) ×min(RT,SJ))) + (WINQ − max(APR,WINQ)))
Rule #8 – Objectives(697.64,6306.54,0.06,114.89,4488.11)
Abs((((RJ/SJ)/PR)/PR)/min((min(PR,(RJ/SJ)) − (SJ/rJ)),(DD − WINQ))) × Abs(((PR/(min(PR,RM) + WINQ)) ×min(RT,SJ))/min((RT × min(PR,(RJ/SJ))),RT))
Rule #9 – Objectives(845.94,2261.88,0.27,150.63,1074.42)
((((−rJ − (−rJ%(SJ − RT))) − (WINQ + (0.559 + PR))) − max(Abs(−RT + rJ),−rJ)) − ((SJ − max(Abs(SJ −RT),SJ)) − (−rJ/min((RM + WT),DD)))) + RT − 1 − SJ + DD − APR × PR − WINQ − 3 × (0.559 + PR)
Rule #10 – Objectives(737.34,3790.15,0.13,84.86,1118.69)
max((WT − 2 × SJ + 0.563716 − APR × PR),((Abs(IF(PR + SJ,RM/PR,2 × SJ − WT))/max(PR + 0.024362229, max(SJ,RT)))/Abs(max(−APR + WINQ,APR)/(RM/PR))))

*IF(a,b,c) will return b if a $\geq$ 0; otherwise it will return c.

based on very basic mathematical operations, and therefore it is possible to simplify these rules or to understand how they can effectively solve the DJSS problems.

The performance of the example rules and some benchmark rules on two unseen simulation scenarios $\langle 50,95,4 \rangle$ and $\langle 50,95,6 \rangle$ are shown in Table 7. It is easy to realise that most benchmark rules are dominated, regarding all objectives, by some example evolved rules. For instance, sophisticated rules such as RR and COVERT are greatly dominated by rules #3 and #4 in the two testing simulation scenarios. PT+WINQ+SL is the only benchmark rule that is not dominated by our example rules, based on the average objective values shown in the table. This is not surprising since PT+WINQ+SL is one of the most competitive rules, but there are still several evolved rules that can dominate PT+WINQ+SL as shown in Fig. 8 and Fig. 9. Rules #1 and #7 are two rules with results quite similar to those from PT+WINQ+SL and only slightly slightly worse than PT+WINQ+SL in some objectives. In $\langle 50,95,4 \rangle$, rule #1 is only worse than PT+WINQ+SL for T. However, it is noted that rule #1 can achieve much better $\%T$ and T_{max}.

Table 7 Performance of example evolved rules

$\langle 50,95,4 \rangle$	F	F_{max}	%T	T	T_{max}
PT+WINQ+SL	2991.03	7551.84	88.47	1431.15	4491.71
Slack/OPN	4532.45	13977.43	97.45	2930.66	11409.89
RR	2943.40	15050.27	85.46	1374.69	12255.53
COVERT	2744.83	37158.13	77.18	1166.01	34645.73
2PT+WINQ+NPT	2355.21	31321.28	45.65	1010.55	28785.17
PT+WINQ+NPT+WSL	3356.03	7695.30	94.06	1771.03	4481.94
SPT+PW+FDD	3710.73	6769.10	96.94	2115.04	5268.29
OPFSLK/PT	3108.31	7648.78	91.88	1530.30	5909.04
FDD	3740.53	6810.86	96.98	2144.74	5322.64
SL	3638.32	7967.81	98.59	2038.43	4732.04
Rule #1	2933.19	7096.85	68.44	1446.81	4193.42
Rule #2	3381.54	6018.19	85.10	1812.70	5486.42
Rule #3	2548.54	9075.00	46.06	1126.13	6047.45
Rule #4	2472.12	9615.42	51.21	1018.41	6983.91
Rule #5	2527.99	8182.37	56.04	1081.19	5192.21
Rule #6	2229.99	13394.98	37.33	928.19	10625.92
Rule #7	3049.49	7492.19	90.03	1462.91	4406.31
Rule #8	2323.42	16446.96	24.61	1021.91	13515.90
Rule #9	3042.21	6800.97	86.39	1477.09	4645.73
Rule #10	2629.42	7945.21	63.78	1124.64	4772.82

$\langle 50,95,6 \rangle$	F	F_{max}	%T	T	T_{max}
PT+WINQ+SL	2714.00	7982.00	61.49	657.95	3148.09
Slack/OPN	3941.20	12989.18	77.26	1581.59	8974.86
RR	2811.91	11882.35	53.49	611.18	7107.35
COVERT	2999.75	31445.59	56.98	676.51	27970.88
2PT+WINQ+NPT	2355.21	31321.28	26.12	701.61	27540.38
PT+WINQ+NPT+WSL	2932.42	8021.92	68.06	771.68	3037.79
SPT+PW+FDD	3710.73	6769.10	81.53	1421.97	5064.32
OPFSLK/PT	3108.31	7648.78	68.09	927.30	5441.89
FDD	3740.53	6810.86	82.00	1448.29	5118.74
SL	3524.23	8647.86	91.95	1163.81	3672.67
Rule #1	2768.33	8172.39	40.60	817.59	3702.13
Rule #2	3264.87	6243.41	57.39	1081.99	5428.79
Rule #3	2444.66	9914.98	26.03	576.57	5190.51
Rule #4	2419.77	9898.71	28.31	520.51	5564.72
Rule #5	2426.92	9221.52	34.36	585.90	4464.28
Rule #6	2213.12	14116.06	23.18	651.85	9846.80
Rule #7	2761.50	7893.13	51.89	588.13	3061.45
Rule #8	2280.87	17798.84	14.04	627.54	13237.53
Rule #9	2934.48	7009.52	60.37	798.31	3879.64
Rule #10	2487.80	9041.86	34.35	514.90	4150.00

6 Conclusions

Most of the dispatching rules for DJSS problems proposed in the literature are designed for minimising a specific objective. However, the choice of a suitable dispatching rule has to depend on the performance of the rule across multiple conflicting objectives. In this work, we show how we can use GP to handle this issue. The proposed MO-GPHH method aims at exploring the Pareto front of evolved rules which can be used to support the decision making process. Extensive experiments have been performed and the results show that the evolved Pareto front contains superior rules as compared with rules reported in the literature when both single and multiple objectives are considered. Moreover, it has been shown that the obtained Pareto front can provide valuable insights on how trade-offs should be made.

We have also discussed and implemented different analyses on the experimental results, which help us confirm the effectiveness of the evolved dispatching rules. In these analyses, we focus on two issues. First, we try to define a standard procedure in order to properly compare the performance of rules within the multi-objective stochastic environments. Second, we need to find a way to assess the robustness of the evolved rules under different simulation scenarios. Although they are two very important issues, there have been no existing guidelines on how they should be done. In this work, we proposed different approaches to handle these issues. Even though there are still some limitations with these approaches, they can nevertheless be used as a good way to assess the performance of rules in such a complicated problem. Certainly, these two issues can also be interesting issues for future studies.

Acknowledgements. This work is supported in part by the Marsden Fund of New Zealand Government (VUW0806 and 12-VUW-134), administrated by the Royal Society of New Zealand, and the University Research Fund (200457/3230) at Victoria University of Wellington.

References

1. Atlan, L., Bonnet, J., Naillon, M.: Learning distributed reactive strategies by genetic programming for the general job shop problem. In: Proceedings of the 7th Annual Florida Artificial Intelligence Research Symposium (1994)
2. Baker, K.R.: Sequencing rules and due-date assignments in a job shop. Management Science 30, 1093–1104 (1984)
3. Banzhaf, W., Nordin, P., Keller, R., Francone, F.: Genetic Programming: An Introduction. Morgan Kaufmann, San Francisco (1998)
4. Bhowan, U., Johnston, M., Zhang, M., Yao, X.: Evolving diverse ensembles using genetic programming for classification with unbalanced data. IEEE Transactions on Evolutionary Computation (2012), doi:10.1109/TEVC.2012.2199119
5. Burke, E.K., Hyde, M.R., Kendall, G., Ochoa, G., Ozcan, E., Woodward, J.R.: Exploring hyper-heuristic methodologies with genetic programming. Artificial Evolution 1, 177–201 (2009)

6. Deb, K., Pratap, A., Agarwal, S., Meyarivan, T.: A fast and elitist multiobjective genetic algorithm: NSGA-II. IEEE Transactions on Evolutionary Computation 6(2), 182–197 (2002)
7. Dimopoulos, C., Zalzala, A.M.S.: Investigating the use of genetic programming for a classic one-machine scheduling problem. Advances in Engineering Software 32(6), 489–498 (2001)
8. Geiger, C.D., Uzsoy, R.: Learning effective dispatching rules for batch processor scheduling. International Journal of Production Research 46, 1431–1454 (2008)
9. Geiger, C.D., Uzsoy, R., Aytug, H.: Rapid modeling and discovery of priority dispatching rules: An autonomous learning approach. Journal of Heuristics 9(1), 7–34 (2006), doi: http://dx.doi.org/10.1007/s10951-006-5591-8
10. Hildebrandt, T., Heger, J., Scholz-Reiter, B.: Towards improved dispatching rules for complex shop floor scenarios: a genetic programming approach. In: GECCO 2010: Proceedings of the 12th Annual Conference on Genetic and Evolutionary Computation, pp. 257–264. ACM, New York (2010)
11. Holthaus, O., Rajendran, C.: Efficient jobshop dispatching rules: Further developments. Production Planning & Control 11(2), 171–178 (2000)
12. Ingimundardottir, H., Runarsson, T.P.: Supervised learning linear priority dispatch rules for job-shop scheduling. In: Coello, C.A.C. (ed.) LION 2011. LNCS, vol. 6683, pp. 263–277. Springer, Heidelberg (2011)
13. Jakobović, D., Budin, L.: Dynamic scheduling with genetic programming. In: Collet, P., Tomassini, M., Ebner, M., Gustafson, S., Ekárt, A. (eds.) EuroGP 2006. LNCS, vol. 3905, pp. 73–84. Springer, Heidelberg (2006)
14. Jakobović, D., Jelenković, L., Budin, L.: Genetic programming heuristics for multiple machine scheduling. In: Ebner, M., O'Neill, M., Ekárt, A., Vanneschi, L., Esparcia-Alcázar, A.I. (eds.) EuroGP 2007. LNCS, vol. 4445, pp. 321–330. Springer, Heidelberg (2007)
15. Jayamohan, M.S., Rajendran, C.: New dispatching rules for shop scheduling: a step forward. International Journal of Production Research 38, 563–586 (2000)
16. Jones, A., Rabelo, L.C.: Survey of job shop scheduling techniques. Tech. rep., NISTIR, National Institute of Standards and Technology, Gaithersburg, US (1998)
17. Koza, J.R.: Genetic Programming: On the Programming of Computers by Means of Natural Selection. MIT Press (1992)
18. Li, X., Olafsson, S.: Discovering dispatching rules using data mining. Journal of Scheduling 8, 515–527 (2005)
19. Miyashita, K.: Job-shop scheduling with GP. In: GECCO 2000: Proceedings of the Genetic and Evolutionary Computation Conference, pp. 505–512 (2000)
20. Montgomery, D.C.: Design and Analysis of Experiments. John Wiley & Sons (2001)
21. Nguyen, S., Zhang, M., Johnston, M., Tan, K.C.: A coevolution genetic programming method to evolve scheduling policies for dynamic multi-objective job shop scheduling problems. In: CEC 2012: IEEE Congress on Evolutionary Computation, pp. 3332–3339 (2012)
22. Nguyen, S., Zhang, M., Johnston, M., Tan, K.C.: Evolving reusable operation-based due-date assignment models for job shop scheduling with genetic programming. In: Moraglio, A., Silva, S., Krawiec, K., Machado, P., Cotta, C. (eds.) EuroGP 2012. LNCS, vol. 7244, pp. 121–133. Springer, Heidelberg (2012)
23. Nie, L., Shao, X., Gao, L., Li, W.: Evolving scheduling rules with gene expression programming for dynamic single-machine scheduling problems. The International Journal of Advanced Manufacturing Technology 50, 729–747 (2010)

24. Panwalkar, S.S., Iskander, W.: A survey of scheduling rules. Operations Research 25, 45–61 (1977)
25. Pinedo, M.L.: Scheduling: Theory, Algorithms, and Systems, 3rd edn. Springer (2008)
26. Rafter, J.A., Abell, M.L., Braselton, J.P.: Multiple comparison methods for means. SIAM Review 44(2), 259–278 (2002)
27. Rajendran, C., Holthaus, O.: A comparative study of dispatching rules in dynamic flowshops and jobshops. European Journal of Operational Research 116(1), 156–170 (1999)
28. Sels, V., Gheysen, N., Vanhoucke, M.: A comparison of priority rules for the job shop scheduling problem under different flow time- and tardiness-related objective functions. International Journal of Production Research (2011)
29. Tay, J.C., Ho, N.B.: Evolving dispatching rules using genetic programming for solving multi-objective flexible job-shop problems. Computer and Industrial Engineering 54, 453–473 (2008)
30. Vepsalainen, A.P.J., Morton, T.E.: Priority rules for job shops with weighted tardiness costs. Management Science 33, 1035–1047 (1987)
31. Wang, Z., Tang, K., Yao, X.: Multi-objective approaches to optimal testing resource allocation in modular software systems. IEEE Transactions on Reliability 59(3), 563–575 (2010)
32. Yin, W.J., Liu, M., Wu, C.: Learning single-machine scheduling heuristics subject to machine breakdowns with genetic programming. In: CEC 2003: IEEE Congress on Evolutionary Computation, pp. 1050–1055 (2003)

Dynamic Vehicle Routing: A Memetic Ant Colony Optimization Approach

Michalis Mavrovouniotis and Shengxiang Yang

Abstract. Over the years, several variations of the dynamic vehicle routing problem (DVRP) have been considered due to its similarities with many real-world applications. Several methods have been applied to address DVRPs, in which ant colony optimization (ACO) has shown promising results due to its adaptation capabilities. In this chapter, we generate another variation of the DVRP with traffic factor and propose a memetic algorithm based on the ACO framework to address it. Multiple local search operators are used to improve the exploitation capacity and a diversity scheme based on random immigrants is used to improve the exploration capacity of the algorithm. The proposed memetic ACO algorithm is applied on different test cases of the DVRP with traffic factors and is compared with other peer ACO algorithms. The experimental results show that the proposed memetic ACO algorithm shows promising results.

1 Introduction

The vehicle routing problem (VRP) is a classical combinatorial optimization problem, in which a fleet of vehicles need to satisfy the demand of customers while minimizing the overall routing cost, e.g., the distance travelled, starting from and ending to the depot [11]. It has wide applications in the real world, including three main categories: services [6, 29], transport of goods [2, 49], and transport of persons [18, 20]. Moreover, the arc routing problem, which is the arc routing counterpart of the VRP, has received a lot of attention recently, due to its importance in many real-world applications [12, 44, 47].

Michalis Mavrovouniotis · Shengxiang Yang
Centre for Computational Intelligence (CCI),
School of Computer Science and Informatics,
De Montfort University,
The Gateway, Leicester LE1 9BH, U.K.
e-mail: {mmavrovouniotis, syang}@dmu.ac.uk

A.Ş. Etaner-Uyar et al. (eds.), *Automated Scheduling and Planning*,
Studies in Computational Intelligence 505,
DOI: 10.1007/978-3-642-39304-4_11,

The VRP has been proven to be *NP*-hard [28]. A number of variations of the traditional VRP have been studied, including the capacitated VRP (CVRP) with service times, where each vehicle has a limited capacity and service time; the multiple depot VRP, where multiple depots exist; the VRP with time windows (VRPTW), where each customer must be visited during a specific time slot; the VRP with pick-up and delivery, where goods have to be picked-up and delivered in specific amounts at the customers; and the heterogeneous fleet VRP, where vehicles have different capacities.

In contrast to the classical definition of the VRP where the inputs are known beforehand, in real-world applications, the information available may change during the execution of routes. In such cases, we have a dynamic environment. Over the years, many variations of the dynamic VRP (DVRP) have been studied, including the VRP with stochastic demands, the dynamic VRPTW, and many others (see [42] for a comprehensive review). In this book chapter, we consider and generate our own DVRP with traffic factors where the cost between two locations (or customers) varies depending on the period of the day.

A simple solution to address the DVRP is to consider the arrival of new traffic information as a problem that needs to be solved from scratch. However, such solution requires an extensive computation time to re-optimize while in real-world applications the time available is limited [26]. Ant colony optimization (ACO) algorithms have proved to be good meta-heuristics to the simplest version of the DVRP with traffic factors, which is the dynamic travelling salesman problem (DTSP) [35, 36] and to other stationary or DVRP variations [1, 45]

Since ACO algorithms have been designed for static optimization problems, they lose their adaptation capabilities quickly because of the stagnation behaviour, where all ants follow the same path from the early stages of the execution [9, 15]. Recently, several approaches have been proposed to avoid the stagnation behaviour and address the single- or multi-vehicle DVRP, which includes: (1) local and global restart strategies [23]; (2) pheromone manipulation schemes to maintain diversity [17]; (3) increasing diversity via immigrants schemes [35, 36]; (4) memory-based approaches [21, 24]; and (5) memetic algorithms (MAs) [33].

Among the approaches, the memetic ACO (M-ACO) algorithm has shown promising results, which is a hybridization of an ACO algorithm with a local search (LS) operator. Every iteration of the algorithm, the best ant is selected for local improvement by a LS operator. The MAs based on ACO are not so common, since most MAs are based on the evolutionary computation framework [33]. Some MAs based on ACO have been proposed for some problems under static environments [31, 30, 53], because of the strong exploitation an LS scheme provides. However, in dynamic optimization problems (DOPs), exploration needs to be increased in order to address the stagnation behaviour. Therefore, in M-ACO a diversity scheme based on random immigrants is integrated to achieve a balance between exploitation and exploration.

The remaining of this chapter is outlined as follows. Section 2 describes the proposed DVRP with traffic factors used in the experiments. Section 3 describes one of the best performing ACO algorithms applied to the DVRP, whereas Section 4 gives details of the proposed MA based on ACO. Section 5 presents the experimental results and analysis. Finally, Section 6 concludes this contribution and points out the future work.

2 Problem Description

2.1 *Static VRP*

The VRP became one of the most popular combinatorial optimization problems, due to its similarities with many real-world applications. The VRP is classified as *NP*-hard [28]. The basic VRP is the capacitied VRP, where a number of vehicles with a fixed capacity need to satisfy the demand of all the customers, starting from and finishing to the depot. A VRP without the capacity constraint or with one vehicle can be seen as a travelling salesman problem (TSP). There are many variations and extensions of the VRP, such as the multiple depot VRP, the VRP with pickup and delivery, the VRP with time windows and combinations of different variations (for more details see [50]). In this paper the basic VRP is considered.

Usually, the VRP is represented by a complete weighted graph $G = (V,E)$, with $n+1$ nodes, where $V = \{u_0,\ldots,u_n\}$ is a set of vertices corresponding to the customers (or delivery points) u_i $(i = 1,\cdots,n)$ and the depot u_0 and $E = \{(u_i,u_j) : i \neq j\}$ is a set of edges. Each edge (u_i,u_j) is associated with a non-negative d_{ij} which represents the distance (or travel time) between u_i and u_j. For each customer u_i, a non-negative demand δ_j is given. For the depot u_0, a zero demand is associated, i.e., $\delta_0 = 0$.

The aim of the VRP is to find the route (or a set of routes) with the lowest cost without violating the following constraints:

- Every customer is visited exactly once by only one vehicle.
- Every vehicle starts and finishes at the depot.
- The total demand of every vehicle route must not exceed the vehicle capacity Q.

Formally, the VRP can be described as follows:

$$Minimize \sum_{i=0}^{n}\sum_{j=0}^{n} d_{ij} \sum_{k=1}^{v} x_{ij}^{k}, \tag{1}$$

subject to:

$$\sum_{j=0}^{n} \delta_j \left(\sum_{i=0}^{n} x_{ij}^{k} \right) \leq Q, \forall k \in \{1,\ldots,v\}, \tag{2}$$

$$x_{ij}^{k} = \begin{cases} 1, & \text{if } (u_i,u_j) \text{ is covered by vehicle } k, \\ 0, & \text{otherwise,} \end{cases} \tag{3}$$

where $x_{ij}^{k} \in \{0,1\}$, n is the number of customers, v is the number of vehicles, which is not fixed but chosen by the algorithm during execution, d_{ij} is the distance between customers u_i and u_j, δ_j is the demand of customer u_j, and Q is the capacity of vehicle k. The objective function in Equation (1) is to minimize the distance travelled by all vehicles that are used subject to the capacity constraint in Equation (2).

A lot of algorithms have been proposed to solve small instances of different variations of the VRP, either exact or approximation algorithms [41, 50]. Although exact algorithms guarantee to provide the global optimum solution, an exponential time is required in the worst case scenario, because the VRP is *NP*-hard [28]. On the other hand, approximation algorithms, i.e., evolutionary algorithms or ACO algorithms, can provide a good solution efficiently but cannot guarantee the global one [25, 39].

2.2 Dynamic VRP

The VRP becomes more challenging if it is subject to a dynamic environment, since the moving optimum needs to be tracked [26]. From the stationary VRP described above we generate a dynamic variation where the inputs of the problem change dynamically during the execution of the algorithm. There are many variations of the DVRP, such as the DVRP with dynamic demand [27, 43].

In this paper, we generate a DVRP with traffic factors, where each edge (u_i, u_j) is associated with a traffic factor t_{ij}. Therefore, the cost to travel from u_i to u_j is $c_{ij} = d_{ij} \times t_{ij}$. Note that the cost to travel from u_j to u_i may differ due to different traffic factors. For example, one road may have more traffic in one direction and no traffic in the opposite direction.

Every f iterations, a random number $R \in [F_L, F_U]$ is generated to represent potential traffic jams, where F_L and F_U are the lower and upper bounds of the traffic factor, respectively. Each edge has a probability m to have a traffic factor, by generating a different R to represent high and low traffic jams in different roads, i.e., $t_{ij} = 1 + R$ where the remaining edges are set to have a traffic factor $t_{ij} = 1$, which indicates no traffic. Note that f and m represent the frequency and magnitude of changes in the DVRP, respectively.

Depending on the period of the day, environments with different traffic factors can be generated. For example, during the rush hour periods, a higher probability is given to generate R closer to F_U, whereas during evening hour periods, a higher probability is given to generate R closer to F_L.

3 Ant Colony Optimization for the DVRP

3.1 Ant Colony System (ACS)

An ACO algorithm consists of a population of μ ants where they construct solutions and share information with each other via their pheromone trails [3]. Ants "read" pheromone from others and "write" pheromone to their trails. The first ACO algorithm developed is the Ant System (AS) [13]. Many variations and extensions of the AS have been developed over the years and applied to different optimization problems [4, 14, 10, 32, 46].

The best performing ACO algorithm for the VRP is the ACS [14, 16] and it has a wide application in the real-world application [45]. There is a multi-colony variation of this algorithm applied to the VRPTW [16]. In this paper, the single colony variation is considered, which has been applied to the DVRP with stochastic demands [37, 38]. Initially, all the ants are placed on the depot and all pheromone trails are initialized with an equal amount. With a probability $1-q_0$, where $0 \leq q_0 \leq 1$ is a parameter of the *pseudo-random* proportional decision rule (usually set to 0.9 for ACS), an ant, say ant k, chooses the next customer j from customer i, as follows:

$$p_{ij}^k = \begin{cases} \frac{[\tau_{ij}]^{\alpha}[\eta_{ij}]^{\beta}}{\sum_{l \in N_i^k}[\tau_{il}]^{\alpha}[\eta_{il}]^{\beta}}, \text{if } j \in N_i^k, \\ 0, \qquad \text{otherwise}, \end{cases} \tag{4}$$

where τ_{ij} is the existing pheromone trail between customers i and j, η_{ij} is the heuristic information available a priori, which is defined as $1/c_{ij}$, where c_{ij} is the distance travelled (including t_{ij}) between customers i and j, N_i^k denotes the neighbourhood of unvisited customers of ant k when its current customer is i, and α and β are the two parameters that determine the relative influence of pheromone trail and heuristic information, respectively. With the probability q_0, ant k chooses the next city, i.e., z, with the maximum probability, which satisfies the following formula:

$$z = \underset{j \in N_i^k}{\text{argmax}} \, [\tau_{ij}]^{\alpha}[\eta_{ij}]^{\beta}. \tag{5}$$

However, if the choice of the next customer will lead to an infeasible solution, i.e., exceeding the maximum capacity of the vehicle, the depot is chosen and a new vehicle route starts.

When all ants construct their solutions, the best ant retraces the solution and deposits pheromone globally according to its solution quality on the corresponding trails, as follows:

$$\tau_{ij} \leftarrow (1-\rho)\tau_{ij} + \rho\Delta\tau_{ij}^{best}, \forall (i,j) \in T^{best}, \tag{6}$$

where $0 < \rho \leq 1$ is the pheromone evaporation rate and $\Delta\tau_{ij}^{best} = 1/C^{best}$, where C^{best} is the total cost of the best tour T^{best}. Moreover, a local pheromone update is performed every time an ant chooses another customer j from customer i as follows:

$$\tau_{ij} \leftarrow (1-\rho)\tau_{ij} + \rho\tau_0, \tag{7}$$

where ρ is defined as in Equation (6) and τ_0 is the initial pheromone value.

The pheromone evaporation is the mechanism that helps the population to forget useless solutions constructed in previous environments and adapt to the new environment. The recovery time depends on the size of the problem and magnitude of change.

3.2 React to Dynamic Changes

In Bonabeau *et al.* [5], it was discussed that traditional ACO algorithms may have good performance in DOPs, since they are very robust algorithms. The mechanism which enables ACO algorithms to adapt in DOPs is the pheromone evaporation. Lowering the pheromone values enables the algorithm to forget bad decisions made in previous iterations. Moreover, when a dynamic change occurs, it will eliminate the pheromone trails of the previous environment that are not useful, or not visited frequently, in the new environment, where the ants may be biased and can not adapt well.

The ACS algorithm, which follows the traditional ACO framework, can be applied directly to the proposed DVRPs with traffic factors, without any modifications, apart from the heuristic information where the traffic factor needs to be considered. Further special measures when a dynamic change occurs are not required, due to the pheromone evaporation. On the other hand, the ACS algorithm with a complete re-initialization of pheromone trails when a dynamic change occurs may look a better choice instead of relying just to the pheromone evaporation. However, such actions can be a sufficient choice in DOPs where the frequency of change is available beforehand, which usually is not the case in real-world applications.

4 Proposed Approach for the DVRP

4.1 Framework of ACO-Based Memetic Algorithm

The M-ACO algorithm has been previously applied to the DTSP and showed some promising results [33]. It is based on the P-ACO algorithm framework which is the memory-based version of traditional ACO and has been developed especially for DOPs, due to to the knowledge transferred directly from the previous environment to a new one using the solutions stored in the population list. In this chapter, we apply the M-ACO algorithm to the DVRP.

The framework of M-ACO differs from the traditional ACO framework since it maintains a memory (population-list) of limited size, which is used to store the best ants. The population-list is used every iteration to update the pheromone, instead of the whole population. The pheromone trails on each iteration depend on the ants stored in the population-list and pheromone evaporation is not applied.

The construction of solutions, the initial phase and the first iterations of the M-ACO algorithm work in the same way as in the traditional ACO algorithm; see Equation (4). The pheromone trails are initialized with an equal amount of pheromone and the population-list M of size K is empty. For the first K iterations, the best-so-far ant is selected to be improved by a LS operator and deposits a constant amount of pheromone, which is defined as follows:

$$\tau_{ij} \leftarrow \tau_{ij} + \Delta\tau_{ij}^{k}, \forall\, (i,j) \in T^{k}, \tag{8}$$

where $\Delta\tau_{ij}^k = (\tau_{max} - \tau_{init})/K$. Moreover, τ_{max} and τ_{init} denote the maximum and initial pheromone amount, respectively. This positive update procedure of Equation (8) is performed whenever an ant enters the population-list.

On iteration $K+1$, the best-so-far ant, which has been improved by a LS operator previously, enters the population-list and updates its pheromone trails positively. However, the ant that entered the population-list first needs to be removed in order to make room for the iteration-best ant, and thus, a negative constant update to its corresponding pheromone trails is done, which is defined as follows:

$$\tau_{ij} \leftarrow \tau_{ij} - \Delta\tau_{ij}^k, \forall\,(i,j) \in T^k, \tag{9}$$

where $\Delta\tau_{ij}^k$ is defined as in Equation (8). This mechanism keeps the pheromone trails between a certain minimum value τ_{min}, which is equal to τ_{init}, and a maximum value τ_{max}, which can be calculated by $\tau_{init} + \sum_{k=1}^{K} \Delta\tau_{ij}^k$. We have seen the importance of keeping the pheromone trails within certain bound from the ACS (implicitly) and Max-Min AS (explicitly) [46], which are the two of the best performing ACO algorithms for stationary problems.

This population-list update policy is based on the *Age* of ants. However, other strategies have also been proposed by researchers, such as *Quality* and *Prob* [22]. From the experimental results in [22], the default *Age* strategy is more consistent and performs better than the others, since other strategies have more chances to maintain identical ants into the population-list, which leads the algorithm to the stagnation behaviour. This is due to the fact that high levels of pheromone will be generated into a single trail and dominate the search space.

Furthermore, a diversity scheme based on random immigrants is applied with the M-ACO, because of the strong exploitation the LS operator will provide. Even if the *Age* strategy is used, the population-list may store identical solutions and lead the algorithm to stagnation behaviour and degrade the performance in DOPs. Therefore, the diversity scheme checks whether the ants in the population-list keeps a certain diversity. If not, a random immigrant is generated to replace an ant in the population-list. The M-ACO framework is presented in Algorithm 1.

4.2 Swap Local Search Operators

Within M-ACO, LS operators are applied to improve the solution represented by ants in a local area. The main concern of ACO-based MAs is which ant should be selected for local improvement. Usually, LS is applied to all the ants in the current population to improve their solution quality [46]. However, this may be infeasible for DOPs considering that the evaluations per iteration is limited [52]. A good choice is to select only the best ant for local improvement for several LS steps.

In M-ACO, multiple LS operators are used, i.e., the simple swap and adaptive swap operators [34]. In the simple swap operator, two customers, say customers i and j, are randomly selected and swapped from ant p. Differently, in the adaptive swap operator, a customer i is randomly selected from ant p and the same customer i is located in another ant q, which is randomly selected. Then, the predecessor

Algorithm 1. M-ACO for DVRP

```
1: Initialize parameters
2: Initialize pheromone trails τ_init
3: M := empty
4: while (termination condition not satisfied) do
5:     for (k := 1 to μ) do
6:         Construct a solution by ant k
7:         Update statistics
8:     end for
9:     best := find the best ant
10:    Swaps(best) using Algorithm 2
11:    M.enQueue(best)
12:    Add pheromone using Equation (8)
13:    if (M is full) then
14:        M.deQueue()
15:        Remove pheromone using Equation (9)
16:    end if
17:    if (Div of M is zero using Equation (15)) then
18:        temp := generate a random immigrant
19:        Replace a randomly selected ant in M by temp
20:    end if
21: end while
```

customer $i+1$ from ant q is located in ant p. Finally, the swap of the customers in p is performed between the predecessor $i+1$ and the city adapted from the predecessor of ant q. Note that all the swaps are allowed whenever the capacity constraint is still satisfied, in order to represent a feasible solution.

The use of multiple LS operators has been found beneficial in many MAs [48, 52]. This is because different LS operators may improve the solution quality on different problem instances, due to the problem dependency. Moreover, different LS operator may perform better on different periods of the optimization process [8, 40]. For example, in M-ACO for the DTSP [33] simple and adaptive inversions are used as LS in the DTSP. The simple inversion has been found more effective on later stages of the optimization process, and the adaptive inversion on early stages. The two LS operators are selected probabilistically to promote both competition and cooperation of different LS operators and activate them on different periods of the optimization process.

Similarly, in M-ACO for the DVRP, both the simple and adaptive swap operators are activated probabilistically at every step of an LS operation on every iteration of the algorithm as follows. Let p_s and p_a denote the probability of applying simple and adaptive swaps to the individual selected for LS, respectively, where $p_s + p_a = 1$. Initially, the probabilities are both set to 0.5 in order to promote a fair competition between the two operators. The probabilities are adjusted according to the improvement each inversion operator has achieved on every LS step. The probability of the operator with the higher improvement will be increased using a mechanism, which is similar to the one introduced in [33, 52] and is presented in Algorithm 2.

Algorithm 2. Swaps(*best*)

1: **if** (environmental change is detected) **then**
2: $p_a := p_s := 0.5$;
3: **end if**
4: $\xi_a := \xi_s := 0$
5: **for** ($i := 1$ to ls) **do**
6: **if** ($rand() \leq p_s$) **then**
7: Perform simple swap
8: Update ξ_s
9: **else**
10: Perform adaptive awap
11: Update ξ_a
12: **end if**
13: **end for**
14: Update p_a and p_s

Let ξ denote the degree of improvement of the selected ant after an LS step, which is calculated as follows:

$$\xi = \frac{\left|C^{best'} - C^{best}\right|}{C^{best}}, \tag{10}$$

where $C^{best'}$ is the tour cost of the best ant after applying an LS step, using the simple or adaptive swap operator, and C^{best} is the tour cost of the best ant before applying the LS step. When the number of LS steps reaches the pre-set step size, denoted as ls, the degree of improvement regarding simple swap and adaptive swap operators, denoted as ξ_s and ξ_a, respectively, is calculated and used to adjust the probabilities of selecting simple and adaptive swap operators in the next iteration, $p_s(t+1)$ and $p_a(t+1)$, as follows:

$$p_s(t+1) = p_s(t) + \xi_s(t), \tag{11}$$

$$p_a(t+1) = p_a(t) + \xi_a(t), \tag{12}$$

$$p_s(t+1) = \frac{p_s(t+1)}{p_s(t+1) + p_a(t+1)}, \tag{13}$$

$$p_a(t+1) = 1 - p_s(t+1), \tag{14}$$

where $\xi_s(t)$ and $\xi_a(t)$ are the total degree of improvement achieved by simple and adaptive swap operators at iteration t, respectively.

4.3 Increasing the Population Diversity

The main problem of ACO algorithms when applied to DOPs is the premature convergence. Once the algorithm has converged to an optimum, it cannot adapt well to the new environment when a dynamic change occurs. This is because high intensity of pheromone trails are generated to the solution before the change and may bias the population after the change until they are eliminated. As a result, the population of ants will lose their adaptation capabilities to explore the promising areas in the search space.

When an LS operator is applied to ACO, the premature convergence becomes even worse because of the strong exploitation the LS operator provides. Immigrants schemes have been found effective when integrated to ACO algorithms [35, 36], since they maintain a certain level of diversity within the population. The general idea of immigrants schemes is to generate immigrant ants, in every iteration of the algorithm, that represent random solutions, to replace a small portion of ants in the current population [19, 51].

In the M-ACO algorithm, there is no need to generate immigrant ants in every iteration in order not to disturb the optimization process of the LS operator. Therefore, to increase the diversity without disturbing the optimization process of the LS, the random immigrants are generated only when all the ants currently stored in the population-list are identical. To check whether the ants in the population-list are identical a diversity metric is used on every iteration i of run j, as follows:

$$Div = \frac{1}{K(K-1)} \sum_{p=1}^{K} \sum_{q \neq p}^{K} S(p,q), \tag{15}$$

where K is the size of the population-list and $S(p,q)$ is a similarity metric between ant p and ant q, which is calculated as follows:

$$S(p,q) = 1 - \frac{CE_{pq}}{n + avg(NV_p, NV_q)}. \tag{16}$$

where CE_{pq} are the common edges of ants p and q, n is the number of customers, and NV_p and NV_q are the number of vehicles of ants q and p, respectively. This diversity metric is based on the genotype of the solution and it is more accurate, than a metric based on the phenotype. This is because for a permutation problem, two different solutions may have identical phenotype. However, a method based on the genotype is much more computationally expensive than the phenotype one, but considering that it is applied to the population-list of small size (usually $K = 4$), it is sufficient to use it.

If Div is 0.0, it means that all the ants are identical. As a result, a high intensity of pheromone will be generated into one trail, forcing the population to converge into one path only. Therefore, every time the population reaches $Div = 0.0$, a random immigrant ant is generated to replace an ant in the population-list.

4.4 Behaviour in Dynamic Environments

The M-ACO algorithm may not have pheromone evaporation, but it has a more aggressive mechanism to eliminate pheromone trails that may limit the adaptation capabilities of ACO. The corresponding pheromone trails of the ant replaced from the population-list are removed.

In dynamic environments, the worst solution in the previous environment may be fit in a new environment. The traditional ACO framework considers the quality of solution to deposit pheromone, i.e., the better the solution, the higher the pheromone is deposited. In M-ACO, a constant amount of pheromone is deposited for all solutions stored in the population-list giving an equal chance to less fitted solutions, but possibly useful to the new environment, to be considered.

Finally, both p_a and p_s from Equations (11) and (12) are set to their initial value, i.e., 0.5, when an environmental change occurs in order to re-start the cooperation and competition when a new environment arrives. An environmental change can be detected by re-evaluating the population-list in every iteration t. Therefore, whenever there is a change to the solution quality of any ant stored in the population-list from iteration t in the next iteration $t+1$, it means that a dynamic change has occurred.

5 Experimental Study

5.1 Experimental Setup

The proposed M-ACO algorithm is tested on the DVRP instances that are constructed from three static benchmark VRP problem instances[1]. Using the method described in Section 2, several dynamic test cases of DVRP with traffic factors are generated with $F_L = 0$ and $F_U = 5$. The value of f was set to 10 and 100, indicating fast and slow environmental changes, respectively. The value of m was set to 0.1, 0.25, 0.5, and 0.75, indicating the degree of environmental changes from small, to medium, and large, respectively. As a result, eight dynamic test DVRPs, i.e., two values of $f \times$ four values of m, were generated from each static VRP instance.

In order to analyze and investigate the performance of the M-ACO algorithm, several ACO algorithms taken from the literature are considered as the peer algorithms for comparison as follows:

- ACS [38]: a traditional ACO algorithm which is described in Section 3
- P-ACO [24]: the memory-based version of ACO algorithm that has been applied to the DTSP. In this chapter, we apply P-ACO to the DVRP since our proposed algorithm uses its framework.
- Random immigrants ACO (RIACO) [35]: it has a different framework from both ACO and P-ACO, since it maintains a short-term memory where no ants can survive in more than one iteration. Random immigrants are generated to increase diversity.

[1] Taken from the Fisher benchmark instances available at
`http://neo.lcc.uma.es/vrp/`

The parameters of the investigated algorithms are chosen from our preliminary experiments and some of them are taken from the literature [35, 36]. For all algorithms $\alpha = 1$, $\beta = 5$, $q_0 = 0.0$ (expect in ACS which uses the pseudo-random proportional rule; therefore $q_0 = 0.9$).

The number of ants μ in the population for each algorithm varies in order to have the same number of evaluations every iteration. For M-ACO and P-ACO $K = 4$ and for M-ACO the number of LS steps is set to $ls = 15$. The population size for ACS and RIACO was set to $\mu = 50$, for P-ACO was set to $\mu = 50 - K$, and for M-ACO was set to to $\mu = 50 - K - ls$.

For each algorithm on a DVRP, 30 independent runs were executed on the same environmental changes. The algorithms were executed for 1000 iterations and one observation was taken every iteration. The overall performance of an algorithm on a DVRP instance is defined as follows:

$$\bar{P}^{best} = \frac{1}{G} \sum_{i=1}^{G} \left(\frac{1}{R} \sum_{j=1}^{R} P_{ij}^{best} \right), \tag{17}$$

where G is the total number of iterations, R is the number of runs, and P_{ij}^{best} is the tour cost of the best-so-far ant, after a change, for iteration i of run j, respectively.

5.2 *Effect of LS Operators*

The experimental results regarding the effect of the LS operators are presented in Fig. 1 for different static VRP instances. We use the framework of M-ACO described in Section 4, without LS, with simple swap, with adaptive swap, and with both simple and adaptive swaps activated as in Equations (11) and (12). From the experimental results, several observation can be drawn.

First, it is obvious that all algorithms with LS operators outperform the one where LS is not used in all problems. This is as expected since LS operators promote the exploitation and help the algorithm converge to a much better solution.

Second, different LS operators work better on different instances as expected. Simple swap performs better on `F-n45-k4` over the other two LS operators. On the other hand, the combination of simple and adaptive swaps performs better on the remaining problem instances. This result indicates that LS operators are problem-dependent, which is natural since it is almost impossible to develop an algorithm to outperform the remaining algorithms on all problem instances.

Third, the combination of the two LS operators usually performs better. The adaptive swap in `F-n72-k4` and `F-n135-k7` converges faster than the simple swap. On the other hand, the simple swap is able to converge to a better optimum than the adaptive one. This is because the adaptive swap after several iterations is more likely to have similar ants in the population. As a result, two customers obtained from similar (or even identical) ants will not have an effect for the swap.

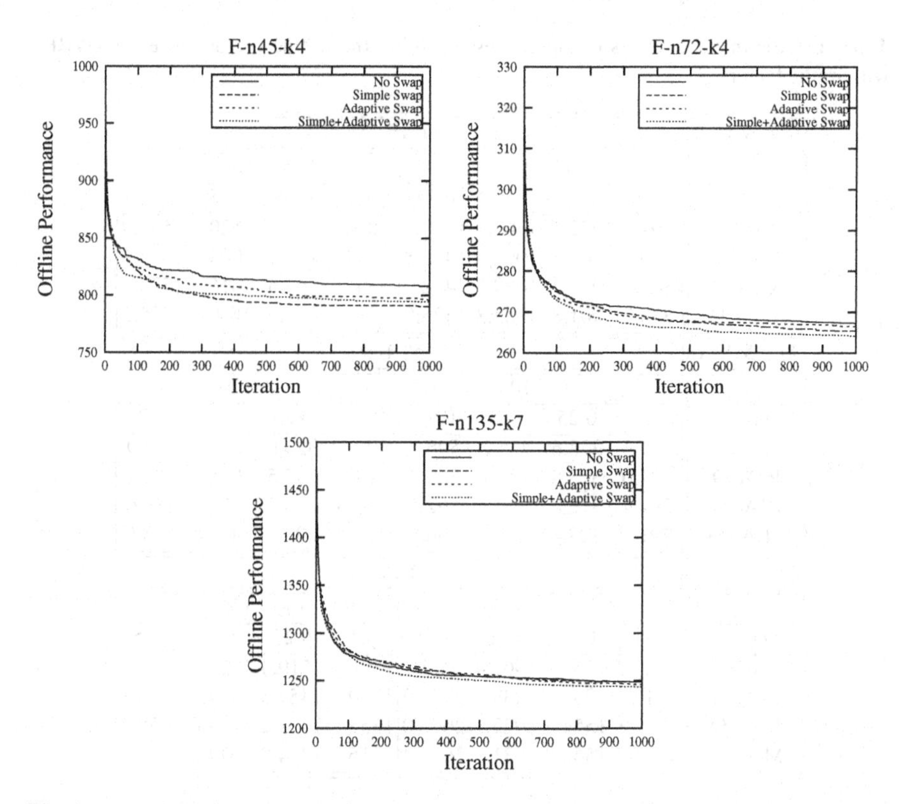

Fig. 1 Dynamic behaviour of the P-ACO with different LS operators on the stationary VRP instances.

5.3 Comparing M-ACO with Other Peer ACO Algorithms

The experimental results regarding the offline performance of the investigated algorithms for different dynamic test cases of DVRP with traffic factors are presented in Table 1. The corresponding Wilcoxon rank-sum test results at a 0.05 level of significance are presented in Table 2. In the comparisons, "+" or "−" indicates that the first algorithm is significantly better or the second algorithm is significantly better, respectively, and "∼" indicates no significance between the algorithms. Moreover, to better understand the dynamic behaviour of algorithms, the offline performance against the first 500 iterations is plotted in Fig. 2 for DVRPs with $f = 100$ and $m = 0.1$ and $m = 0.75$. From the experimental results, several observations can be made by comparing the behaviour of the algorithms.

First, ACS is outperformed by its competitors in all problem instances for all dynamic cases; see the comparisons of ACS ⇔ P-ACO, ACS ⇔ RIACO and M-ACO ⇔ ACS in Table 2. This is because ACS depends on pheromone evaporation to eliminate pheromone trails that represent solutions of the previous environment that are not useful to the new environment. As a result, the population of ants may be

Table 1 Experimental results of algorithms regarding the offline performance for DVRPs with traffic factors

Alg. & Inst.	F-n45-k4							
	$f = 10$				$f = 100$			
$m \Rightarrow$	0.1	0.25	0.5	0.75	0.1	0.25	0.5	0.75
ACS	897.5	972.5	1205.6	1648.0	883.4	929.1	1120.2	1536.9
P-ACO	839.7	903.3	1092.9	1486.8	836.2	862.1	1003.5	1356.6
RIACO	841.2	902.4	1089.5	1482.9	834.9	867.5	1016.1	1375.1
M-ACO	838.2	901.8	1085.3	1478.0	823.8	855.2	994.4	1336.2
Alg. & Inst.	F-n72-k4							
	$f = 10$				$f = 100$			
$m \Rightarrow$	0.1	0.25	0.5	0.75	0.1	0.25	0.5	0.75
ACS	305.3	338.6	426.2	596.2	297.3	324.6	412.7	547.9
P-ACO	291.0	323.8	406.1	568.2	274.2	297.5	367.5	478.4
RIACO	294.4	322.8	401.7	562.5	280.6	303.5	375.2	489.6
M-ACO	291.0	323.4	405.1	566.0	273.2	296.7	365.4	476.6
Alg. & Inst.	F-n135-k7							
	$f = 10$				$f = 100$			
$m \Rightarrow$	0.1	0.25	0.5	0.75	0.1	0.25	0.5	0.75
ACS	1427.7	1567.3	1967.4	2745.7	1383.7	1519.4	1820.5	2536.2
P-ACO	1412.1	1565.7	1939.5	2705.5	1319.3	1452.5	1674.4	2280.8
RIACO	1417.8	1554.2	1922.1	2676.0	1353.1	1457.2	1698.6	2358.4
M-ACO	1411.8	1563.8	1931.9	2695.4	1316.7	1444.9	1674.3	2277.0

Table 2 Statistical test results regarding the offline performance of the algorithms for DVRPs with traffic factors

Alg. & Inst.	F-n45-k4				F-n72-k4				F-n135-k7			
$f = 10$, $m \Rightarrow$	0.1	0.25	0.5	0.75	0.1	0.25	0.5	0.75	0.1	0.25	0.5	0.75
ACS⇔P-ACO	−	−	−	−	−	−	−	−	−	−	−	−
ACS⇔RIACO	−	−	−	−	−	−	−	−	−	−	−	−
P-ACO⇔RIACO	+	∼	−	∼	+	−	−	−	+	−	−	−
M-ACO⇔ACS	+	+	+	+	+	+	+	+	+	+	+	+
M-ACO⇔P-ACO	+	∼	+	+	∼	∼	+	+	∼	∼	+	+
M-ACO⇔RIACO	+	∼	+	+	+	−	−	−	+	−	−	−
$f = 100$, $m \Rightarrow$	0.1	0.25	0.5	0.75	0.1	0.25	0.5	0.75	0.1	0.25	0.5	0.75
ACS⇔P-ACO	−	−	−	−	−	−	−	−	−	−	−	−
ACS⇔RIACO	−	−	−	−	−	−	−	−	−	−	−	−
P-ACO⇔RIACO	∼	+	+	+	+	+	+	+	+	+	+	+
M-ACO⇔ACS	+	+	+	+	+	+	+	+	+	+	+	+
M-ACO⇔P-ACO	+	+	+	+	+	∼	+	∼	∼	+	∼	∼
M-ACO⇔RIACO	+	+	+	+	+	+	+	+	+	+	+	+

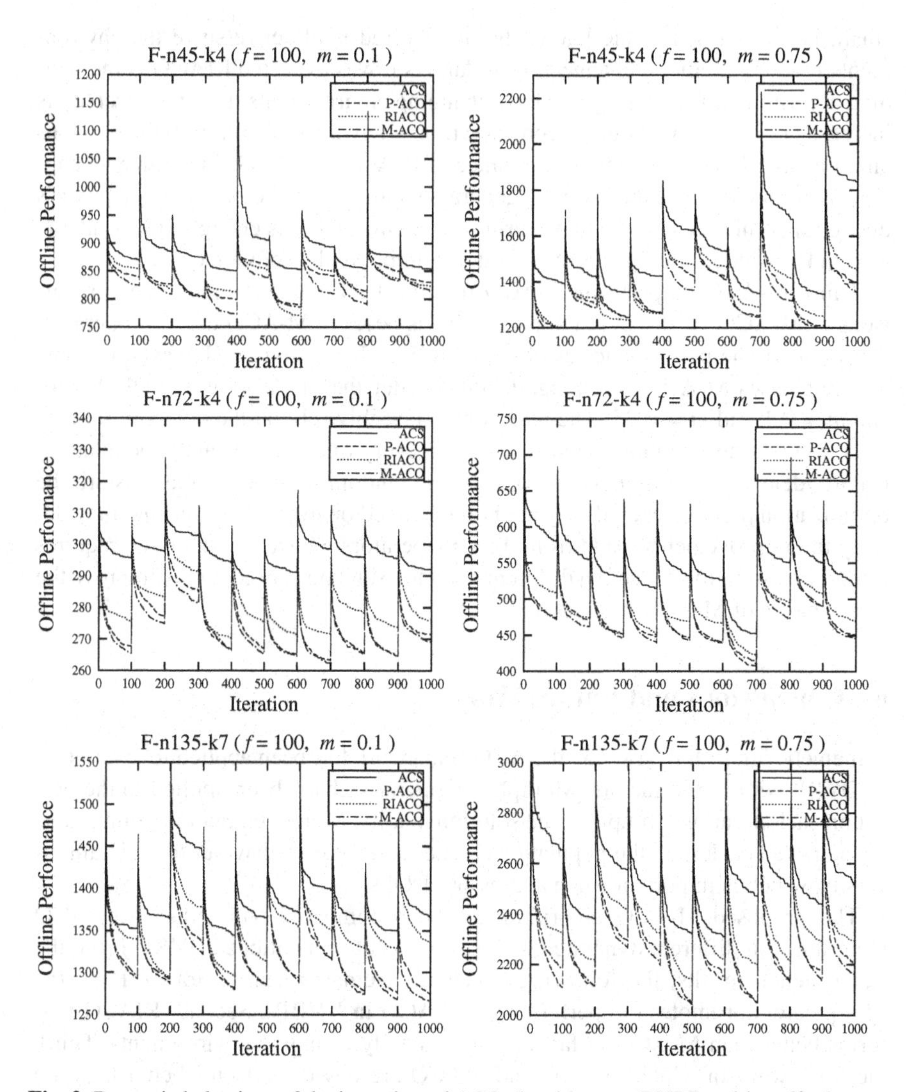

Fig. 2 Dynamic behaviour of the investigated ACO algorithms on DVRPs with traffic factors

biased and explore areas away from the optimum and the performance is degraded. On the other hand, RIACO increases the diversity using random immigrants, and in P-ACO and M-ACO, knowledge is transferred directly using the stored solutions in the population-list.

Second, M-ACO outperforms RIACO in almost all dynamic cases with $f = 10$ and $m = 0.1$, and all the dynamic cases with $f = 100$; see the comparisons of M-ACO $\Leftrightarrow$ RIACO in Table 2. In fast changing environments, i.e., when $f = 10$, there is no enough time to gain knowledge and transfer it to the new environment. However, in cases where the environment changes slightly or when the search space is small, i.e., F-n45-k4, the knowledge transferred is effective since the environments before and after a change are similar. As it was expected RIACO is the winning algorithm in fast and significantly changing environments since the diversity is increased randomly, without the consideration of any information from the previous environment. In contrast, the performance of RIACO is degraded in slowly changing environments since the diversity generated may disturb the optimization process due to randomization. M-ACO performs better since there is enough time available to gain knowledge and for the LS operator to improve the solution quality.

Third, M-ACO outperforms P-ACO in almost all dynamic cases; see comparisons of M-ACO $\Leftrightarrow$ P-ACO in Table 2. This is because M-ACO uses LS operators that improve the performance as observed in Fig. 1, previously. However, in some dynamic cases M-ACO is not significantly better than P-ACO, e.g., with $f = 10$ and $m = 0.1$ and $m = 0.25$. The reason may possibly rely on the diversity scheme used, where random immigrants are activated every time the ants in the population-list are identical. As discussed previously, the random immigrants may disturb the optimization process, even if they are not generated on every iteration, and may destroy the improvements made from the LS operators. However, this issue requires further investigation, since if a different diversity scheme is used it may improve the performance of M-ACO even more.

6 Conclusions and Future Work

A memetic algorithm based on the ACO framework has been applied to a variation of DVRP with traffic factors. Multiple LS operators have been applied to the best ant on each iteration to improve the solution quality. Moreover, random immigrants are activated each time the population reaches stagnation behaviour to maintain diversity, which is important when addressing DOPs.

The proposed M-ACO algorithm has been compared with other peer ACO algorithms on different dynamic test cases of the aforementioned DVRP. From the experimental results, the following concluding remarks can be drawn. First, the LS operators promotes the performance of ACO in DVRPs. Second, RIACO performs better than M-ACO in fast and significantly changing environments. Third, the framework in which M-ACO and P-ACO are based, performs better than the traditional ACO framework in DVRPs. Finally, the knowledge transferred from M-ACO and P-ACO is effective in slowly changing environments and some fast changing environments that change slightly.

In general, the performance of M-ACO is good from our preliminary results, but it can be furthermore improved if a better balance between exploration and exploitation can be achieved. The exploitation is achieved using the LS operators, whereas the exploration using the random immigrants. However, in some dynamic cases the diversity scheme used may destroy the improvement gained from the LS operator. A similar observation has been found when M-ACO has been applied to the DTSP [33].

Therefore, for future work it would be interesting to apply other diversity scheme, i.e., activate elitism-based immigrants, or even multiple diversity schemes to M-ACO and investigate if an appropriate balance between exploration and exploitation can be achieved for M-ACO in DOPs. Another future work is to apply other more specialized LS operators to M-ACO that may improve the performance even more [7].

Acknowledgements. This work was supported by the Engineering and Physical Sciences Research Council (EPSRC) of U.K. under Grant EP/K001310/1.

References

1. Bell, J.E., McMullen, P.R.: Ant colony optimization techniques for the vehicle routing problem. Advanced Engineering Informatics 18, 41–48 (2004)
2. Bielding, T., Görtz, S., Klose, A.: On-line routing per mobile phone: a case on subsequence deliveries of newspapers. In: Beckmann, M., et al. (eds.) Innovations in Distribution Logistics. LNEMS, vol. 619, pp. 29–51. Springer, Heidelberg (2009)
3. Bullnheimer, B., Haïti, R., Strauss, C.: An improved ant system algorithm for the vehicle routing problem. Annals of Operations Research 89, 319–328 (1999)
4. Bullnheimer, B., Hartl, R.F., Strauss, C.: A new rank based version of the ant system - a computational study. Central European Journal for Operations Research and Economics 7(1), 25–38 (1999)
5. Bonabeau, E., Dorigo, M., Theraulaz, G.: Swarm Intelligence: From Natural to Artificial Systems. Oxford University Press, New York (1999)
6. Borenstein, Y., Shah, N., Tsang, E., Dorne, R., Alsheddy, A., Voudouris, C.: On the partitioning of dynamic workforce scheduling problems. Journal of Scheduling 13(4), 411–425 (2010)
7. Bräysy, O., Gendreau, M.: VRPTW, Part I: Route construction and local search algorithms. Transportation Science 39, 104–118 (2005)
8. Caponio, A., Cascella, G.L., Neri, F., Salvatore, N., Summer, M.: A fast adaptive memetic algorithm for online and offline control design of PMSM drives. IEEE Transactions on Systems, Man and Cybernetics, Part B: Cybernetics 37, 28–41 (2007)
9. Colorni, A., Dorigo, M., Maniezzo, V.: Distributed optimization by ant colonies. In: Proceedings of the 1st European Conference on Artificial Life, pp. 134–142 (1992)
10. Cordón, O., de Viana, I.F., Herrera, F., Moreno, L.: A new ACO model integrating evolutionary computation concepts: The best worst Ant System. In: Proceedings of the 2nd International Workshop on Ant Algorithms, pp. 22–29 (2000)
11. Dantzig, G., Ramser, J.: The truck dispatching problem. Management science 6(1), 80–91 (1959)
12. De Rosa, B., Improta, G., Ghiani, G., Musmanno, R.: The arc routing and scheduling problem with transshipment. Transportation Science 36(3), 301–313 (2002)
13. Dorigo, M., Maniezzo, V., Colorni, A.: Ant system: optimization by a colony of cooperating agents. IEEE Transactions Systems, Man and Cybernetics, Part B: Cybernetics 26(1), 29–41 (1996)
14. Dorigo, M., Gambardella, L.M.: Ant colony system: a cooperative learning approach to the travelling salesman problem. IEEE Transactions on Evolutionary Computation 1(1), 53–66 (1997)
15. Dorigo, M., Stützle, T.: Ant Colony Optimization. The MIT Press, London (2004)

16. Gambardella, L.M., Taillard, E., Agazzi, G.: MACS-VRPTW: A multiple ant colony system for vehicle routing problems with time windows. In: Corne, D., et al. (eds.) New Ideas in Optimization, pp. 63–76 (1999)
17. Eyckelhof, C.J., Snoek, M.: Ant Systems for a Dynamic TSP. In: Dorigo, M., Di Caro, G.A., Sampels, M. (eds.) Ant Algorithms 2002. LNCS, vol. 2463, pp. 88–99. Springer, Heidelberg (2002)
18. Fabri, A., Recht, P.: On dynamic pickup and delivery vehicle rouyting with several time windows and waiting times. Transportation Research Part B: Methodological 40(4), 279–291 (2006)
19. Grefenestette, J.J.: Genetic algorithms for changing environments. In: Proceedings of the 2nd International Conference on Parallel Problem Solving from Nature, pp. 137–144 (1992)
20. Gribkovskaia, I., Laporte, G., Shlopak, A.: A tabu search heuristic for a routing problem arising in servicing of offshore oil and gas platforms. Journal of the Operational Research Society 59(11), 1449–1459 (2008)
21. Guntsch, M., Middendorf, M.: Applying population based ACO to dynamic optimization problems. In: Dorigo, M., Di Caro, G.A., Sampels, M. (eds.) Ant Algorithms 2002. LNCS, vol. 2463, pp. 111–122. Springer, Heidelberg (2002)
22. Guntsch, M., Middendorf, M.: A population based approach for ACO. In: Cagnoni, S., Gottlieb, J., Hart, E., Middendorf, M., Raidl, G.R. (eds.) EvoIASP 2002, EvoWorkshops 2002, EvoSTIM 2002, EvoCOP 2002, and EvoPlan 2002. LNCS, vol. 2279, pp. 72–81. Springer, Heidelberg (2002)
23. Guntsch, M., Middendorf, M.: Pheromone modification strategies for ant algorithms applied to dynamic TSP. In: Boers, E.J.W., Gottlieb, J., Lanzi, P.L., Smith, R.E., Cagnoni, S., Hart, E., Raidl, G.R., Tijink, H. (eds.) EvoIASP 2001, EvoWorkshops 2001, EvoFlight 2001, EvoSTIM 2001, EvoCOP 2001, and EvoLearn 2001. LNCS, vol. 2037, pp. 213–222. Springer, Heidelberg (2001)
24. Guntsch, M., Middendorf, M., Schmeck, H.: An ant colony optimization approach to dynamic TSP. In: Proceedings of the 2001 Genetic and Evolutionary Computation Conference, pp. 860–867 (2001)
25. He, J., Yao, X.: From an individual to a population: An analysis of the first hitting time of population-based evolutionary algorithms. IEEE Transactions on Evolutionary Computation 6(5), 495–511 (2002)
26. Jin, Y., Branke, J.: Evolutionary optimization in uncertain environments - a survey. IEEE Transactions on Evolutionary Computation 9(3), 303–317 (2005)
27. Kilby, P., Prosser, P., Shaw, P.: Dynamic VRPs: A study of scenarios, Technical Report APES-06-1998, University of Strathclyde, U.K. (1998)
28. Labbe, M., Laporte, G., Mercure, H.: Capacitated vehicle routing on trees. Operations Research 39(4), 61–622 (1991)
29. Larsen, A., Madsen, O.B.G., Solomon, M.M.: The priori dynamic travelling salesman problem with time windows. Transportation Sciences 38(4), 459–472 (2004)
30. Lee, Z.-J., Su, S.-F., Chuang, C.-C., Liu, K.-H.: Genetic algorithm with ant colony optimization for multiple sequence alignment. Applied Soft Computing 8(1), 55–78 (2006)
31. Lim, K.K., Ong, Y.-S., Lim, M.H., Chen, X., Agarwal, A.: Hybrid ant colony algorithms for path planning in sparse graphs. Soft Computing 12(10), 981–994 (2008)
32. Maniezzo, V., Colorni, A.: The ant system applied to the quadratic assignment problem. IEEE Transactions on Knowledge and Data Engineering 9(5), 769–778 (1999)
33. Mavrovouniotis, M., Yang, S.: A memetic ant colony optimization algorithm for the dynamic travelling salesman problem. Soft Computing 15(7), 1405–1425 (2011)

34. Mavrovouniotis, M., Yang, S.: An ant system with direct communication for the capacitated vehicle routing problem. In: Proceedings of the 2011 Workshop on Computational Intelligence, pp. 14–19 (2011)
35. Mavrovouniotis, M., Yang, S.: Ant colony optimization with immigrants schemes for the dynamic vehicle routing problem. In: Di Chio, C., et al. (eds.) EvoApplications 2012. LNCS, vol. 7248, pp. 519–528. Springer, Heidelberg (2012)
36. Mavrovouniotis, M., Yang, S.: Ant colony optimization with memory-based immigrants for the dynamic vehicle routing problem. In: Proceedings of the 2012 IEEE Congress on Evolutionary Computation, pp. 2645–2652 (2012)
37. Montemanni, R., Gambardella, L., Rizzoli, A., Donati, A.: A new algorithm for a dynamic vehicle routing problem based on ant colony system. In: Proceedings of the 2nd International Workshop on Freight Transportation and Logistics, pp. 27–30 (2003)
38. Montemanni, R., Gambardella, L., Rizzoli, A., Donati, A.: Ant colony system for a dynamic vehicle routing problem. Journal of Combinatorial Optimization 10(4), 327–343 (2005)
39. Neumann, F., Witt, C.: Runtime analysis of a simple ant colony optimization algorithm. Algorithmica 54(2), 243–255 (2009)
40. Neri, F., Toivanen, J., Cascella, G.L., Ong, Y.-S.: An adaptive multimeme algorithm for designing HIV multidrug therapies. IEEE/ACM Transactions on Computational Biology and Bioinformatics 4(2), 264–278 (2007)
41. Osman, I.: Metastrategy simulated annealing and tabu search algorithms for the vehicle routing problem. Annals of Operations Research 41, 421–451 (1993)
42. Pillac, V., Gendreau, M., Guèret, C., Medaglia, A.L.: A review of dynamic vehicle routing problems. Technical Report, CIRRELET-2011-62 (2011)
43. Psaraftis, H.: Dynamic vehicle routing: status and prospects. Annals of Operations Research 61, 143–164 (1995)
44. Polacek, M., Doerner, K., Hartl, R., Maniezzo, V.: A variable neighborhood search for the capacitated arc routing problem with intermediate facilities. Journal of Heuristics 14(5), 405–423 (2008)
45. Rizzoli, A.E., Montemanni, R., Lucibello, E., Gambardella, L.M.: Ant colony optimization for real-world vehicle routing problems - from theory to applications. Swarm Intelligence 1(2), 135–151 (2007)
46. Stützle, T., Hoos, H.: The MAX-MIN ant system and local search for the traveling salesman problem. In: Proceedings of the 1997 IEEE International Conference on Evolutionary Computation, pp. 309–314 (1997)
47. Tagmouti, M., Gendreau, M., Potvin, J.: Arc routing problems with time- dependent service costs. European Journal of Operational Research 181(1), 30–39 (2007)
48. Talbi, E.G., Bachelet, V.: Cosearch: a parallel cooperative metaheuristic. Journal of Math. Model Algorithms 5(1), 5–22 (2006)
49. Taniguchi, E., Thompson, R.: Modelling city logistics. Transportation Research Record: Journal of the Transportation Research Board 1790(1), 45–51 (2002)
50. Toth, P., Vigo, D.: Branch-and-bound algorithms for the capacitated VRP. In: Toth, P., Vigo, D. (eds.) The Vehicle Routing Problem, pp. 29–51 (2001)
51. Yang, S.: Genetic algorithms with memory and elitism based immigrants in dynamic environments. Evolutionary Computing 16(3), 385–416 (2008)
52. Wang, H., Wang, D., Yang, S.: A memetic algorithm with adaptive hill climbing strategy for dynamic optimization problems. Soft Computing 13(8-9), 763–780 (2009)
53. Zhang, X., Tang, L.: A new hybrid ant colony optimization algorithm for the vehicle routing problem. Pattern Recognition Letters 30(9), 848–855 (2008)

Author Index